Carbon Accounting and Savanna Fire Management

Carbon Accounting and Savanna Fire Management

Editors: Brett P. Murphy, Andrew C. Edwards, Mick Meyer and Jeremy Russell-Smith

CSIRO PUBLISHING

National Library of Australia Cataloguing-in-Publication entry

Carbon accounting and savanna fire management / edited by
Brett P Murphy, Andrew C Edwards, CP (Mick) Meyer and Jeremy Russell-Smith.

9780643108516 (paperback)
9780643108523 (epdf)
9780643108530 (epub)

Includes bibliographical references and index.

Greenhouse gas mitigation – Australia, Northern.
Savanna ecology – Australia, Northern.
Wilderness areas – Fire management – Australia, Northern.
Carbon dioxide mitigation – Australia, Northern.

Murphy, Brett P., editor.
Edwards, Andrew C., editor.
Meyer, C. P. (Mick), editor.
Russell-Smith, Jeremy editor.

363.738740994

Published by

CSIRO Publishing
36 Gardiner Road, Clayton VIC 3168
Private Bag 10, Clayton South VIC 3169
Australia

Telephone: [+613] 9545 8555
Email: csiropublishing@csiro.au
Website: www.publishing.csiro.au

Front cover: (Top) Low-intensity early dry season fire in central Arnhem Land. Credit: Brett Murphy. (Bottom) Frequency of late dry season fires from 1997 to 2007. Credit: Darwin Centre for Bushfire Research.
Back cover: (Top) Early dry season fire in Kakadu National Park. Credit: Clay Trauernicht. (Bottom) Burnt area mapping, 2013. Credit: Darwin Centre for Bushfire Research.

Set in 10.5/12 Minion & Stone Sans
Edited by Elaine Cochrane
Cover design by James Kelly
Typeset by Thomson Digital
Printed by Ingram Lightning Source

Feb26_RP_ILS

Contents

Preface and acknowledgements

This book charts the ongoing development of methodologies for greenhouse gas emissions abatement and carbon sequestration through savanna fire management for application in remote fire-prone northern Australian settings. In 2009, a precursor to the present volume was published. There we described the cultural, ecological and economic context behind the development of an emissions abatement methodology under higher rainfall (>1000 mm per annum [p.a.]) conditions. A key impetus for that earlier methodology was to afford an opportunity for Indigenous (Aboriginal) land managers to derive economic and enterprise development benefits from implementing strategic prescribed fire management based on a customary model. Since then, that methodology has been formally accredited as part of Australia's National Greenhouse Gas Inventory, and applied increasingly in economically marginal pastoral enterprise settings, both Indigenous and non-Indigenous.

This book describes the technical infrastructure and ongoing research underpinning the development of two further savanna burning methodologies: (1) an extension of the higher rainfall (>1000 mm p.a.) savanna burning abatement methodology to lower rainfall (600–1000 mm p.a.) conditions; and (2) work towards a new methodology addressing carbon sequestration in woody biomass (live and dead) associated with enhanced fire management under higher rainfall conditions. Despite substantial advances towards the development and practical application of savanna burning methodologies, described herein, that development continues to be a work-in-progress.

The research described here has been funded principally through the Indigenous Carbon Farming Fund administered by the Australian Government Department of the Environment (DoE; formerly the Department of Climate and Energy Efficiency), the Indigenous Land Corporation (ILC), and The Nature Conservancy (TNC). Overall administration of project components described here was undertaken by the North Australian Indigenous Land and Sea Management Alliance (NAILSMA), under the guidance of a Steering Committee involving DoE (Dr Rochelle Christian, Dr Karen King, Dr Brendan Pippen, Allyn White, Hilton Taylor, Allan Cassidy), ILC (Nerissa Walton), TNC (Dr Geoff Lipsett-Moore, Dr James Fitzsimons, Fergus McDonald, David Hinchley), NAILSMA (Glenn James, Christine Michael, Angela Shima), Charles Darwin University (Prof Lindsay Hutley, Dr Peter Jacklyn).

Research partners included in this enterprise have been: the Darwin Centre for Bushfire Research (DCBR), Research Institute for Environment and Livelihoods (RIEL), Charles Darwin University (CDU), Darwin; CSIRO Land and Water Flagship, Darwin; CSIRO Oceans and Atmosphere Flagship, Aspendale; School of Botany, The University of Melbourne, Parkville; Australian Wildlife Conservancy, Mornington Station. The assistance of Indigenous Ranger Groups in the undertaking of the field program is gratefully acknowledged: Mimal Rangers, Bulman; Yugal Mangi Rangers, Ngukurr; Waanyi Garawa

Rangers, Borroloola; Carpentaria Land Council Rangers, Doomagee and Burketown; Kimberley Land Council Rangers, Balanggarra and Wunggur.

As well as the authors of the various chapters (see pp. xi–xiv), we would like to give special acknowledgement to chapter reviewers:

Dr John Armston (Department of Science, Information Technology, Innovation and the Arts, Queensland Government)

Prof Andrew Campbell (Charles Darwin University)

Dr Ron Craig (Landgate, Western Australian Government)

Dr Rod Fensham (University of Queensland)

Ian Fox (The University of Adelaide, South Australia)

Dr Andy Hudak (USDA Forest Service; USA)

Prof Lindsay Hutley (Charles Darwin University)

Dr Ida Kubiszewski (Australian National University)

Dr Karen King (Department of Environment, Australian Government)

Mitchell Lendrum (Department of Environment, Australian Government)

Dr Clare Murphy (University of Wollongong)

Dr Sofia Oliveira (Charles Darwin University)

Dr Steve Roxburgh (Land and Water Flagship, CSIRO)

Dr Hilary Smith (Latitude Forest Services Pty Ltd)

Dr Ron Steffen (Green Mountain College, Vermont, USA)

Steve Sutton (Charles Darwin University)

Associate Prof Richard Thackway (University of Queensland)

Dan Tindall (Department of Science, Information Technology, Innovation and the Arts, Queensland Government)

Prof Peter Whitehead (Charles Darwin University; North Australian Indigenous Land and Sea Management Alliance)

Additionally, we note that Chapters 2 and 12 were reviewed through the peer review processes of *The Rangelands Journal* and the *International Journal of Wildland Fire*, respectively. We thank the publishers of both journals, CSIRO Publishing, for permission to reproduce those papers as chapters here.

The Editors
Dr Brett P Murphy
Dr Andrew C Edwards
Dr CP (Mick) Meyer
Prof Jeremy Russell-Smith

List of contributors

Shaun Ansell
Warddeken Land Management Limited
PO Box 785, Nightcliff, NT 0814, Australia

Mila Bristow
Research Institute for Environment and Livelihoods
Charles Darwin University, Darwin, NT 0909, Australia

Peter Brocklehurst
Department of Land Resource Management, Northern Territory Government
PO Box 496, Palmerston, NT 0831, Australia

Susan Campbell
CSIRO Oceans and Atmosphere Flagship
GPO Box 664, Canberra, ACT 2601, Australia

Garry D. Cook
CSIRO Land and Water Flagship
PMB 44, Winnellie, NT 0822, Australia

Martin Cope
CSIRO Oceans and Atmosphere Flagship
PMB 1, Aspendale, VIC 3195, Australia

Aaron Crosbie
Cape York Sustainable Futures
4/275 McCormack Street, Cairns, QLD 4870, Australia

Nicholas Cuff
Department of Land Resource Management, Northern Territory Government
PO Box 496, Palmerston, NT 0831, Australia

Mark Desailly
Bushfires NT
PO Box 37346, Winnellie, NT 0821, Australia

Andrew C Edwards
Darwin Centre for Bushfires Research
Research Institute for Environment and Livelihoods, Charles Darwin University,
Darwin, NT 0909, Australia

Jay Evans
Darwin Centre for Bushfires Research
Research Institute for Environment and Livelihoods, Charles Darwin University, Darwin, NT 0909, Australia

Rohan Fisher
Darwin Centre for Bushfires Research
Research Institute for Environment and Livelihoods, Charles Darwin University, Darwin, NT 0909, Australia

Jarrad Holmes
Kimberley Land Council
PO Box 2145, Broome, WA 6725, Australia

Lindsay B Hutley
Research Institute for Environment and Livelihoods
Charles Darwin University, Darwin, NT 0909, Australia

Peter Jacklyn
Darwin Centre for Bushfires Research
Research Institute for Environment and Livelihoods, Charles Darwin University, Darwin, NT 0909, Australia

Glenn James
North Australian Indigenous Land and Sea Management Alliance
Charles Darwin University, Darwin, NT 0909, Australia

Aidan Joseph
Cape York Sustainable Futures
4/275 McCormack Street, Cairns, Qld 4870, Australia

Sunhee Lee
CSIRO Oceans and Atmosphere Flagship
PMB 1, Aspendale, VIC 3195, Australia

Sarah Legge
Australian Wildlife Conservancy
PO Box 8070, Subiaco East, WA 6008, Australia

Ben Lewis
Firestick and Associates
Katherine, NT 0850, Australia

Fay Lewis
Australian Wildlife Conservancy
PO Box 8070, Subiaco East, WA 6008, Australia

Adam C. Liedloff
CSIRO Land and Water Flagship
PMB 44, Winnellie, NT 0822, Australia

Ashok Luhar
CSIRO Oceans and Atmosphere Flagship
PMB 1, Aspendale, VIC 3195, Australia

Dominique Lynch
Darwin Centre for Bushfires Research
Research Institute for Environment and Livelihoods, Charles Darwin University, Darwin, NT 0909, Australia

Stefan W. Maier
Darwin Centre for Bushfires Research
Research Institute for Environment and Livelihoods, Charles Darwin University, Darwin, NT 0909, Australia

CP (Mick) Meyer
CSIRO Oceans and Atmosphere Flagship
PMB 1, Aspendale, VIC 3195, Australia

Ross Mitchell
CSIRO Oceans and Atmosphere Flagship
GPO Box 664, Canberra, ACT 2601, Australia

Catherine Monagle
Institute of the Advanced Study of Sustainability
United Nations University, Yokohama, Japan

Brett P Murphy
NERP Environmental Decisions Hub
School of Botany, The University of Melbourne, VIC 3010, Australia

Sofia Oliveira
Research Institute for Environment and Livelihoods
Charles Darwin University, Darwin, NT 0909, Australia

Anna Pickworth
Kimberley Land Council
PO Box 2145, Broome, WA 6725, Australia

Jeremy Russell-Smith
Darwin Centre for Bushfires Research, Research Institute for Environment and Livelihoods, Charles Darwin University, Darwin, NT 0909, Australia; and North Australian Indigenous Land and Sea Management Alliance, Charles Darwin University, Darwin, NT 0909, Australia

Tom Vigilante
Wunambal Gaambera Aboriginal Corporation
Kalumburu, WA 6740, Australia

Jake Weigl
Warddeken Land Management Limited
PO Box 785, Nightcliff, NT 0814, Australia

Peter J Whitehead
Darwin Centre for Bushfires Research, Research Institute for Environment and Livelihoods, Charles Darwin University, Darwin, NT 0909, Australia; and North Australian Indigenous Land and Sea Management Alliance, Charles Darwin University, Darwin, NT 0909, Australia

Cameron P Yates
Darwin Centre for Bushfires Research
Research Institute for Environment and Livelihoods, Charles Darwin University, Darwin, NT 0909, Australia

Stuart Young
CSIRO Oceans and Atmosphere Flagship
PMB 1, Aspendale, VIC 3195, Australia

1

Reimagining fire management in fire-prone northern Australia*

Jeremy Russell-Smith and Peter J Whitehead

Summary

The northern savannas, occupying a quarter of the Australian land mass, constitute the most fire-prone landscapes of a fiery land – in recent years, on average, 18% of the savannas have been burnt each year. Land use is predominantly given over to extensive beef cattle pastoralism – although over much of the rangelands this is at best an economically marginal activity, especially in frequently burnt, higher rainfall northern regions. Outside of urban centres most of the savanna human population is Indigenous, increasingly rapidly, and remains impoverished. Much of the northern cadastre is Indigenously owned outright, or is subject to continued rights of access and use under Native Title arrangements. In this paper we articulate a distinctly northern Australian understanding of and approach to the management of fire, matched to the biophysical and social realities of the north. Despite many obstacles to effective management associated with remoteness, sparse population and limited transport and other infrastructure, management of fire can be improved for positive ecological and social outcomes over a large part of the Australian land mass. Taking advantage of emerging carbon and biodiversity environmental services developments, skilled fire management offers a culturally apt enterprise opportunity where few others exist.

Introduction

Australia is rightly recognised as a fire-prone continent. Fuelled by public media and official enquiries (e.g. Ellis *et al.* 2004; Teague *et al.* 2010), the dominant popular perception is that fire is a particularly southern Australian phenomenon, characterised by life- and property-threatening conflagrations under extreme summer fire-weather conditions. Indeed, under envisaged climate change scenarios of marked increase in number of days of extreme temperature and diminishing regional water availability (CSIRO and Bureau of Meteorology

* This chapter is an updated and substantially reworked version of a chapter previously published as Russell-Smith J, Whitehead PJ, Cooke PM, Yates CP (2009) Challenges and opportunities for fire management in fire-prone northern Australia. In *Culture, Ecology and Economy of Fire Management in North Australian Savannas: Rekindling the* Wurrk *Tradition* (Eds J Russell-Smith, PJ Whitehead, P Cooke) pp. 1–22. CSIRO Publishing: Melbourne.

2007; Garnaut 2008; Indian Ocean Climate Initiative 2012), wildfire activity in densely populated south-eastern and south-western Australia is likely to increase in decades to come. Given contemporary patterns of human settlement, it is entirely reasonable therefore that such preoccupation with southern landscape fire management issues will continue.

However, a broader understanding of the geographic scale and frequency of fire patterning in Australia reveals that fire incidence in the 'deep south' is, in fact, scant by comparison with periodic fire occurrence under typically low rainfall conditions in central Australia, and especially compared with annual–biennial fire recurrence under monsoonal conditions in north Australian savannas (Russell-Smith *et al.* 2007; Murphy *et al.* 2013). Contrary to the narrow views of Australia's only ostensibly 'national' bushfire enquiry (Ellis *et al.* 2004), such fire regimes in 'regional Australia' are exerting significant impacts on a range of regional cultural, production, and biodiversity values (e.g. Dyer *et al.* 2001; Allan and Southgate 2002; Williams *et al.* 2002; Russell-Smith *et al.* 2003; Woinarski *et al.* 2007a). The greenhouse gas and particulate emissions resulting from these fire regimes, and the radiative forcing, associated especially with savanna fires, have global-scale climate implications (e.g. Luhar *et al.* 2008; Meyer *et al.* 2008; ANGA 2011).

In response to this generic national myopia concerning landscape fire issues in northern Australia, here we set out to (1) describe contemporary fire patterning across the savannas, (2) explore the particular biophysical and societal features associated with and giving rise to those patterns, and especially (3) make the case that savanna fire, rather than being seen simply as an aberrant, problematic landscape property, in fact can afford savanna land managers – and especially disadvantaged Indigenous communities – significant employment opportunities and cultural benefits in a re-imagined, diversified savanna economy.

Fire mapping and contemporary burning patterns

Remotely sensed data sources and website portals have started to transform widely held misperceptions that bushfire events, and their social, environmental and economic implications, are a particularly southern Australian phenomenon. On the basis of continental mapping of large fire-affected areas (>~2–4 km^2) for the period 1997–2004 derived from AVHRR imagery, Russell-Smith *et al.* (2007) observed that 76% of total mean fire-affected area (508 000 km^2 p.a.) occurred in the northern savannas. Expressed as a proportion of continental land area defined by rainfall classes, a mean of 0.6% of southern Australia (53% of continental land area) was fire-affected each year, 5% of central Australia (25% of continent), and 23% of northern Australia (22% of continent). These general patterns are reflected also in updated, 1997–2011, continental mapping of the frequency of large fires derived from AVHRR imagery (Fig. 1.1).

At a continental scale, fire extent is substantially explained by rainfall seasonality, i.e. the relative amount of rainfall received in the wettest compared with the driest quarter (Russell-Smith *et al.* 2007; Murphy *et al.* 2013). In the monsoonal north, intense bursts of high summer rainfall separated by an extended annual dry season (southern winter and spring) 'drought', during which next to no rain may fall for 6 months or more, drive an annual cycle of elevated fire risk alternating with periods of little or no fire risk. In the arid centre, erratically variable periods of above average rainfall and subsequent high rates of plant growth may alternate with dry periods extending over several years to decades. Susceptibility of landscapes to fire therefore changes over longer than annual cycles, but very large wildfires occur with some regularity, fuelled mostly by grassy fuels like spinifex (*Triodia* spp.). In parts of mesic southern Australia (especially forested regions),

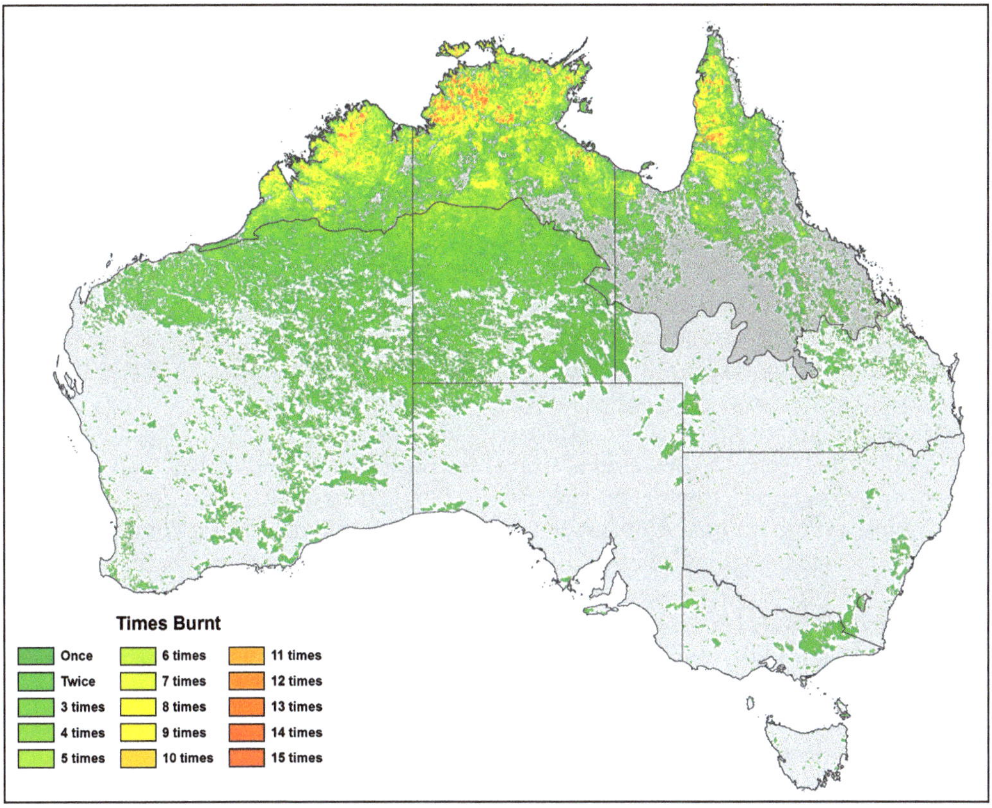

Figure 1.1. Frequency of large fire-affected areas (>~2–4 km^2) derived from AVHRR imagery, 1997–2011. North of line indicates tropical savannas region as defined by the Tropical Savannas Cooperative Research Centre. (Data from: Western Australia Department of Land Information – Landgate).

vulnerabilities increase with fuel accumulation over many years, so that when the mostly summer fires occur they can be extraordinarily intense.

For the tropical savannas, mean annual rainfall declines rapidly from >2000 mm in parts of the far north to ~500 mm in southern regions. The annual mean extent of large fires follows this general trend (Chapter 2). Significantly, modelling of fire extent with a range of biophysical variables (e.g. antecedent rainfall, or fire, in previous year(s); land use) suggests that rainfall is sufficiently annually reliable in the north to support annually recurrent burning (Russell-Smith *et al.* 2007; Chapter 2). With declining, less annually reliable rainfall, fire propagation relies on cumulative antecedent rainfall for the development of adequate fuel loads and fuel continuity (Allan and Southgate 2002).

A salient feature of contemporary savanna fire regimes concerns the predominance of fires occurring in the latter part of the dry season, typically under severe fire weather conditions (periodically strong south-easterly winds, high temperatures, low humidities, fully cured fuels). For example, of the annual mean 338 000 km^2 of the tropical savannas region affected by large fires over the period 1997–2011, 69% occurred in the late dry season months August–November. Fire regimes dominated by frequent, large, late dry season fires are commonplace in many regions of northern Australia, especially the Kimberley region in the north-west, the Top End of the Northern Territory, and western Cape

York Peninsula (Fig. 1.1). Contemporary north Australian fire regimes have significant impacts on regional biodiversity values (e.g. Russell-Smith and Bowman 1992; Woinarski *et al.* 2001, 2010, 2011; Franklin *et al.* 2005; Russell-Smith *et al.* 2010, 2012), and have significant implications for greenhouse gas emission estimates and related carbon dynamics (Cook and Meyer 2009; Murphy *et al.* 2009, 2010; Russell-Smith *et al.* 2009a; Meyer *et al.* 2012).

Of particular relevance here, 'Prescribed burning of savannas' is listed as an accountable activity under the provisions of the Kyoto Protocol. While the protocol was only ratified by Australia relatively recently, Australia's National Greenhouse Gas Inventory (NGGI) has reported on savanna burning emissions from 1990. Typically, accountable greenhouse gas emissions contribute between 1–3% of Australia's NGGI (ANGA 2011). However, following international convention, Australia's NGGI considers only emissions of the accountable greenhouse gases methane (CH_4) and nitrous oxide (N_2O). Carbon dioxide itself is not accounted for since it is assumed (often incorrectly, see Cook *et al.* 2005; Cook and Meyer 2009) that CO_2 emissions in one burning season are negated by vegetation growth in subsequent growing seasons (IPCC 1997). In fact CO_2, together with more reactive species (e.g. the ozone precursors comprising CO, volatile organic compounds and oxides of nitrogen) that are released into the atmosphere over a typically long burning season, is likely to have a substantial impact on regional atmospheric composition and its inter-annual variability.

In the past few years very substantial developments have taken place in the north, providing opportunities for land managers to undertake strategic landscape-scale fire management to deliver commercial greenhouse gas emission offset projects, particularly under the mantle of the Commonwealth Government's Carbon Farming Initiative (CFI; Commonwealth of Australia 2013). Leading the way has been the Western Arnhem Land Fire Abatement (WALFA) project, operating over 28 000 km^2 of Aboriginal-owned land under a 17-year voluntary (*sensu* Bayon *et al.* 2006) offset arrangement with a multinational energy corporate (Russell-Smith *et al.* 2009b; Heckbert *et al.* 2011, 2012; Cook *et al.* 2012). In its first 7 years of operation WALFA has successfully reduced greenhouse gas emissions by 37.7% (or 127 000 t CO_2-e yr^{-1}) relative to the 10-year pre-project baseline period (Russell-Smith *et al.* 2013). In 2013, the first savanna burning project was declared under CFI law on Fish River Station, a 1700 km^2 property managed on behalf of Aboriginal landowners by the Indigenous Land Corporation (see www.ilc.gov.au/site/page.cfm?u = 335) and is now eligible to trade in Australia's compliance market. At the present time, many more projects across the north are already, or in the process of being, CFI-registered. While the present federal government's proposals will, if brought into law, alter the way credits are bought and sold, key features of the CFI will be retained, including eligibility of savanna fire projects.

Achieving significant greenhouse gas emissions abatement, and more effective savanna landscape fire management generally, are thus components of the same fundamental problem – how do we practically and economically reduce the incidence and extent of contemporary late dry season wildfires?

Ecological and cultural landscapes

We approach this question by first considering why it has proved so difficult to manage Australia's savanna landscapes to avert unwelcome environmental and social change, despite the absence of intense population pressures.

Ecology

North Australian savanna environments are relatively intact structurally and many areas are recognised for their important biodiversity values, including major centres of endemism (with many plants and animals that occur nowhere else; Woinarski *et al.* 2007a). But there have been losses of ecological function – chiefly evidenced in biodiversity declines – at several levels and across large areas (Woinarski *et al.* 2001, 2011; Franklin *et al.* 2005). Scattered richer patches of the landscape are asked to maintain natural production but at the same time support economic (agricultural and pastoral) production. More resilient and reliably productive environments (like wetlands) are valued highly by many different interests or sectors (Jonauskas 1996). Rivers are mostly unregulated. Most rivers carrying high wet season flows cease flowing during parts of the annual dry season, so that many water-dependent ecosystems are maintained by springs or other near-surface groundwater. Primary industries (agriculture and mining) rely heavily on groundwater extraction, even in the high rainfall areas, and so can impact these systems.

Temperatures are uniformly high, and often very high. Rainfall is intensely seasonal with no equivalent to the more equable wet tropics of the east coast of Australia. In the seasonal tropics, timing of the onset and cessation of wet conditions is highly unpredictable, and evaporation rates greatly exceed precipitation for most of the year. Hence the length of the growing season is variable but often short. Extreme weather events are common (storm, cyclone) and recur on a range of spatial and temporal scales. Landscapes are old, frequently reworked, and highly weathered. Soils are often of low fertility or low water-holding capacity and in many settings highly erodible, and so limiting for agriculture (Woinarski and Dawson 2002). These features have contributed to the problematic history, including repeated failures, of agriculture in northern Australia (Cook 2009).

Many areas of northern Australia contain exploitable concentrations of minerals and fossil energy, some of which are already in production and many others are under development. Mineral extraction often involves relatively low-grade deposits requiring movement of large quantities of overlying or intervening rock, which build long-term problems of acid formation during the subsequent oxidation of waste rock and drainage from it (e.g. Harries 1997). On-site processing can add to pressures on water availability and also compromise water quality. The Northern Territory Government has recently acknowledged weaknesses in environmental management of 'legacy' mines by imposing a levy on new ventures to fund remediation to reduce long-term impacts (see http://newsroom.nt.gov.au/index.cfm?fuseaction = viewRelease&id = 10733&d = 5).

Society

It is difficult to present a comprehensive statement of the particular social and economic character of the savannas. Most routinely produced demographic and socioeconomic data are presented to jurisdictional or regional boundaries that do not align with biophysical domains. The profile to follow is therefore based on several individual studies and one-off estimates made over the last several years drawing on the 2006 Census, supplemented with some preliminary observations by the Australian Bureau of Statistics and others using the 2011 Census.

The northern savanna population, including the major towns of Cairns, Townsville, Mt Isa (Queensland), Darwin, Palmerston, Katherine (Northern Territory), and Broome (Western Australia), is relatively small (751 000 in 2011), widely dispersed, and approaching 17% Indigenous. Average population density of ~0.3 persons km^{-2} is low by Australian (3.0 persons km^{-2}) and global standards, and lower still away from major centres.

Excluding the major town centres, there is ~1 person for each 700 ha (0.14 persons km^{-2}). By any standard, the tropical savannas of Australia are sparsely inhabited (Taylor *et al.* 2006) and the scope to marshal human resources for intensive land management is correspondingly weak.

Outside the towns, the savanna population was around 500 000 based on 2011 census figures, with Indigenous people comprising a much greater proportion of the population. In the Kimberley and Top End savannas, about half of the population is Indigenous, and in very remote regions generally, it is more than 90%. Nationally, 45.4% of the population living in areas classified by the Australian Bureau of Statistics as very remote is Indigenous (Taylor 2006). In the Northern Territory, most Indigenous people (70%) live on lands held under Aboriginal communal title (Taylor 2003). As highlighted by Taylor (2006), this means that Indigenous people and their institutions predominate over most of the Australian land mass and the northern savannas in particular.

The savanna population is growing, with an increase of 8% between the 1996 and 2001 censuses. Rate of increase is higher among the Indigenous population, which grew by 14.9% compared with 6.5% in the non-Indigenous community. Non-Indigenous population growth is confined largely to the major centres. Projections to 2021 see a continuation of higher rates of growth in the Indigenous population (25.7%) over this period than in the non-Indigenous population (14.8%) (Taylor *et al.* 2006). Preliminary analyses of the 2011 Census indicate Indigenous population growth rates higher than projected (Biddle 2013).

At regional scales, rates of population growth are highly variable, both through time and by location, being strongly influenced by shifts in immigration and emigration of non-Indigenous people tracking employment and other economic opportunities. Sites of high local Indigenous population growth are often poorly matched to areas of likely job growth (Taylor 2003). Rates of Indigenous mobility may be increasing, but emigration from remote and very remote regions does not appear to have kept pace with natural population increase, so the populations of these regions mostly continue to rise (Biddle 2013; Biddle and Markham 2013).

Despite some recent improvements, morbidity and mortality rates remain unacceptably high among the Indigenous population (Burgess *et al.* 2005; SCRGSP 2011). Educational systems have broken down in many remote areas (Collins 1999), so that many Indigenous people suffer from poor literacy and numeracy and experience difficulty in taking advantage of such mainstream employment opportunities as may be available in the regions.

Compounding this educational deficit, the savanna is subject to other knowledge deficits. The formal scientific understanding of regional environments and natural resources is weaker than in more densely settled jurisdictions due to historical and contemporary limitations on technical capacity and research investment. Dependence on census data aggregated to inappropriate boundaries (Griffith 2010) limits understanding of regional human demography. Regional economies are weakly characterised and their dynamics poorly understood.

In contrast to these formal knowledge gaps, many Indigenous people have very detailed knowledge of the landscapes and resources for which they are responsible (e.g. Garde *et al.* 2009), but mechanisms for applying this and other local knowledge are poorly developed and often contested (e.g. see Brook and McLachlan 2005; Gilchrist *et al.* 2005).

Economy

Like its ecology, a key defining attribute of the mainstream savanna economy is its geographic patchiness and variability through time. Volatility is associated with low enterprise diversity, dependence on variable commodity prices and extractive industries, and relatively small size. A 2006 estimate of gross incomes from mineral extraction in the savannas (excluding coal) was about $6 billion p.a. and this is likely to have increased (Brereton *et al.* 2009). Estimates of the annual value of cattle production achieved on 90% of the north's land area are something over $1 billion p.a. (NALWTF 2009). Another estimate for gross value from both production and processing put the northern pastoral contribution to the Australian economy at $5 billion in 2009–2010 (Gleeson *et al.* 2012). Estimates vary in their structure and assumptions, meaning some figures may not be directly comparable within or across sectors. Recent constraints on access to the live export market in Indonesia have highlighted the fragility of the industry as presently structured (see, for example, www.abc.net.au/news/rural/specials/live-export).

Tourism, which is highly dependent on access to natural environments and related activities like recreational fishing, generated expenditures exceeding $2.8 billion p.a. in 2006/7 and is an important source of employment (Clark *et al.* 2009), although in some locations costs of servicing tourists may be high compared with benefits captured locally (Greiner *et al.* 2004). Approximately 8% of the savanna population is engaged in tourism or hospitality industries, but employees are mostly non-Indigenous.

The formal savanna economy is very open, reflected in a very high ratio of imports and exports to gross product, with little local production of manufactured or otherwise highly processed items and limited value-adding to exports. Correlation between unemployment and employment rates is weak, reflecting gaps in local and regional labour markets and consequent importation of skilled and semi-skilled labour during economic upswings (Gerritsen 2010). Many savanna residents enjoy well above average incomes, but a mostly unemployed and growing regional Indigenous population is on chronically low to very low incomes (e.g. Taylor 2003).

It follows that at the regional scale, multipliers (measures of the extent to which money circulates and stimulates additional activity in the economy) from activities like pastoralism and mining are low, because local communities are often poorly placed to provide the goods or services sought by these industries, including labour (Welters 2010). Multipliers from public sector investments in services like health are substantially greater (Stoeckl *et al.* 2007).

Defence facilities and their personnel make a large contribution to economic activity. There is a strong general dependence on public expenditure well above national per capita averages. Low capacity to raise revenue locally, combined with the costs imposed by remoteness, mean that maintaining services, especially in the regions, depends on national fiscal equalisation policies implemented through the Commonwealth Grants Commission (Morris 2003). Total subventions are based predominantly on the size of the population requiring services. It follows that funds needed to manage issues influenced strongly by land area and biophysical pressures are often inadequate (see Altman *et al.* 2007).

Operating mostly separately from the mainstream economy is an Indigenous customary economy based on direct non-commercial exploitation of natural resources (Altman 1987). That economy is stronger in the remoter areas, and there intersects with the mainstream economy chiefly through production of arts and crafts, which often draws on continued associations with country and use of natural products (Altman 2003). In many parts of the savannas, Indigenous society has been substantially supported by welfare payments and Australian Government job creation and community development programs. That support – particularly the Community Development Employment Program (CDEP) – has been restructured to emphasise 'real' jobs and hence engagement with the mainstream economy (see Jordan 2012), despite the weakness or absence of mainstream job markets in remote locations.

In these circumstances, confronting the north's conservation and other land management goals through employment of Indigenous residents with the necessary skills and commitment appears an obvious step. The opportunity has been taken up by national, state and territory governments through programs like Working on Country (Allen Consulting Group 2011; Urbis Pty Ltd 2012), by conservation NGOs through support for Indigenous Protected Areas (e.g. www.karrkad-kandji.org.au/index.php/supporters), and industry through offset programs, including those based on emissions abatement (Russell-Smith *et al.* 2009b).

Land use

The greater part of the northern savannas (about 90%) is used ostensibly for pastoral production, especially the grazing of cattle (*Bos taurus* and *Bos indicus*), and also sheep in parts of western Queensland (Fig. 1.2). Most of this land is public land leased from state and Northern Territory governments for pastoral use. Given the very large areas used and the associated costs of infrastructure, returns on assets from pastoralism can be very low (0.3 to 2.0% average over a decade: McCosker *et al.* 2010). Parlous assessments like these, which showed property owners spending more than they earned in 6 of 7 years, have been used to justify resistance to increases in returns to the public from leases. Rents are often set at peppercorn levels (a few cents per hectare) that may not cover administrative costs, let alone provide a commercial return (Productivity Commission 2002).

Despite the very small area used for mining purposes, such land use constitutes the greatest economic return to the regional economy (see above). Parks and reserves, including Indigenous protected areas, support much of the tourist activity in the north, which generates more direct employment than either pastoralism or mining (e.g. 6.1% for tourism

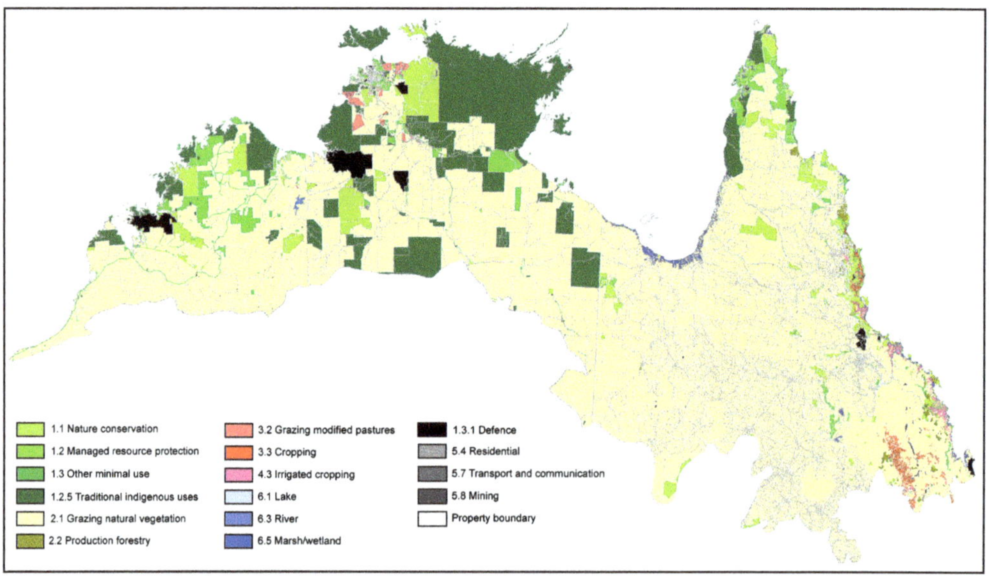

Figure 1.2. Generalised land use of the tropical savannas region (data from: Land Use of Australia Version 3, Bureau of Rural Sciences, Commonwealth of Australia. 2008).

versus ~1.6% and 3.8% respectively in the Northern Territory (TRA 2012; www.nt.gov.au/d/Primary_Industry/Content/File/publications/books_reports/indigenous_pastoral_program.pdf; http://www.nt.gov.au/d/Minerals_Energy/index.cfm?header = AboutMines andEnergy).

The predominance of Indigenous people living outside the major towns and associated rapid population growth is reflected increasingly in changing patterns of legal ownership of and interests in land. Recent data indicate that around 19% of the tropical savannas region is owned or managed by Indigenous people (Fig. 1.3a), ranging from 36% of savannas in the Northern Territory to 6% in Queensland. Additionally, Indigenous interests in land, as expressed through determinations of and applications made for Native Title under the Commonwealth of Australia's *Native Title Act 1993*, indicate that, as of May 2013: (1) determinations of Native Title have been granted for a further 22%, predominantly in Western Australia (Fig. 1.3b); and Registered or Scheduled Native Title applications (i.e. still to be determined) cover more than 43% of the tropical savannas region, ranging from 52% of Western Australian savannas to 40% in Queensland (Fig. 1.3c).

Summary

These observations illustrate some important features of land and resource management in northern Australia. Low human population density and weak infrastructure compromise capacity to manage pervasive damaging processes like wildfire that require active intervention to manage their impacts (Whitehead 2000; Whitehead *et al.* 2002). Because the spatially dominant land uses (pastoralism, Indigenous customary use and tourism) produce low or dispersed incomes, local people often lack the resources to respond effectively without public support. The north Australian state/territory jurisdictions have historically found it hard to fund interventions demanding coordinated action over large areas, because their funding sources and models for allocating funds respond most strongly to population. Regional land managers are therefore increasingly reliant on Australian Government funding programs that vary idiosyncratically in their targets and timelines, both of which are, moreover, often a poor fit to northern environments. And all forms of support may contract, given governments' present emphasis on fiscal austerity.

Mainstream economic developments have rarely been designed to complement or substitute for government support for land management capability. Developments like mines are patchy in space and time, and local or regional interests often find it difficult to connect to them. Immigrants responding successfully to such development opportunities may stay for relatively short periods and export benefits as savings rather than spending or reinvesting locally. The economic and related social benefits and environmental costs of savanna development can be inequitably distributed, with many Indigenous people remaining severely disadvantaged despite being 'land rich', are and obliged to accept the environmental costs of development.

Weak access to benefits may compound disadvantage when distinct 'cultural' values are not easily communicated and so receive inadequate consideration in areas subject to development. In most parts of the savannas it is possible to identify people with particular ties to and obligations to care for sites proposed for development. It is nearly always necessary to deal with complex cross-cultural issues and competing interests. Despite the sparseness of population, there is no *terra nullius*. Approaches to development that ignore this reality risk creating conflict and related social obstacles to orderly and genuinely sustainable development.

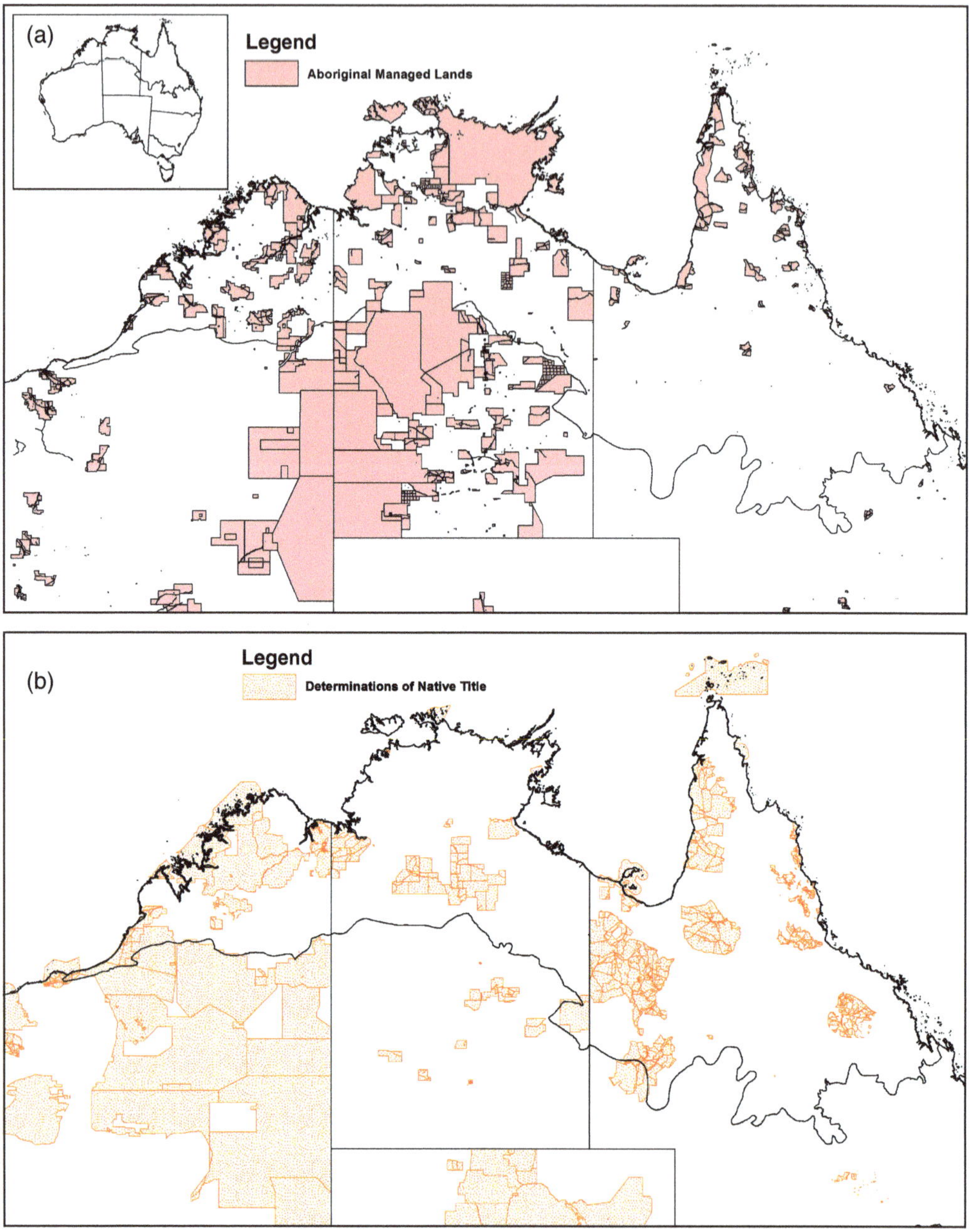

Figure 1.3. Current status of Aboriginal interests in land. (a) Aboriginal-owned or managed lands (data from: Indigenous Land Corporation. 2013). (b) Determinations of Native Title (data from: National Native Title Tribunal, June 2013). (c) Native Title Applications (data from: National Native Title Tribunal, June 2013).

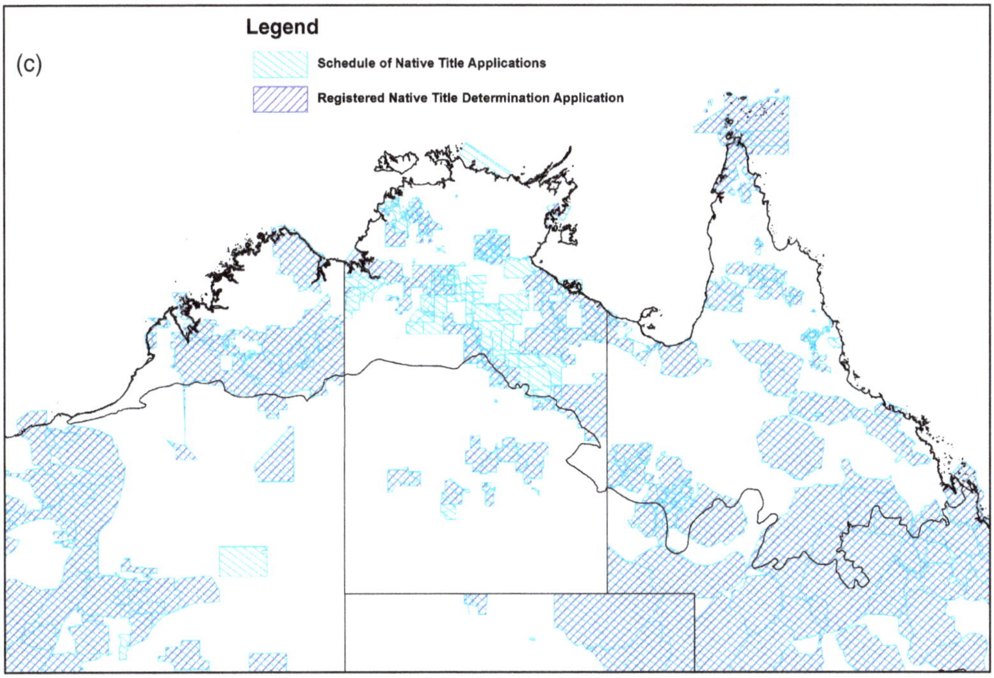

Figure 1.3. (Continued).

A reimagined future for managing the savannas

The Australian Government recently resuscitated the notion of accelerated northern development, with a particular emphasis on feeding Asia from broad-scale agriculture, established through investment from China (Australian Government 2012). This Asian Century vision identifies environmental sustainability and Indigenous interests as apparently peripheral, albeit contentious, issues to be managed (DFAT 2012: p. 44), rather than as fundamental drivers of policy. However, as major landholders and key stakeholders, Indigenous people have direct interests in the potential to use their lands for appropriate development. In addition, they have less direct but nonetheless critical interests in developments on other lands that affect landscape and ecological processes and so have substantial off-site impacts. Water use is of particular concern (NAILSMA 2009a, 2009b).

Government and industry views of the roles of Indigenous landowners and land managers need to change. Indigenous interests should be recognised and treated as important potential investors – of skills, knowledge, and land and associated assets – in appropriate northern development, and as full participants in all related planning processes, rather than marginalised as sources of difficulties requiring management by and for others.

Today's Asian Century version of the inevitability of agricultural development in the north contrasts with a long history of agriculture failure (Cook 2009), and considered advice provided by the Northern Australia Land and Water Task Force (2009). That group of Indigenous and industry leaders and technocrats reviewed lessons from lengthy experience and messages from contemporary science about problems with present land use and holes in the northern food bowl mythology. The Task Force promoted a substantial role for agriculture in relatively small-scale mosaics, but it also recognised important land use and livelihood alternatives that would collectively transform northern Australia's economy and society over the next few decades. In part those alternatives flow from acceptance of the need for change in land management use and practice towards genuine sustainability.

Holmes (2002, 2010) argues that transformation is underway, echoing changes in rural landscapes in other affluent nations. Over the last several decades, land ownership has shifted to include more Indigenous and conservation lands. But extensive pastoralism still dominates, with often 'flimsy ... occupance' (Holmes 2010) of most (90%) northern lands generating a small portion of regional incomes and employment. Many properties in marginal country are incapable of producing adequate returns on capital from the ostensible purpose of granting access to these public lands – commercially viable cattle production – and so rely on increasing valuation of leases in the good times to capture adequate returns over the longer term. Even if such sites remain under ownership of committed pastoralists, diversification of use may reduce pressures to 'force' incomes from grazing that some lands are unable to sustain. But shifts in the mix of uses are inhibited by laws governing pastoral leases as well as sectoral resistance to change (Cribb *et al.* 2009).

As a central component of the post-productivist transition (Holmes 2002), transfer of land to Indigenous people has greatly outpaced access to the resources needed to support use or management, or even to resume residence, so adverse impacts from fire, weeds and feral animals are hard to manage. Ongoing degradation of savannas by poor fire and grazing management (Whitehead 2000; Franklin *et al.* 2005; Woinarski *et al.* 2007a) requires active correction, whatever the mix of future developments.

On top of these costs, entrenched socioeconomic disadvantage demands urgent attention, so Indigenous landowners may feel obligated to extract incomes from their lands. Traditional landowners are under great pressure to make important decisions about future use; now rather than later. Orthodox uses offer starkly contrasting options. Among the more accessible is inclusion of lands in protected lands systems. Joint management systems, under which lands are formally declared as reserves and often held by the state under long-term leases, place the greatest constraints on future land use, in exchange for long-term commitments to employ community members in park management. Indigenous protected areas (IPAs) place fewer restrictions on use and have proved highly attractive to landowners, even though government financial support is usually modest, and certainly less enduring, relative to declared jointly managed areas. Partnerships with conservation NGOs may also be proposed, and funding from non-government sources is increasingly common.

Another distinct class of options derives from external investors gaining cheap access to large areas of land. Proposals for marginal uses may depend for their viability on attributing low or no value to the land on which they take place, but they may generate some employment or other benefit attractive to communities. For example, pastoral lessees pay a few cents per hectare (Productivity Commission 2002). Commercial valuation of a forestry venture on the Tiwi lands which required clearing 30 000 ha of native forest realised an annual rental of \$3 ha^{-1} yr^{-1} (cited in SECARC 2009). That project collapsed in 2009 and attempts to resurrect it remain in doubt. Economically fragile ventures and their after-effects may restrict future uses. Marginal uses with such low rates of return can be considered economically sustainable only when ecosystem services and natural capital (*sensu* Costanza *et al.* 1997) are ascribed little or no value and are accessed effectively free of charge.

If savanna landscapes and their owners and managers are to benefit from a rural transformation, landowners facing stark and potentially irreversible choices like these require comprehensive, high quality, unbiased, non-ideological advice that weighs up costs and benefits and openly acknowledges risks – including consideration of the less orthodox, emerging opportunities. An expanded palette of options is needed, the most important of which are discussed below.

Valuing ecosystem services

As indicated above, the full costs and benefits of contemporary savanna management activities and emerging options are yet to be evaluated properly. Benefits in water quality and quantity, landscape and soil stability, biodiversity, aesthetics, and climate change mitigation activities (e.g. savanna burning emissions abatement, biomass sequestration) need to be valued formally so that landholders have stronger incentives to provide them and some segments of society can no longer free-ride on the efforts of others to care for environments. Environmental problems require expensive active correction and will not be managed by simply adding irrigated agriculture or continuing with business-as-usual pastoral management activities.

A salutory case is provided by emerging understanding of the importance of, and opportunity for, carbon conservation management in more-or-less structurally intact north Australian savanna systems (Garnaut 2008). The Carbon Farming Initiative (CFI) is the first national-scale example of a market in ecosystem services (carbon credits) that must be reliably quantifiable and meet other rigorous standards (e.g. 'additionality') specified in law. The first methodology approved was for abatement of emissions from savanna burning in relatively high rainfall (>1000 mm p.a.) savannas (Commonwealth of Australia 2013). That methodology is currently being extended to encompass a larger climatic / geographic envelope (Chapter 2), and to include significant carbon sequestration opportunities associated with enhanced savanna fire management (DCCEE 2013). While inconsistent state / territory regulatory frameworks provide a level of uncertainty for undertaking savanna burning projects in some jurisdictions (Hepburn 2009; Dore *et al.* 2014), the current CFI legislation opens the door to allowing all land managers to undertake savanna burning emissions abatement activities as long as they can demonstrate the legal right to undertake such projects on nominated tracts of land.

Enhancing sources of support for providers of environmental services

Federal and state governments have environmental offsets policies backed by laws that require residual detriment from major developments to be compensated. In the north Australian environment where management challenges focus on broad landscape function rather than the uniqueness of particular sites, two important opportunities are created for the design of offsets.

First, rather than seek like-for-like protection of a site similar to the one impacted, offset expenditures can generate greater conservation returns by being directed to the widespread pressures like wildfire that are inherently difficult to manage. Direction of funds in infrastructure-poor, remote regions can provide the relatively modest but critical resources needed to drive coordinated management responses. Work by the North Australian Indigenous Land and Sea Management Alliance (NAILSMA) and its partners has shown that startup expenditures of about $7 million could establish several regional fire projects of scale equivalent to or greater than WALFA to deliver biodiversity benefits over 10 million ha, as well as greenhouse gas emissions abatement and carbon sequestration in the millions of tonnes of CO_2-e. Sustaining such projects over the long term will depend on a reasonable carbon price being achievable under the Emissions Reduction Fund or voluntary markets (NAILSMA 2014).

Second, these sorts of projects can ensure that benefits of economic development reach people who need them most. Well designed offsets can create benefits in sites where other forms of employment are absent, and in this way contribute to major national social goals such as 'Closing the Gap' on Indigenous disadvantage. Governments can encourage Indigenous engagement in offset markets by design of their environmental offsets policies to favour or at least not to discriminate against Indigenous providers in remote regions. As argued by Stoeckl (2010), increasing the quantity of goods and services produced in northern Australia will not overcome regional and Indigenous disadvantage. Structural change or altering the nature of goods and services and the way in which they are produced will also be necessary.

Valuing Indigenous knowledge and skills

Management of fire is demonstrably an area of special interest and relevance to Indigenous north Australians. Given the failure of formal education systems in many parts of north Australia, while the educational gap is being closed it makes sense to develop forms of employment that emphasise highly developed customary skills in land and resource management. Well-designed offset systems provide an important option to create remote labour markets matched to the present skills and interests of unemployed or underemployed residents, contributing to Closing the Gap goals in both employment and enterprise development.

The federal government's Working on Country program funds Indigenous ranger groups who contract to undertake specified kinds of work in nominated locations. The Indigenous Protected Area (IPA) program supports traditional owners of lands of conservation value to deliver on objectives set out in an agreed management plan for a defined site. The IPA arrangements are regarded as sufficiently robust to warrant recognition in Australia's national reserve system. State and territory governments as well as some commercial entities, conservation NGOs and research agencies support Indigenous Ranger groups to undertake specified tasks. These systems implicitly place a value on ecosystem services provided by well-managed Indigenous lands. In the north Australian context such programs typically explicitly value and support development and maintenance of physical and social infrastructure needed for realistic pursuit of projects in landscape-scale fire management.

Encouraging Indigenous enterprise/business development

Experience gained in delivery of environmental services to government and other contractors can be applied by regional and remote organisations to develop businesses in commercial environmental services. It will be important to support the design of offset programs that encourage local investments in business systems and skills to establish continuity of supply, and that will reassure purchasers that the quality of environmental and social offsets will reflect well on developers. This is a key step in transitioning local community land management structures from 'green welfare'-based dependency to building autonomous economic independence for regional community-based enterprises (Altman and Kerins 2012).

Many resource extraction companies, for example, recognise that the effects of their presence on local people may be too often negative because while the local people struggle to take up direct employment in increasingly specialised resource extraction methods or related services, they bear environmental and social costs. An important goal for such companies is to support local people to create enduring businesses – even if they operate entirely outside the mining, petroleum and gas industries – and so leave a positive legacy (MCA 2011).

Pastoral diversification

Use of large areas of northern landscapes for marginal pastoralism that causes long-term environmental damage for little return is clearly unsustainable. In more productive areas, practice should be encouraged to make better use of favourable sites and withdraw stock permanently from unfavourable parts of larger properties. Often these coincide with rugged terrain with high biodiversity values. In both of these situations it will be necessary to replace the withdrawn management effort to prevent ungrazed areas becoming sites of regular wildfire. In marginal situations especially, economic diversification of allowable activities on pastoral leases is a critical issue facing the northern pastoral industry (e.g. WA Government 2009; Walsh *et al.* 2014; Walton *et al.* 2014). At the time of writing, several pastoral leases in the Northern Territory have developed proposals for managing components of their estate for emissions abatement through improved fire management. One property, Fish River Station, owned by the Indigenous Land Corporation, recently has announced the sale of ~26 000 carbon credits to a commercial buyer as part of a CFI-approved project (see www.abc.net.au/news/2013-06-06/caltex-carbon-credits/4736858). Further feasible options include exploring biodiversity credits and stewardship arrangements (Douglass *et al.* 2011). Access to incomes from such ecosystem services may support landholders to improve the environmental outcomes from other forms of use. The Northern Territory has recently changed pastoral laws to simplify approvals for non-pastoral uses. Unfortunately promotional material makes no reference to environmental services or related opportunities (Northern Territory Government 2014), even though these would be least likely to conflict with coexisting native title rights.

Fire management in a transformed savanna economy

As we have seen, Australian regional perspectives on and approaches to the management of fire are strongly influenced by interplay among the frequency of fire, its intensity, the assets that are put at risk, and the resources that are available to control fire regimes. In densely settled southern Australia, the dominant objective is to protect human life and property from the infrequent but very intense wildfires that can invade urban or peri-urban landscapes or smaller rural settlements. In central Australia, occasional huge fires sweep through the region, putting at risk built assets, pastures and the biota. In the last decade, mainland extinction of an endangered macropod, the mala or rufous hare-wallaby (*Lagorchestes hirsutus*) was attributed to such a fire (Woinarski *et al.* 2007b).

In the tropical savannas of northern Australia, encompassing about a quarter of the nation's land area (or more than 1.9 million km^2), fire is frequent, occurs over more than 6 months of each year, affects large areas, and varies greatly in intensity. Instead of treating occurrence of fire as aberrant or as evidence of breakdown of flawed management systems, responses to fire risks need to focus on the reasons for fire and the ways in which active application of fire can produce and, indeed, may be essential for delivering conservation and social benefits of local, regional, and national significance.

Unfortunately, regional diversity is not always reflected in fire policy and law imposed from mostly southern Australian capital cities, where approaches are dictated by an ongoing antagonism between threats to life and property on the one hand, and demands to forgo regular fuel reduction burning in the interests of aesthetics, avoiding health effects on local human populations, and biodiversity conservation on the other. Because fires are viewed as aberrations, too little attention is paid to opportunities afforded through emerging economic paradigms, particularly the potential to deploy skilful fire management to influence carbon dynamics at the national scale and address global warming.

In this chapter we have articulated a distinctly northern Australian understanding of and approach to the management of fire, matched to the biophysical and social realities of the north. We contend that, despite the many obstacles to effective management associated with remoteness, sparse population and limited transport and other infrastructure, management of fire can be improved for positive ecological and social outcomes over a large part of the Australian land mass. Skilled fire management offers a culturally apt enterprise opportunity where few others exist.

Acknowledgements

Cameron Yates provided updated versions of maps. Dionne Walsh and Neil MacDonald (both Northern Territory Department of Primary Industries and Fisheries) alerted us to recent publications addressing north Australian pastoral land management issues. Joel Dillon (Indigenous Land Corporation) provided updated spatial data for Indigenous lands.

References

Allan GE, Southgate IR (2002) Fire regimes in the spinifex landscapes of Australia. In *Flammable Australia: The Fire Regimes and Biodiversity of a Continent*. (Eds RA Bradstock, JE Williams and AM Gill) pp. 145–176. Cambridge University Press: Cambridge, UK.

Allen Consulting Group (2011) 'Assessment of the economic and employment outcomes of the Working on Country program, October 2011.' Report to Australian Government Department of Sustainability, Environment, Water, Population and Communities: Canberra.

Altman JC (1987) *Hunter-Gatherers Today*. Australian Institute of Aboriginal & Torres Strait Islander Studies: Canberra.

Altman JC (2003) People on country, healthy landscapes and sustainable Indigenous economic futures: The Arnhem Land case. *The Drawing Board: An Australian Review of Public Affairs* **4**(2), 65–82.

Altman J, Kerins S (2012) *People on Country: Vital Landscapes, Indigenous Futures*. The Federation Press: Sydney.

Altman JC, Buchanan GJ, Larsen L (2007) 'The environmental significance of the Indigenous estate: Natural resource management as economic development in remote Australia.' CAEPR Discussion Paper No. 286/2007. Centre for Aboriginal Economic Policy Research, Australian National University: Canberra.

ANGA (Australian National Greenhouse Accounts) (2011) 'National Inventory Report 2009, Vol. 1.' Australian Government Department of Climate Change & Energy Efficiency: Canberra.

Australian Government (2012) 'Australia in the Asian Century: White Paper.' Australian Government Department of Prime Minister and Cabinet: Canberra.

Bayon R, Hawn A, Hamilton K (2006) *Voluntary Carbon Markets: An International Business Guide, What They Are and How They Work*. Earthscan: London.

Biddle N (2013) 'CAEPR Indigenous population project – 2011 census papers. Paper 5: population and age structure.' Centre for Aboriginal Economic Policy Research, Australian National University: Canberra.

Biddle N, Markham F (2013) 'CAEPR Indigenous population project – 2011 census papers. Paper 9: mobility.' Centre for Aboriginal Economic Policy Research, Australian National University: Canberra.

Brereton D, Klimenko V, Cote C, Evans R (2009) The minerals industry and land and water development in northern Australia. In 'Northern Australia Land and Water Science Review: Full Report.' Chapter 8: 8.1–8.50. CSIRO Sustainable Agriculture Flagship and Northern Australia Land and Water Taskforce: Canberra.

Brook RK, McLachlan SM (2005) On using expert-based science to 'test' local ecological knowledge. *Ecology and Society* **10**(2): r3. www.ecologyandsociety.org/vol10/iss2/resp3

Burgess C, Johnston F, Bowman D, Whitehead PJ, O'Dea K (2005) Healthy Land: Healthy People? Exploring the health benefits of Indigenous Land Management. *Australian and New Zealand Journal of Public Health* **29**, 117–122. doi:10.1111/j.1467-842X.2005.tb00060.x

Clark E, Abel N, Measham T, Morison J, Rippin L (2009) Water-based tourism and recreation in northern Australia. In 'Northern Australia Land and Water Science Review.' pp. 11.1–11.54. Australian Government Department of Infrastructure, Transport, Regional Development and Local Government: Canberra.

Commonwealth of Australia (2013) Carbon Credits (Carbon Farming Initiative) Reduction of Greenhouse Gas Emissions through Early Dry Season Savanna Burning—1.1) Methodology Determination 2013. www.comlaw.gov.au/Series/F2013L01165

Collins R (1999) 'Learning Lessons: An independent review of Indigenous Education in the Northern Territory.' Northern Territory Government Department of Education: Darwin.

Cook G (2009) Historical perspectives on land use development in northern Australia: with emphasis on the Northern Territory. In 'Northern Australia Land and Water Science Review: Full Report.' Chapter 6: 6.1–6.30. CSIRO Sustainable Agriculture Flagship and Northern Australia Land and Water Taskforce: Canberra.

Cook GD, Meyer CP (2009) Fire, fuels and greenhouse gases. In *Culture, Ecology and Economy of Fire Management in North Australian Savannas: Rekindling the* Wurrk *Tradition*. (Eds J Russell-Smith, PJ Whitehead, PM Cook) pp. 313–327. CSIRO Publishing: Melbourne.

Cook GD, Liedloff AC, Eager RW, Chen X, Williams RJ, O'Grady AP, Hutley LB (2005) The estimation of carbon budgets of frequently burnt tree stands in savannas of northern Australia using allometric analysis and isotopic discrimination. *Australian Journal of Botany* **53**, 621–630. doi:10.1071/BT04150

Cook GD, Jackson S, Williams RJ (2012) A revolution in northern Australia fire management: recognition of Indigenous knowledge, practice and management. In *Flammable Australia: Fire Regimes, Biodiversity and Ecosystems in a Changing World*. (Eds RA Bradstock, AM Gill, RJ Williams) pp. 293–306. CSIRO Publishing: Melbourne.

Costanza R, d'Arge R, de Groot R, Farber S, Grasso M, Hannon B, Limburg K, Naeem S, O'Neill RV, Paruelo J, Raskin RG, Sutton P, van den Belt M (1997) The value of the world's ecosystem services and natural capital. *Nature* **387**, 253–260. doi:10.1038/387253a0

Cribb J, Harper G, Stone P (2009) Sustaining growth of the northern beef industry. In 'Northern Australia Land and Water Science Review: Full Report.' Chapter 5: pp. 5.1–5.30. CSIRO Sustainable Agriculture Flagship and Northern Australia Land and Water Taskforce: Canberra.

CSIRO, Bureau of Meteorology (2007) 'Climate change in Australia: technical report 2007'. CSIRO, Australia. www.climatechangeinaustralia.gov.au

DCCEE (Australian Government Department of Climate Change & Energy Efficiency) (2013) Indigenous Carbon Farming Fund. Australian Government Department of Climate

Change & Energy Efficiency, Canberra. www.climatechange.gov.au/reducing-carbon/grants/indigenous-carbon-farming-fund/round-one-2012-13

DFAT (Australian Government Department of Foreign Affairs and Trade) (2012) 'Feeding the future: a joint Australia–China study on strengthening agricultural investment and technological cooperation to improve food security.' Australia–China Joint Working Group, Department of Foreign Affairs and Trade (Australia) and Ministry of Commerce (China): Canberra.

Dore J, Michael C, Russell-Smith J, Tehan M, Caripes M (2014) Carbon projects and Indigenous land tenure in northern Australia. *The Rangelands Journal* **36**, 389–402.

Douglass LL, Possingham HP, Carwardine J, Klein CJ, Roxburgh SH, Russell-Smith J, Wilson KA (2011) The effect of carbon credits on savanna land management and priorities for biodiversity conservation. *PLoS ONE* **6**(9), e23843. doi:10.1371/journal.pone.0023843

Dyer R, Jacklyn P, Partridge I, Russell-Smith J, Williams RJ (2001) 'Savanna burning: understanding and using fire in northern Australia.' Tropical Savannas Cooperative Research Centre: Darwin.

Ellis S, Kanowski P, Whelan R (2004) 'National Inquiry on Bushfire Mitigation and Management'. Commonwealth of Australia: Canberra. www.coag.gov.au

Franklin DC, Whitehead PJ, Pardon G, Matthews J, McMahon P, McIntyre D (2005) Geographic patterns and correlates of the decline of granivorous birds in northern Australia. *Wildlife Research* **32**(5), 399–408. doi:10.1071/WR05052

Garde M, Nadjamerrek B, Kolkiwarra M, Kalarriya J, Djandjomerr J, Birriyabbirriya B, Bilindja R, Kubarkku M, Biless P (2009) The language of fire: seasonality, resources and landscape burning on the Arnhem Land Plateau. In *Culture, Ecology and Economy of Fire Management in North Australian Savannas: Rekindling the* Wurrk *Tradition*. (Eds J Russell-Smith, PJ Whitehead, PM Cook) pp. 86–164. CSIRO Publishing: Melbourne.

Garnaut R (2008) 'The Garnaut climate change review: final report'. Cambridge University Press: Melbourne.

Gerritsen R (2010) Post-colonial model for north Australian political economy. In *North Australian Political Economy: Issues and Agendas*. (Ed. R Gerritsen) pp. 18–40. Charles Darwin University Press: Darwin.

Gilchrist G, Mallory M, Merkel F (2005) Can local ecological knowledge contribute to wildlife management? Case studies of migratory birds. *Ecology and Society* **10**(1), 20. www.ecologyandsociety.org/vol10/iss1/art20

Gleeson T, Martin P, Mifsud C (2012) 'North Australian beef industry: assessment of risks and opportunities. May 2012.' Australian Bureau of Agricultural and Resource Economics and Sciences: Canberra.

Greiner R, Mayocchi C, Larson S, Stoeckl N, Schweigert R (2004) 'Benefits and costs of tourism for remote communities: Case study: Carpentaria Shire, north-west Queensland.' Tropical Savannas Cooperative Research Centre: Darwin.

Griffith D (2010) How policy development methods – together with inadequate statistics and flawed classifications – have done and still do distort policy implementation in northern Australia. In *North Australian Political Economy: Issues and Agendas*. (Ed. R Gerritsen) pp. 41–54. Charles Darwin University Press: Darwin.

Harries J (1997) 'Acid mine drainage in Australia: its extent and potential future liability.' Supervising Scientist Report 125. Office of the Supervising Scientist: Canberra.

Heckbert S, Russell-Smith J, Reeson A, James G (2011) Traditional solutions: savanna burning, climate change, and Indigenous disadvantage. *Solutions* **2**, 50–56.

Heckbert S, Russell-Smith J, Reeson A, Davies J, James G, Meyer C (2012) Spatially explicit benefit–cost analysis of fire management for greenhouse gas abatement. *Austral Ecology* **37**, 724–732. doi:10.1111/j.1442-9993.2012.02408.x

Hepburn S (2009) Carbon rights as new property: the benefits of statutory verification. *The Sydney Law Review* **31**, 239–271.

Holmes J (2002) Diversity and change in Australia's rangelands: a post-productivist transition with a difference. *Transactions of the Institute of British Geographers* **27**, 362–384. doi:10.1111/1475-5661.00059

Holmes J (2010) Divergent Regional Trajectories in Australia's Tropical Savannas: Indicators of a Multifunctional Rural Transition. *Geographical Research* **48**, 342–358. doi:10.1111/j.1745-5871.2009.00630.x

Indian Ocean Climate Initiative (2012) 'Western Australia's Weather and Climate: A Synthesis of Indian Ocean Climate Initiative Stage 3 Research.' CSIRO and Bureau of Meteorology: Australia.

IPCC (1997) *Revised 1996 Intergovernmental Panel on Climate Change (IPCC) Guidelines for National Greenhouse Gas Inventories.* 3 vols. IPCC/OECD/IEA: Paris, France.

Jonauskas P (Ed.) (1996) Making multiple use work. In 'Proceedings of the Wetland Workshop, 6–7 December 1994.' Northern Territory Government Department of Lands, Planning and Environment: Darwin.

Jordan K (2012) Closing the employment gap through work for the dole? Indigenous employment and the CDEP scheme. *Journal of Australian Political Economy* **69**, 29–58.

Luhar AK, Mitchell RM, Meyer CP, Qin Y, Campbell S, Gras JL, Parry D (2008) Biomass burning emissions over northern Australia constrained by aerosol measurements: II – Model validation, and impacts on air quality and radiative forcing. *Atmospheric Environment* **42**, 1647–1664. doi:10.1016/j.atmosenv.2007.12.040

MCA (Minerals Council of Australia) (2011) 'Minerals industry: Indigenous economic development strategy.' Minerals Council of Australia: Canberra. www.minerals.org.au/news/minerals_industry_indigenous_economic_development_strategy

McCosker T, McLean D, Holmes P (2010) 'Northern beef situation analysis 2009.' Meat and Livestock Australia: Sydney.

Meyer CP, Luhar AK, Mitchell RM (2008) Biomass burning emissions over northern Australia constrained by aerosol measurements: I – Modelling the distribution of hourly emissions. *Atmospheric Environment* **42**, 1629–1646. doi:10.1016/j.atmosenv.2007.10.089

Meyer CP, Cook GD, Reisen F, Smith TEL, Tattaris M, Russell-Smith J, Maier SW, Yates CP, Wooster MJ (2012) Direct measurements of the seasonality of emission factors from savanna fires in northern Australia. *Journal of Geophysical Research, D, Atmospheres* **117**, D20305. doi:10.1029/2012JD017671

Morris A (2003) Powerhouse or Mendicant? Is the Territory an Engine of Growth or a 'drag' on the Federation? Paper presented to the Symposium Beyond the Frontier: Sustainable Futures for Northern Australia, 17–18 July 2003, Charles Darwin University, Darwin.

Murphy BP, Russell-Smith J, Watt F, Cook GD (2009) Fire management and woody biomass carbon stocks in mesic savannas. In *Culture, Ecology and Economy of Fire Management in North Australian Savannas: Rekindling the* Wurrk *Tradition*. (Eds J Russell-Smith, PJ Whitehead, PM Cook) pp. 361–394. CSIRO Publishing: Melbourne.

Murphy BP, Russell-Smith J, Prior LD (2010) Frequent fires reduce tree growth in north Australian savannas: implications for tree demography and carbon sequestration. *Global Change Biology* **16**, 331–343. doi:10.1111/j.1365-2486.2009.01933.x

Murphy BP, Bradstock RA, Boer MM, Carter J, Cary GJ, Cochrane MA, Fensham RJ, Russell-Smith J, Williamson GJ, Bowman DMJS (2013) Fire regimes of Australia: a pyrogeographic model system. *Journal of Biogeography* **40**, 1048–1058. doi:10.1111/jbi.12065

NAILSMA (2009a) 'A policy statement on north Australian Indigenous water rights.' Indigenous Water Policy Group, North Australian Indigenous Land and Sea Management Alliance: Darwin.

NAILSMA (2009b) 'Standing together for water rights: North Australian Indigenous Experts Water Futures Forum Report.' Workshop Proceedings, Mary River, 5–6 August 2009. North Australia Indigenous Land and Sea Management Alliance: Darwin.

NAILSMA (2014) 'Emissions Reduction Fund: Submission on the Green Paper.' Submission to the Department of Environment. North Australian Indigenous Land and Sea Management Alliance: Darwin.

NALWTF (Northern Australia Land and Water Task Force) (2009) 'Sustainable development of northern Australia.' A report to government from the Northern Australia Land and Water Task Force. December 2009. Australian Government Department of Infrastructure, Transport, Regional Development and Local Government: Canberra.

Northern Territory Government (2014) Pastoral lease diversification: Where are the opportunities? Northern Territory Government, Darwin. http://lrm.nt.gov.au/__data/assets/pdf_file/0003/351624/FS-3-Where-are-the-opportunities.pdf Accessed 14 May 2014.

Productivity Commission (2002) 'Pastoral Leases and Non-Pastoral Land Use.' Productivity Commission: Canberra.

Russell-Smith J, Bowman DMJS (1992) Conservation of monsoon rainforest isolates in the Northern Territory, Australia. *Biological Conservation* **59**, 51–63. doi:10.1016/0006-3207(92)90713-W

Russell-Smith J, Yates CP, Edwards A, Allan GE, Cook GD, Cooke P, Craig R, Heath B, Smith R (2003) Contemporary fire regimes of northern Australia: change since Aboriginal occupancy, challenges for sustainable management. *International Journal of Wildland Fire* **12**, 283–297. doi:10.1071/WF03015

Russell-Smith J, Yates CP, Whitehead PJ, Smith R, Craig R, Allan GE, Thackway R, Frakes I, Cridland S, Meyer CP, Gill AM (2007) Bushfires 'down under': patterns and implications of contemporary Australian landscape burning. *International Journal of Wildland Fire* **16**, 361–377. doi:10.1071/WF07018

Russell-Smith J, Murphy BP, Meyer CP, Cook GD, Maier S, Edwards AC, Schatz J, Brocklehurst P (2009aImproving estimates of savanna burning emissions for greenhouse accounting in northern Australia: limitations, challenges, applications. *International Journal of Wildland Fire* **18**, 1–18. doi:10.1071/WF08009

Russell-Smith J, Whitehead PJ, Cooke PM (Eds) (2009b) *Culture, Ecology and Economy of Fire Management in North Australian Savannas: Rekindling the* Wurrk *Tradition*. CSIRO Publishing: Melbourne.

Russell-Smith J, Yates CP, Brock C, Westcott V (2010) Fire regimes and interval-sensitive vegetation in semi-arid Gregory National Park, Northern Territory. *Australian Journal of Botany* **58**, 300–317.

Russell-Smith J, Edwards AC, Price OF (2012) Simplifying the savanna: the trajectory of fire-sensitive vegetation mosaics in northern Australia. *Journal of Biogeography* **39**, 1303–1317.

Russell-Smith J, Cook GD, Cooke PM, Edwards AC, Lendrum M, Meyer CP, Whitehead PJ (2013) Managing fire regimes in north Australian savannas: applying customary Aboriginal approaches to contemporary global problems. *Frontiers in Ecology and the Environment* **11**, e55–e63. doi:10.1890/120251

SCRGSP (Steering Committee for the Review of Government Service Provision) (2011) 'Overcoming Indigenous Disadvantage: Key Indicators 2011.' Productivity Commission: Canberra.

SECARC (2009) Forestry and mining operations on the Tiwi Islands. The Senate Environment, Communications, and the Arts References Committee: Canberra.

Stoeckl N (2010) Bridging the economic divide: background to, and strategies for bridging the divide between Indigenous and non-Indigenous economies in northern Australia. In *North Australian Political Economy: Issues and Agendas*. (Ed. R Gerritsen) pp. 126–129. Charles Darwin University Press: Darwin.

Stoeckl N, Stanley O, Brown V, Stoeckl W (2007) 'Regional economic multipliers in Australia's tropical savannas.' Report to the Tropical Savannas CRC. James Cook University: Townsville. http://savanna.cdu.edu.au/savanna_web/publications/downloads/Regional-Economic-Multipliers-in-the-Tropical-Savannas.pdf Accessed November 2008.

Taylor J (2003) 'Indigenous economic futures in the Northern Territory: the demographic and socioeconomic background.' CAEPR Discussion Paper 246/2003. Centre for Aboriginal Economic Policy Research, Australian National University: Canberra.

Taylor J (2006) 'Population and diversity: policy implications of emerging Indigenous demographic trends.' CAEPR Discussion Paper No. 283/2006. Centre for Aboriginal Economic Policy Research, Australian National University: Canberra.

Taylor J, Brown D, Bell M (2006) 'Scoping population dynamics and demographic accounting in arid and savanna Australia: methods, issues and outcomes.' Report of a scoping study for the Desert Knowledge and Tropical Savannas Cooperative Research Centres. Desert Knowledge Cooperative Research Centre: Alice Springs.

Teague B, McLeod R, Pascoe S (2010) '2009 Victorian Bushfires Royal Commission Final Report.' Parliament of Victoria: Melbourne.

TRA (Tourism Research Australia) (2012) 'State Tourism Satellite Accounts 2010–11.' Tourism Research Australia: Canberra.

Urbis Pty Ltd (2012) 'Assessment of the social outcomes of the Working on Country program. May 2012.' Report to Australian Government Department of Sustainability, Environment, Water, Population and Communities, Canberra. Urbis Pty Ltd: Canberra.

Walsh D, Russell-Smith J, Cowley R (2014) Fire and carbon management in a diversified north Australian rangelands economy: research, policy and implementation challenges. *The Rangeland Journal* **36**, 313–322.

Walton N, Smith H, Bowen L, Mitchell P, Pethybridge E, Hayes T, O'Ryan M (2014) Opportunities for fire and carbon on pastoral properties in the savanna rangelands: Perspectives from the Indigenous Land Corporation and the Northern Territory Cattlemen's Association. *The Rangeland Journal* **36**, 403–409.

Welters R (2010) Atypical labour markets require atypical policy solutions. In *North Australian Political Economy: Issues and Agendas*. (Ed. R Gerritsen) pp. 55–69. Charles Darwin University Press: Darwin.

Western Australian Government (2009) 'A review of the process to permit diversification on pastoral leasehold land in Western Australia.' Report to the Minister for Food and Agriculture, Western Australia Government. www.agric.wa.gov.au/objtwr/imported_assets/content/past/pastoral_diversification_review_report.pdf

Whitehead PJ (2000) The clever country: where cows manage wildlife. In *Business as Usual? Local Conflicts and Global Challenges*. pp. 155–168. Australian National University: Darwin.

Whitehead PJ, Woinarski JCZ, Franklin D, Price O (2002) Landscape ecology, wildlife management and conservation in northern Australia: matching policy, practice and capability in regional planning. In *Landscape Ecology and Resource Management: Linking Theory and Practice.* (Eds J Bissonette, I Storch) pp. 227–259. Island Press: New York.

Williams RJ, Griffiths AD, Allan GE (2002) Fire regimes and biodiversity in the wet-dry tropical landscapes of northern Australia. In *Flammable Australia: The Fire Regimes and Biodiversity of a Continent.* (Eds RA Bradstock, JE Williams and AM Gill) pp. 281–304. Cambridge University Press: Cambridge, UK.

Woinarski JCZ, Dawson F (2002) Limitless lands and limited knowledge: coping with uncertainty and ignorance in northern Australia. In *Ecology, Uncertainty and Policy: Managing Ecosystems for Sustainability.* (Eds JW Handmer, TW Norton, SR Dovers) pp. 83–115. Prentice-Hall: New York.

Woinarski JCZ, Milne DJ, Wanganeen G (2001) Changes in mammal populations in relatively intact landscapes of Kakadu National Park, Northern Territory, Australia. *Austral Ecology* **26**, 360–370. doi:10.1046/j.1442-9993.2001.01121.x

Woinarski J, MacKey B, Nix H, Traill B (2007a) 'The nature of northern Australia: natural values, ecological processes and future prospects.' Australian National University: Canberra.

Woinarski JCZ, Pavey C, Kerrigan R, Cowie I, Ward S (Eds) (2007b) 'Lost from our landscape: threatened species of the Northern Territory.' Northern Territory Government Department of Natural Resources, Environment and the Arts: Darwin.

Woinarski JCZ, Armstrong M, Brennan K, Fisher A, Griffiths AD, Hill B, Milne DJ, Palmer C, Ward S, Watson M, Winderlich S, Young S (2010) Monitoring indicates rapid and severe decline of native small mammals in Kakadu National Park, northern Australia. *Wildlife Research* **37**, 116–126. doi:10.1071/WR09125

Woinarski JCZ, Legge S, Fitzsimons JA, Traill BJ, Burbidge AA, Fisher A, Firth RSC, Gordon IJ, Griffith AD, Johnson CN, McKenzie NL, Palmer C, Radford I, Rankmore B, Ritchie EG, Ward S, Ziembecki M (2011) The disappearing mammal fauna of Australia: context, cause, and response. *Conservation Letters* **4**, 192–201. doi:10.1111/j.1755-263X.2011.00164.x

2

Fire patterns in north Australian savannas: extending the reach of incentives for savanna fire emissions abatement*

Peter J Whitehead, Jeremy Russell-Smith and Cameron Yates

Summary

Anthropogenic fires in Australia's fire-prone savannas produce up to 3% of the nation's accountable greenhouse gas emissions. Incentives to improve fire management have been created by a nationally accredited savanna burning emissions abatement methodology applying to 483 000 km^2 of relatively high rainfall (>1000 mm p.a.) regions. Drawing on 15 years of fire mapping, this chapter assesses appropriate biophysical boundaries for a savanna burning methodology extended to cover lower rainfall regions. We examine a large random sample of points with at least 300 mm of annual rainfall, to show that: (1) relative fire frequencies (percentage of years with fire) decline from 33.3% in higher rainfall regions (>1000 mm) to straddling ~10% in the range 300–700 mm; (2) there are no marked discontinuities in fire frequency or fire seasonality down the rainfall gradient; (3) at all annual rainfalls, fire frequency is higher when rainfall is more strongly seasonal (very low rainfall in the driest quarter); (4) below 500 mm fire regimes are particularly variable and a large proportion of sampled sites had no fire over the study period; (5) fire is more likely to occur later in the fire season (generating relatively higher emissions) in the 600–700 mm annual rainfall band than in other parts of the rainfall gradient; (6) woodland savannas are most common above and predominantly grassland systems are more common below ~600 mm annual rainfall.

We propose that development of a complementary lower rainfall savanna burning methodology should apply to regions between 600 and 1000 mm annual rainfall and ≤15 mm of rainfall in the driest quarter, adding an area more than 1.5 times the existing methodology's coverage. Given greater variability in biophysical influences on fire regimes and observed levels of fire frequency within this lower rainfall domain, we suggest that criteria for determining baseline (pre-project) periods require estimates of mean annual emissions equivalent in precision to the project on which the higher rainfall methodology was based.

* This chapter is an edited version of an article that was first published in 2014 in *The Rangeland Journal*, Vol. 36(4), pp. 371–388. doi:10.1071/RJ13129

Introduction

One of many challenges confronting sustainable development of Australia's northern savannas is management of landscape-scale fire. On average, 19% of the 1.9 million km^2 tropical savannas region was burned (or, more properly, fire-affected) annually over the period 1997–2005 (Russell-Smith *et al.* 2007). Annual frequencies of burning are higher in high rainfall (>~800 mm) subcoastal regions (Fisher *et al.* 2003; Russell-Smith *et al.* 2003a, 2007; Felderhof and Gillieson 2006; Edwards and Russell-Smith 2009).

Relatively little burning occurs in some locations, especially on more fertile soils in lower rainfall sites under intensive pastoral (beef cattle) production (Craig 1997, 1999; Grice and Slatter 1997; Dyer *et al.* 2001). For the most part, however, contemporary regional fire regimes are characterised by extensive wildfires occurring in the latter part of the ~7 month dry season under relatively severe fire-weather conditions (Williams *et al.* 2002; Yates *et al.* 2008; Russell-Smith *et al.* 2009d). In the absence of adequate management response, these patterns are likely to be exacerbated by climate change, especially by an increase in the number of days of extreme temperature (CSIRO and Bureau of Meteorology 2007).

Implications of contemporary fire regimes for north Australian savanna biodiversity (Williams *et al.* 2002, 2003; Rossiter *et al.* 2003; Russell-Smith *et al.* 2003b; Woinarski *et al.* 2005, 2007), soil conservation (Townsend and Douglas 2000; Russell-Smith *et al.* 2006; Richards *et al.* 2012), greenhouse gas emissions (Meyer 2004; Russell-Smith *et al.* 2009b; DCCEE 2010), carbon bio-sequestration (Cook *et al.* 2005; Williams *et al.* 2009; Murphy *et al.* 2010; Werner 2012), pastoral production (Dyer *et al.* 2001), and broader social implications (Whitehead *et al.* 2003, 2009; Ellis *et al.* 2004; Preece 2007), are increasingly well documented. Key issues include developing economically sustainable landscape-scale fire management, particularly in remote, biodiversity-rich northern regions with low pastoral productivity. Here there are few employment opportunities, especially for Indigenous (Aboriginal) communities (Whitehead *et al.* 2003, 2009; Russell-Smith *et al.* 2009c).

Under the Kyoto Protocol, participating Tier 1 (developed economy) countries like Australia are required, where pertinent, to account for greenhouse gas emissions (specifically methane, nitrous oxide) from 'prescribed burning of savannas' (UNFCCC 1998: Article 3, Annex A). Australia's National Greenhouse Gas Inventory (NGGI) reports on savanna burning emissions that, typically, contribute 1–3% of Australia's NGGI annually (Meyer 2004; DCCEE 2010). The Kyoto Protocol also establishes a framework for developing market-based instruments to address anthropogenic sources and sinks of greenhouse gas emissions (UNFCCC 1998: Article 6). Accountable greenhouse gas emissions from Australian savanna burning are predominantly associated with anthropogenic sources (Russell-Smith *et al.* 2007).

In 2011, Australia enacted a suite of laws creating a domestic carbon market and mechanisms to account for reductions of emissions or carbon storage from improved land management. Credits from savanna fire abatement (reducing emissions) under the *Carbon Credits* (*Carbon Farming Initiative*) *Act* 2011 have already been sold to industry liable under the principal legislation, the *Clean Energy Act* 2011 (Russell-Smith *et al.* 2013). New methodologies presently under development could also allow credits for carbon bio-sequestration in living woody biomass of forests and woodlands, which can be substantially increased by reducing frequency of more intense late-season fires (Murphy *et al.* 2009, 2010). Taken together, the immediate returns from abatement and longer-term benefits from bio-sequestration could offer very significant incentives for addressing core issues in savanna fire management.

At the time of writing (May 2014) the federal government had confirmed its intention to revoke laws for a carbon market and substitute 'direct action': government-run 'reverse auctions' in which offset providers are invited to compete to offer offsets at low prices. Existing carbon farming laws will be retained, allowing the land sector to be a source of offsets paid for from a new, publicly funded Emissions Reduction Fund (Australian Government 2014).

Existing projects

The essential premise of the approved savanna fire methodology (DCCEE 2013) is that, through prescribed early dry season burning, sufficient 'firebreaks' can be developed to arrest extensive wildfires later in the fire season (Russell-Smith *et al.* 2009c, 2013). Emissions are reduced because fewer of the higher intensity fires burn less woody fuels that emit more methane and nitrous oxide than fine (grassy) fuels. Reductions in the total area burned and hence lower total amounts of fine and woody fuels burned also contribute.

The first major savanna burning 'experiment' applying approaches reflected in the now approved methodology began in the late 1990s in 28 000 km^2 of the Western Arnhem Land Fire Abatement (WALFA) project area, operating essentially as a 'voluntary' arrangement (*sensu* Bayon *et al.* 2006) between a multinational energy corporate and Indigenous landowners. Operationally, WALFA is an outstanding success (Russell-Smith *et al.* 2013). Mean annual emissions have been reduced by 37.7% relative to a 10-year baseline (1995–2004) by imposing a distinctly different fire regime. During the baseline period, an average of 7.6% of the area was burned early and 32% late. In the project period (2005–10), an average of 20.9% was burned early, and 10.9% by late-season fires. The total area burned was reduced by 7.9%. WALFA has been emulated on a smaller scale by Indigenous landowners at Fish River, a few hundred kilometres to the south-west. A large non-Indigenous pastoral holding (Dunkheld Pastoral Co.) has also produced savanna burning credits in the Daly River region.

Social benefits from the Indigenous projects have not been formally measured, but at least 175 different individuals have gained some employment, training and incomes (WAFMA partners, unpublished data). Participants are more active on their country, which is thought to foster better physical and mental health (Burgess *et al.* 2005). Incomes provide for a larger paid land management workforce than would otherwise be sustained locally, which contributes to effort on associated Indigenous Protected Areas. At Fish River, carbon incomes are to be reinvested in other enterprises, like tourism. Given the scale and frequency of wildfire in other parts of northern Australia (Russell-Smith *et al.* 2007), there would appear to be scope to capture these sorts of benefits in areas not covered by the existing methodology.

Purpose of study

Here we examine the biophysical potential to implement savanna burning projects in fire-prone regions of northern Australia not covered by the present (>1000 mm) methodology. We critically examine fire regime, climatic and cadastral characteristics which influence opportunities for fire management projects. This assessment extends the more generalised market-based assessment in Heckbert *et al.* (2012). Specifically, we consider the (1) utility of a new methodology for lower rainfall regions (≤1000 mm) to provide incentives for improved fire management, (2) biophysical boundaries for a new methodology, and (3) how a lower rainfall methodology might differ from the existing one.

Methods

The study area is that part of north Australia above the 26th parallel and annual rainfall exceeding 300 mm, excluding the Wet Tropics region on the eastern coast (Fig. 2.1).

A random selection of 10 000 points was intersected with fire maps and other readily available GIS coverages described below. After removing points that intersected landscape features (inland wetlands, bare ground or rock) or vegetation types that are unlikely to be included in a lower rainfall savanna burning methodology (e.g. chenopod and salt marsh, rainforest and vine thicket, mangroves, cleared and non-native vegetation, regrowth) the sample comprised 9132 points, 85.4% of which fell within the lower rainfall (≤1000 mm) part of the study area. The same sample of points was examined for the 15 years of study (i.e. resampling was not done annually). While the points are dimensionless, we treat the attributes at their intersection as representative of areas (sites) within the sampled region, and sometimes use the term sites when referring to the sampled points.

Fire variables

Areas burned were mapped at approximately monthly intervals using AVHRR imagery for each calendar year from 1997 to 2011. Russell-Smith *et al.* (2007) describe methods in detail.

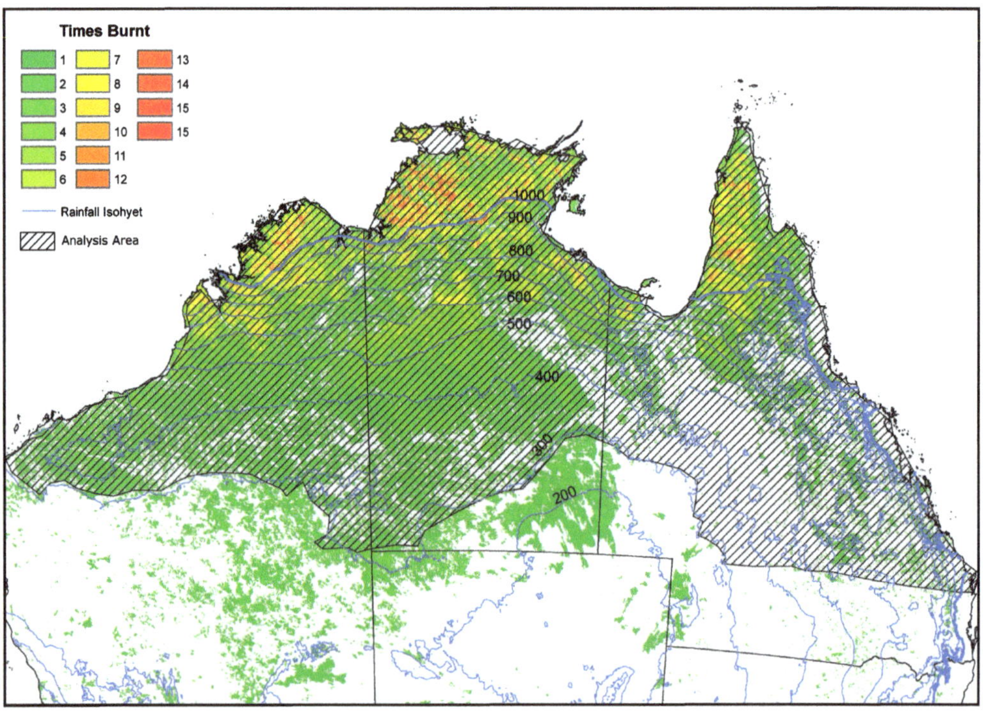

Figure 2.1. Map of the area of northern Australia from which points were randomly sampled for analysis of relationships among fire incidence and biophysical variables, showing fire frequency in 1 × 1 km cells over the period 1997–2011. Annual rainfall isohyets are shown at 100 mm intervals.

Fire frequency and seasonality

Points were intersected with fire incidence maps for each year of the 15-year sequence 1997–2011. Where a point fell within a mapped fire-affected area in any of the monthly maps for a year, it was attributed as burned in that year. For most analyses, these records were aggregated for each point as the *fire frequency* from 0 (no fire recorded in the 15-year period) to 15 (fire recorded in every year). When frequencies were averaged over multiple points they were expressed as percentages of years burned.

The first month in which fire was first recorded (value 1 for January to 12 for December and 0 if no fire) was also recorded for each point for each year. Two variables were used to summarise seasonality: first by the mean of the Julian calendar number of the month of burning (FM) for in each year in which fire occurred (i.e. not including a zero score for years in which fire was not recorded), and second, following existing practice, fires up until and including July were defined as early dry season (EDS) burns and from August as late dry season (LDS) burns.

Antecedent fire

Fire occurrence was related to antecedent fire by lagging fire records (scored 0 in years unburned and 1 when burned) for each site by up to 5 years in increments of 1 year. Antecedent fire was considered for individual years and also aggregated for up to 5 years (i.e. aggregate scores for prior fire ranging from 0 to 5). Mixed effects models were used in analysis to deal with repeated-measures over the 15 years of annual fire records (and 20 years of annual rainfall measures – see below) for the same sites (Crawley 2007; p. 605).

Rainfall variables

Data came from the Bureau of Meteorology as monthly rainfall grids, each of 0.05 × 0.05°, generated using the Barnes successive correction technique (Jones and Weymouth 1997; Jones *et al.* 2006) from irregularly spaced, point-based observations from 6000 rainfall stations, which are sparser in northern Australia where surfaces are correspondingly less robust. This limitation should be recognised in interpreting rainfall quanta and related analyses. The same set of points used to sample fire mapping was intersected with these grids to generate tables of monthly rainfall.

Mean annual rainfall was calculated by summing from July to June (north Australian 'rain-year' encompassing the whole of the summer wet season). The mean of rain-year rainfall for the period July 1970–June 2012 (42 years) was assigned to each point. In some analyses, mean rainfall was divided into classes of 100 mm increments from 300 mm to more than 1500 mm.

Mean wet season rainfall is the sum of monthly rainfall for December to May inclusive within each rain-year. *Mean dry season rainfall* is the sum of monthly rainfall from June to November during each calendar year.

Rainfall seasonality indices

Timing of end of wet season was defined in three ways: as the last month from January to June in each rain-year that received 5 or 8% of total annual rainfall, or 50 mm of rainfall (thresholds), and took values between 1 and 6. Where rainfall patterns did not permit calculation (e.g. if rainfall did not fall below 50 mm during the relevant period), the variable was set to missing. *Start of wet season* was defined as the first month after July that receives an average above threshold, taking values of 7 to 12 in most places. A few sites that recorded a wet season start in January were recorded as 13. *Average length of wet season* was consecutive months with more than threshold rainfall.

Intensity of dry season was indicated for each sampled point by:

- mean rainfall in the driest month of the year (RDM)
- mean rainfall in the driest quarter (RDQ) (lowest sum of monthly means over any consecutive 3-month period)
- the ratio of rainfall in the wettest quarter to the driest quarter (WDR)
- difference between total rainfall during the 6 months of the monsoon season (December–May) and the 6-month dry season (June–November) as a proportion of mean total annual rainfall (WI), which may vary from –1 to +1: positive values indicate greater summer than winter rainfall.

Within-year concentration of rainfall: established indices based on monthly rainfall:

- SI: annual sum of absolute deviations from average monthly rainfall, divided by total annual rainfall (Walsh and Lawler 1981), varying from 0 (rainfall evenly spread across all 12 months) to 1.83 (all rainfall occurs in 1 month)
- PCI: annual sum of monthly rainfall squared divided by annual rainfall squared (Oliver 1980) × 100, which can range in value from 8.3 (uniform) to 100 (highly irregular).

The last two indices differ from others in making no assumptions about the direction of seasonality.

Year-to-year rainfall variability

This was summarised for each point as:

- the standard deviation (s.d.) of annual and other relevant rainfall quanta
- for each year, a rainfall anomaly index (RAI) as the year's rainfall quantum less the mean for the point, divided by the s.d.
- the number of years in which rainfall anomalies exceeding 1.5 (above or below the mean) were observed over periods of up to 5 years.

Vegetation

Vegetation type at each point (VEG) was determined by intersecting points with a GIS coverage from the National Vegetation Information System. Sample points were attributed with level V descriptions, which specify the broad vegetation structure and dominant plant species in the upper-, mid- and under-storeys (ESCAVI 2003).

Land use

Land uses came from the Australian Land Use and Management (ALUM) Classification Version 7 (Bureau of Resource Sciences 2010). In deriving attributes for points we collapsed the classification to a smaller number of one-dimensional categories that we considered most relevant to northern Australia and fire management, namely:

- extensive grazing of natural vegetation (i.e. pastoral lands)
- Indigenous traditional use of natural vegetation

- non-plantation forestry based on harvest from natural vegetation
- conservation parks and reserves
- lands managed for resource protection
- grazing of modified ('improved') pastures, including irrigated pastures
- highly modified lands used for production (cropping, mining, urban, infrastructure)
- wet areas, including marshes, all excluded from analyses.

Greenhouse gas emissions

We used observations of fire frequency and seasonality to calculate an index of emissions intensity using methods described by Russell-Smith *et al.* (2009b). We adjusted for lower productivity and hence less fuel production in lower rainfall (≤1000 mm) regions by:

1. applying to all sites parameters measured from eucalypt woodland (EW) in the high rainfall (>1000 mm) zone, on the grounds that EW is a common vegetation type, especially higher in the 300–1000-mm rainfall gradient, and EW fuel loads and emission factors are at the low end of those measured
2. reducing emissions estimates generated on the EW parameters by one-third given preliminary data indicating lower levels of leaf litter in lower rainfall regions (C Yates, unpublished data).

Indices therefore constitute approximations assumed to be conservative but which are clearly suitable only for broad comparisons among sites. We employed these figures to provide a measure of variance in fire regimes taking account of both fire frequency and seasonality. We calculated the emissions intensity at each random point falling within randomly selected polygons of area 30 000 km^2 (slightly larger than the WALFA area). The mean annual emissions intensity index and the precision of the estimate (standard error of the mean divided by the mean) were calculated for all polygons. The number of random points falling within each polygon varied from 71 to 110.

Relationships between fire regimes and rainfall variables

We derived statistical models to describe relationships between whole of study (15-year) fire regimes and long-term rainfall variables using model selection procedures based on information theory, implemented as described by Burnham and Anderson (2002). We based the set(s) of candidate models on hypotheses that in savannas:

1. repeated fire at the same site requires rapid recovery of fine fuels and reliable annual production driven by sustained moderate to large rainfalls
2. fire is most likely to occur and will spread further under sustained dry conditions, so that irrespective of rainfall quanta and fuel loads, fire frequency will be higher at sites with more intense rainfall seasonality, as measured by the magnitude of rainfall variation between wet and dry seasons
3. consequently, informative analysis of fire frequency should consider both rainfall quanta and rainfall seasonality
4. probability of fire at a given site will increase with the length of the dry season because this will determine the duration of exposure of ignitable fuels to ignitions
5. year-to-year variation in timing of fire will be influenced by the time of cessation of one wet season and the time the next begins.

In general, we accepted the model with the lowest Akaike Information Criterion (AIC) as the best model in the candidate set (Burnham and Anderson 2002). We restricted analysis to simple additive models in most cases, but considered two-way interactions where we identified *a priori* reasons to expect strong interactions. Fire frequency for each point (site) during the 15 years of observations was treated as a binary response variable (years burned versus years unburned) and modelled with binomial errors and default (logit) links (see Crawley 2007; p. 605). Fire seasonality was also treated as response for analyses with the same suite of explanatory variables.

We used similar methods to examine shorter term associations of antecedent rainfall and preceding fire on probability of subsequent fire at individual points in the lower rainfall (≤1000 mm) zone. Specifically we examined associations of fire occurrence and timing with:

- rainfall anomalies (see definitions above) in the immediately preceding wet season and at lags up to 5 years
- aggregations of rainfall anomalies up to 5 years
- wet season timing
- prior incidence of fire and aggregated fire frequency in immediately preceding periods up to 5 years.

Because these analyses involve repeated-measures at the same sites, we used mixed effects models treating sites (point ID) as random variables.

Given difficulties in graphical presentation of all members of large samples of relatively noisy ecological data, we often present trends as means calculated within more or less arbitrary increments of continuous variables. While such means calculated from large samples summarise underlying trends informatively, graphs or formulae do not offer useful predictive tools for local application at smaller scales.

We used the R statistical package (R Development Core Team 2012) for all analyses and graphics.

Results

The study area exceeds 3.27 million km^2 running up to 1400 km inland from the northern coasts. The random selection of points was large enough to sample most of this area. We first summarise the climatic and fire regimes and vegetation types for this region, and then discuss year-to-year dynamics of fire and potential influences on those dynamics.

Rainfall regime

The study area samples the strong north–south/coastal–inland climatic gradient that shapes the ecology and patterns of land use in north Australia (Fig. 2.1). Mean annual rainfall (MAR) in sampled points ranges from 300 mm to more than 2100 mm for individual points (Table 2.1).

Seasonality

Strong seasonality is indicated by generally low winter rainfall compared with the summer wet season rainfall. The region's mean WI of 0.65 is equivalent to mean wet season rainfall 4.7 times mean dry season rainfall. At many points, long-term mean rainfall over the driest 3 months is just a few mm (Table 2.1). The synthetic indices (SI and PCI) and their

Table 2.1. Variables used to summarise the rainfall and fire regimes in the random sample of points ($n = 9132$) from the study area shown in Fig. 2.1

The lowest 42-year average annual rainfall incorporated in the set was deliberately constrained to 300 mm.

Attribute	Acronym	Minimum	Maximum	Mean ± s.d.
Rainfall				
Average annual rainfall (mm)	MAR	300.0	2104.6	640.3 ± 333.9
Average rainfall in wettest month (mm)	RWM	40.4	520.5	154.2 ± 85.6
Average rainfall in driest month (mm)	RDM	0.01	50.4	4.5 ± 5.4
Average rainfall in wettest quarter (mm)	RWQ	114.9	1387.5	399.0 ± 234.9
Average rainfall in driest quarter (mm)	RDQ	0.30	201.1	19.8 ± 17.8
Average wet season (Dec. to May) rainfall (mm)	MWR	190.8	1821.0	540.5 ± 312.4
Average dry season (June to Nov.) rainfall (mm)	MDR	31.0	492.1	99.7 ± 45.0
Rainfall seasonality indices	WI	0.176	0.895	0.649 ± 0.147
	SI	0.493	1.322	1.048 ± 0.100
	PCI	11.6	38.6	24.3 ± 3.3
Fire				
Burned/unburned (individual points in each year)	FYN	0	1	–
Fire frequency (number of times individual points recorded as burned in 15 years)	FF	0	15	2.24 ± 0.03
Fire month (month of fire for individual points in each year)	FM	1	12	–
Mean fire seasonality (mean of fire month for individual points in years in which fire recorded)	FFY	1	12	8.87 ± 0.03
Antecedent fire (fire in the same site 1–5 years prior)	FP1–FP5	0	1	–
Antecedent fire history (frequency of fire at the same site over 5 years prior)	FFP5	0	5	–

associated descriptions indicate conditions ranging from 'rather seasonal' to 'extreme seasonality' (SI; Walsh and Lawler 1981) or from 'moderately seasonal' to 'highly seasonal' (PCI; Oliver 1980). None of the points are in the PCI range (>50) interpreted as indicating 'irregular' rainfall. All parts of the study area are subject to annually recurring seasonal droughts of several months during which there is little or no rainfall.

Inter-annual variability

Over the 42 years of rainfall records, the mean ratio of the standard deviation to the MAR across all points was 0.396. Variance changed with position in the rainfall gradient, with year-to-year variability being much greater in the lower (≤1000 mm) rainfall region, especially in regions averaging less than 600 mm annually. Somewhat different although not inconsistent patterns are evident in the RAI. All sampled points experienced extreme high annual rainfall at least once during the study period (defined here as RAIy >1.5). However, anomalously high annual rainfall occurred somewhat more frequently in sites in the lowest rainfall region (300–400 mm) (generalised linear model on counts, Poisson errors, $P < 0.0001$, explained deviance 4.6%). Probability of a point experiencing anomalously low annual rainfall at least once also varied significantly among rainfall zones ($P < 0.0001$, explained deviance 8.3%), and was most frequent in regions with mean long-term rainfall exceeding 1000 mm ($P \ll 0.0001$, explained deviance 13.2%).

Timing of strongly anomalous years is shown in Fig. 2.2, which shows several peaks of much above average rainfall, some of which occurred during the period 1997–2011, during which fire incidence is analysed. There was also a single less acute spike in anomalously low rainfall during this period (in 2005).

RDQ was mostly very low (Table 2.1) but varied more acutely from year to year relative to annual rainfall at the same sites, with the s.d. averaging 1.8 times the mean. Year-to-year variability of RDQ was relatively lower in drier regions.

Vegetation pattern and rainfall

The random sample of points intersected 20 vegetation descriptions mapped at NVIS Level V (ESCAVI 2003). As well as eliminating obviously irrelevant points from the dataset for

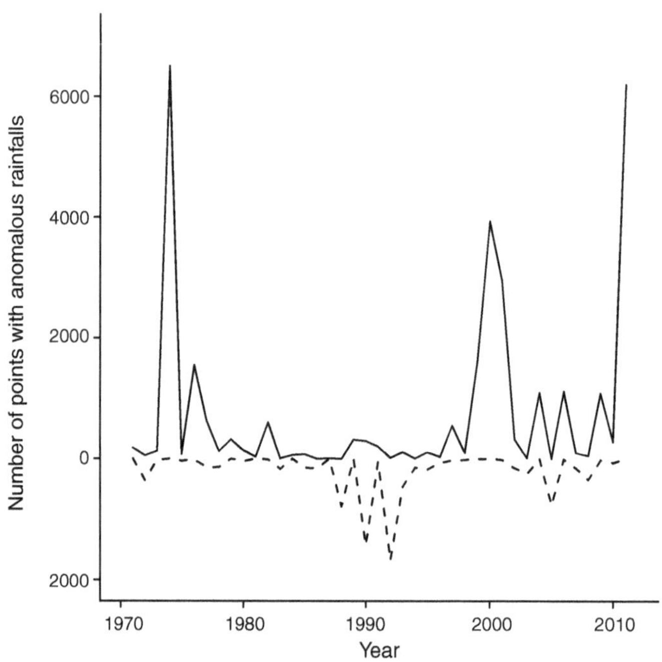

Figure 2.2. Time series of number of sample points for the whole of the study area in which anomalously high (upper plot in solid line) and low annual rainfall (lower plot with dashed line) was reported during the calendar years 1971 to 2011. The criterion for identifying anomalous yearly rainfall was a difference between the annual rainfall and long-term mean annual rainfall exceeding 1.5 standard deviations.

statistical analysis (above), we collapsed several numerically small groups into a category 'Other' (Table 2.2). Vegetation types at sampled points varied in position in the rainfall gradient, with some types concentrated in subsets of the gradient and others being more widely distributed across climatic zones (Table 2.2).

Table 2.2. Distribution of NVIS vegetation types (number of sample points) among rainfall classes

There are often strong associations between rainfall class and vegetation type with, for example, *Acacia*-dominated vegetation being more strongly associated with low rainfall regions and eucalypt woodlands and open forests being more common at the wetter end of the gradient.

	Rainfall interval (in mm)							
Vegetation type	<400	4–500	5–600	6–700	7–800	8–900	9–1000	>1000
Acacia shrublands (ACSH)	456	65	33	56	15	2	0	0
Acacia forests and woodlands (AFW)	54	35	50	58	40	8	13	7
Acacia open woodlands (AOW)	140	41	16	3	0	0	0	0
Casuarina forests and woodlands (CFW)	141	1	0	0	0	0	1	11
Cleared and non-native (CNN)	2	10	134	239	26	25	14	35
Eucalypt open forests (EOF)	0	0	4	21	22	13	23	219
Eucalypt open woodlands (EOW)	135	218	219	233	115	94	33	69
Eucalypt woodlands (EW)	12	16	81	242	229	352	231	604
Hummock grasslands (HG)	1332	798	342	72	37	9	13	7
Melaleuca forests and woodlands (MFW)	13	0	0	19	32	31	53	134
Other (OTH)	53	27	35	73	25	18	9	14
Tropical eucalypt woodlands and grasslands (TEWG)	0	0	1	4	8	22	58	241
Tussock grasslands (TG)	158	427	182	132	50	27	10	28

Note: The 'Other' class is made up relatively poorly represented vegetation types that are sufficiently similar to eligible vegetation types (e.g. include a significant component of woody vegetation) under an existing methodology to suggest they may be considered in a lower rainfall methodology.

Land use and rainfall

Extensive grazing of native pastures was the dominant land use at sample points (55.1%), including the low rainfall (≤1000 mm) zone (57.3%). The next most common was Indigenous use (18.1%). Intensively used points were less than 0.9% of the sample.

Fire regime

Fire was mapped throughout the sampled area, at all positions in the rainfall gradient and in all vegetation types and land uses, but at widely varying frequencies (Fig. 2.1).

Fire frequency

In the region as a whole, 70.6% of sample points had at least one mappable fire during the 15 years (Fig. 2.3). Only two points burned in all 15 years. A sizeable minority of 4.8% experienced fires more than every second year (fire mapped in 8 years or more) and 15.9% burned at least 1 year in 3 (in 5 years or more). There was substantial year-to-year variation in the proportion of sites burned (Fig. 2.4), ranging from 5.9% in 2005 to 24.4% in 2000. There was no evidence of a trend in proportion of sites burned over the region as a whole during the 15-year study period (regression of arcsine-transformed proportion on year: $P = 0.84$).

Fire seasonality

Fires were mapped throughout the year, most occurring from August on (72.8%), treated as late-season fire in the >1000 mm savanna fire methodology. More fires were mapped in

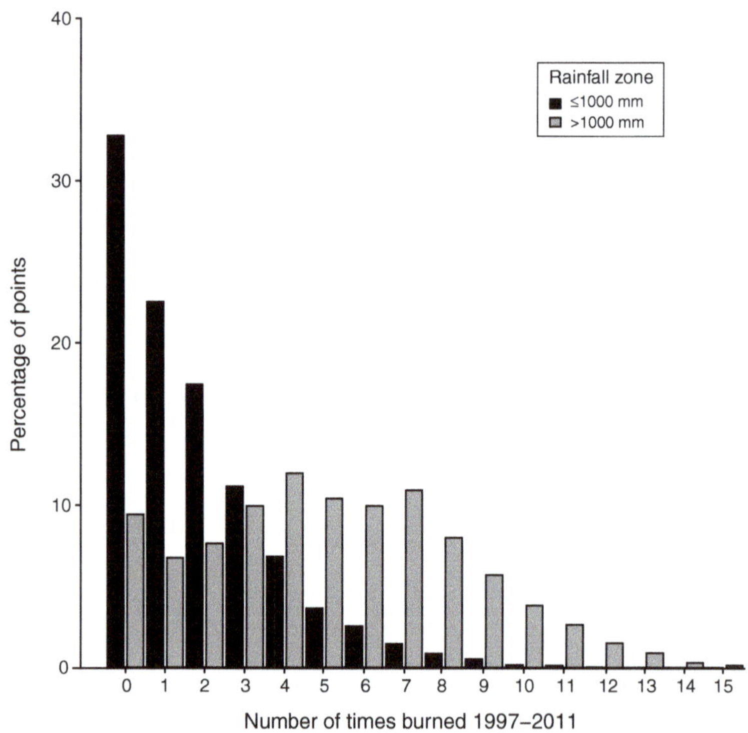

Figure 2.3. Frequency distribution of fires at points from high rainfall (>1000 mm) and lower rainfall (<1000 mm) regions.

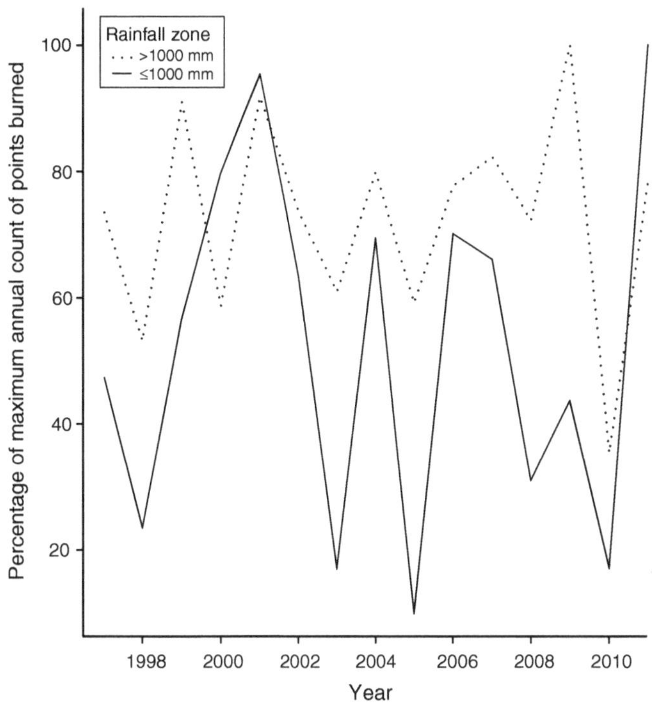

Figure 2.4. Annual time series of number of fires recorded at points in high (>1000 mm) and low (>300 but ≤1000) rainfall regions, shown as a percentage of the maximum number of fires recorded in a year at any time in the series.

October (22.1%) than in any other month (Fig. 2.5). Notably, this peak occurs before lightning ignited fires are likely to occur frequently, indicating predominantly anthropogenic ignitions in all parts of the sampled region (Russell-Smith *et al.* 2007). Annual incidence of LDS fires ranged from 53.5% (2005) to 91.7% (2006).

Within years there was a strong association between the proportion of points burned late and the total points burned during a year. In years with higher LDS fire frequencies, proportionally more points were burned in total (Fig. 2.6: $F_{1,13}$ = 215.8, P <0.0001). However, taking individual points as the scale of analysis, variation in timing of fires with fire frequency was different. Points with higher fire frequencies tended to have more fires early (Fig. 2.7). We suggest that these superficially contradictory observations, associated with aggregation of data in different ways, arise from more sites in total being burned when they are affected by late, often larger, wildfires, while regularly burned sites tend to be ignited earlier, probably through deliberate burning.

Across the region as a whole, there was no apparent trend through time in the annual proportion of points burned late ($F_{1,13}$ = 0.007, P = 0.94).

Fire regime and rainfall quanta

Although there is considerable overlap, distribution of fire frequencies varied substantially between points in the >1000 mm rainfall zone (median fire frequency of 5, maximum 15) and lower rainfall regions (median 1, maximum 12) (Fig. 2.8; χ^2 approximation, P <0.0001). Fires were relatively fewer in the lower rainfall region [13 733 mapped fires at 7796 points

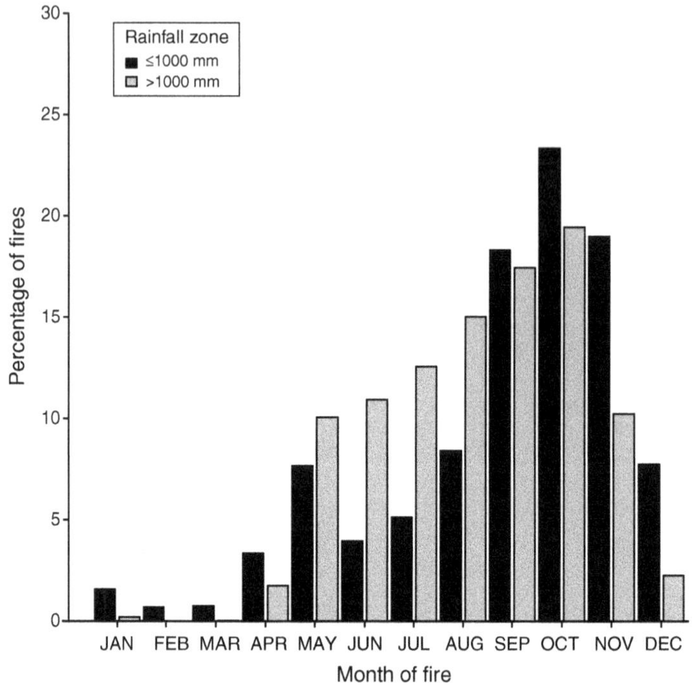

Figure 2.5. Counts of fires recorded in sites in different months pooled across the 15 years of mapping.

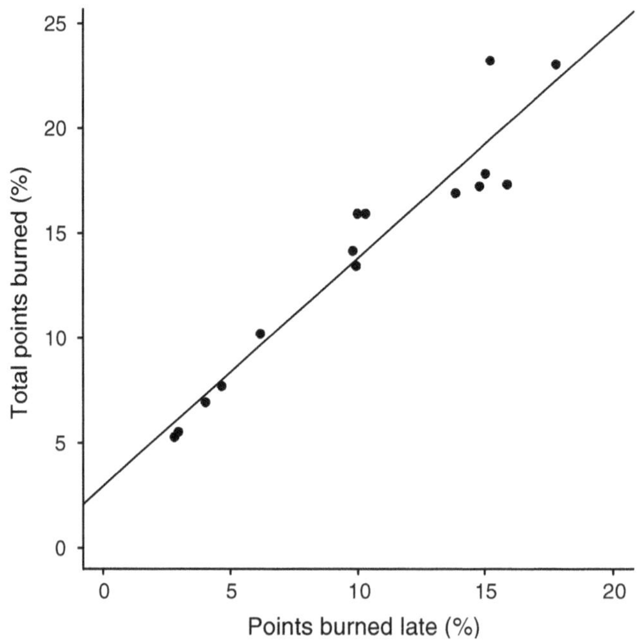

Figure 2.6. Annual association between incidence of fire in the savannas (the whole of the area shown in Fig. 2.1) and early/late seasonality of fire over the period 1997–2011. Line is a simple linear regression.

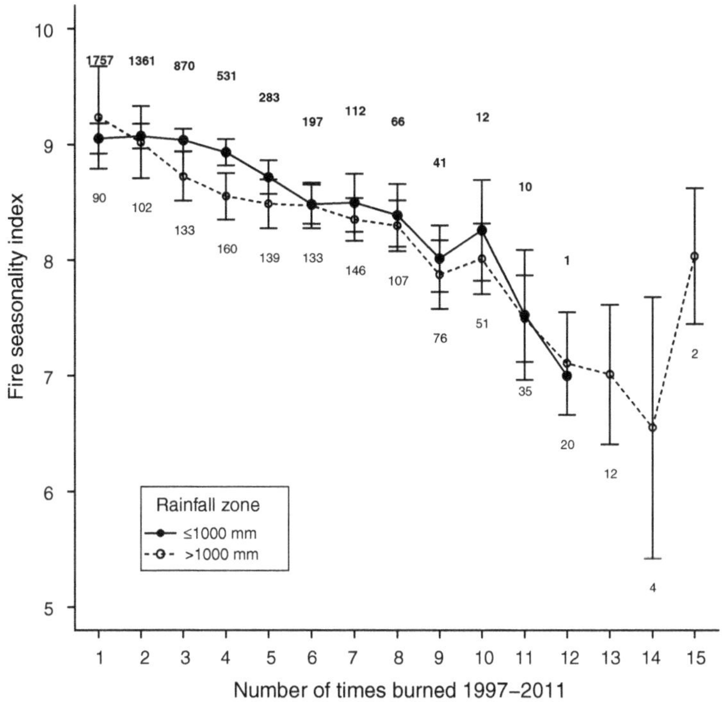

Figure 2.7. Relationship of average seasonality score to fire frequency shown separately for sites in the 1000+ mm rainfall zone and the lower (≤1000 mm) zone.

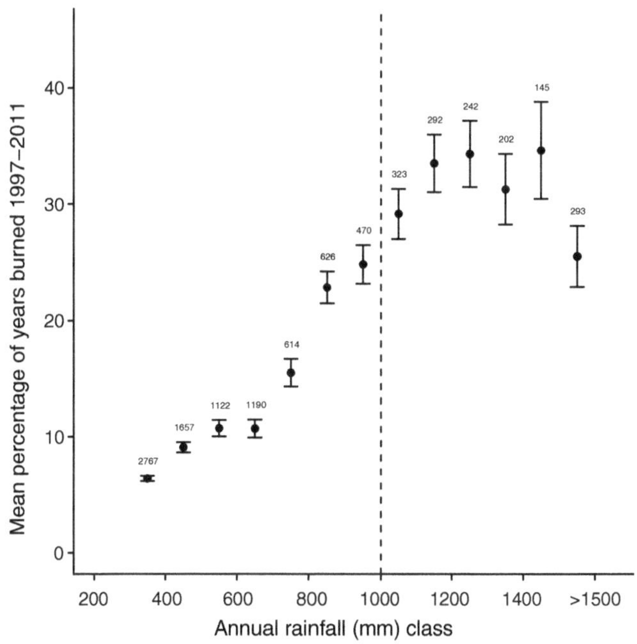

Figure 2.8. Variation in fire frequency along the rainfall gradient. Means are plotted at the midpoint of the increment and bars are 95% confidence intervals. Numbers above bars are number of sites in the rainfall increment. The vertical line separates the low and high rainfall zones.

Table 2.3. Summary of generalised linear models relating single explanatory variables to fire frequency (as binary variable with number of years burned versus years unburned) in the lower-rainfall zone (≤1000 mm), compared by AIC

Variables characterise long-term rainfall regimes at each site and vegetation type and forms of land use. The sample of points after removal of points from the high rainfall zone and to take account of various missing values and irrelevant vegetation types was 7562. Variables were logged to examine departures from linearity. The null deviance for all models was 18 668 and the best model [log (RWQ : RDQ)] explained 32.3% of deviance.

	Model	k	AIC	ΔAIC	Weight	Residual deviance	Explained deviance (%)
1	Rainfall as factor	8	27776	2470	<0.0001	15064	19.3
2	Mean annual rainfall (MAR)	3	27714	2408	<0.0001	15071	19.4
3	Mean rainfall wettest month (RWM)	3	27080	1775	0	14379	23.0
4	Mean rainfall driest month (RDM)	3	26915	1610	<0.0001	14213	23.9
5	Mean rainfall wettest quarter (RWQ)	3	26664	1359	<0.0001	13962	25.2
6	RDQ	3	26836	1531	<0.0001	14134	24.3
7	Ratio RWQ : RDQ	3	27223	1918	<0.0001	14251	22.2
8	Wet season start (5%)	3	30431	5125	<0.0001	17729	5.0
9	Wet season end (5%)	3	28665	3350	<0.0001	15953	14.5
10	Wet season length (5%)	3	28147	2842	<0.0001	15445	17.3
11	Wet season start (8%)	3	30847	5532	<0.0001	18135	2.9
12	Wet season end (8%)	3	31299	5994	<0.0001	18597	0.4
13	Wet season length (8%)	3	30304	4998	<0.0001	17601	5.7
14	Wet season start (50 mm)	3	30010	4704	<0.0001	17308	7.3
15	Wet season end (50 mm)	3	29182	3877	<0.0001	16480	11.7
16	Wet season length (50 mm)	3	29120	3815	<0.0001	16418	12.1
17	Seasonality index WI	3	26543	1238	<0.0001	13841	25.9
18	Seasonality index SI	3	28676	3370	<0.0001	15974	14.4
19	Seasonality index PCI	3	30766	5460	<0.0001	18603	0.3
20	Log (MAR)	3	27773	2428	<0.0001	15031	19.5
21	Log (RWM)	3	26961	1655	<0.0001	14259	23.6
22	Log (RDM)	3	26632	1326	<0.0001	13930	25.4
23	Log (RWQ)	3	26664	1359	<0.0001	13962	25.2
24	Log (RDQ)	3	26465	1160	<0.0001	13763	26.3
25	Log (RWQ : RDQ)	3	25305	0	1.0	12603	32.5
26	Log (WI)	3	26754	1449	<0.0001	14052	24.7
27	Log (SI)	3	28637	3332	<0.0001	15935	14.6
28	Log (PCI)	3	30652	5346	<0.0001	17950	3.8
29	VEG	13	28695	3390	<0.0001	15973	14.4
30	LUS	7	30799	5494	<0.0001	18089	3.1

(average 1.76) versus 6765 fires at 1336 sites (average 5.06 fires/site) in the high rainfall region]. The proportion of points burned in the low rainfall region varied from 2.2% in 2005 to 22.3% in 2011, and from 16.5% (2011) to 46.6% (2009) in the high rainfall region.

In addition to variation in fire frequency between broad rainfall regions (above or below 1000 mm) there was substantial variation along the rainfall gradient within these regions, especially below 1000 mm (Fig. 2.8; generalised linear model, explained deviance of 31.4% for response years burned versus years unburned on rainfall class: $F_{12,9119} = 337.9$, $P < 0.0001$). Relationships of all of the derived long-term rainfall variables with fire frequency considered individually are summarised in Table 2.3.

The distribution of fires across months (Fig. 2.5) pooled across the 15 years of mapping varied significantly between high and lower rainfall regions. The mean FM for all points in which fire was observed (range 1–12) was 9.0 for the low rainfall region and 8.5 for the high rainfall zone ($F_{1,6449} = 65.9$, $P < 0.0001$). The fitted probability of mapped fires occurring in August or later (generalised linear model, binomial response for late versus early on rainfall zone) was 0.77 for the low and 0.64 for the high-rainfall zone. Fires were on average latest in the 600–700 mm rainfall band (Fig. 2.9).

Fire regime and rainfall seasonality

Fire frequency showed statistically significant relationships with all rainfall variables summarised by means and extremes in those (long-term) means, which can be interpreted as relatively stable measures of rainfall regime, including various measures of seasonality. For the low rainfall zone, the single-variable model of this type with the best fit was

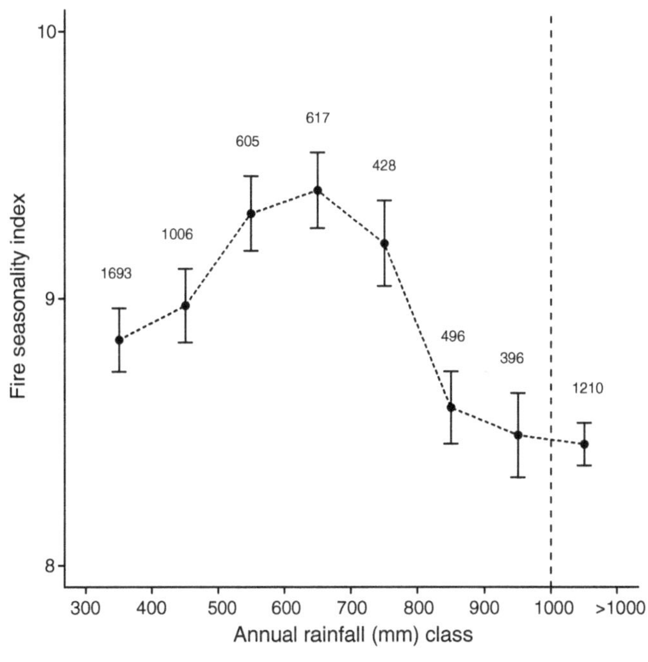

Figure 2.9. Variation in mean seasonality score for points that burned at least once during the period 1997–2011 with position in the rainfall gradient. Observations from the high rainfall zone (>1000 mm) have been pooled. Means are plotted at the midpoint of the increment and bars are 95% confidence intervals. Numbers above bars are number of sites in the increment. Seasonality scores peak (are latest) in the 600–700-mm annual rainfall increment.

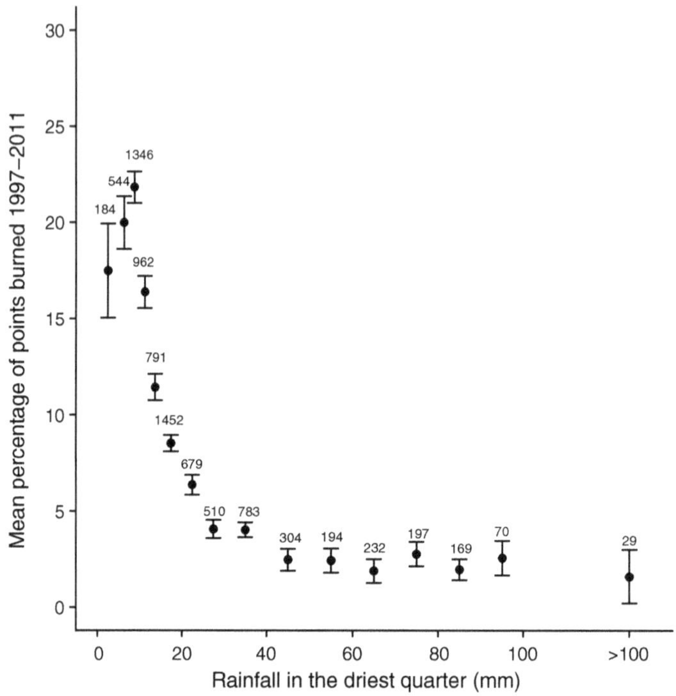

Figure 2.10. Mean fire frequency (as percentage of 15-year sample period) in arbitrary increments of RDQ, presented on a logarithmic scale and plotted at the increment midpoint. Error bars are 95% CI. Numbers above error bars are number of points.

log(RWQ:RDQ) (Table 2.3). Fire frequency was higher in sites with higher ratios. For all variables, levels of explained deviance were modest, being 32.3% for the best model. The next best model (ΔAICc = 1153) was for log(RDQ) at 26.1% (Fig. 2.10). Relative ranking of model fits were unchanged when wet season start, end and length models (which were surprisingly weak) were dropped and sample size was increased accordingly (because the number of sites with missing values for variables included in the candidate set was lower). Associations of fire seasonality (e.g. FFY) with all long-term measures of average rainfall seasonality, including RDQ, were weak (maximum $r^2 = 0.003$) and are not reported in detail.

In sum, these simple analyses suggest that in the lower (≤1000 mm) rainfall zone, fire frequency is associated (1) relatively weakly with coarse measures of timing of wet seasons or their lengths, (2) more strongly and positively with total annual rainfall, and (3) most strongly with measures of within-year rainfall variation that directly contrast wet and dry season rainfall. But even the best models leave much (most) variance unexplained. This is unsurprising given other evidence that ignitions are mostly anthropogenic. Conditions for recurring fire and its spread are established by annual availability of combustible fuels driven by rainfall and seasonal conditions, but social factors determine distribution of ignitions in space and time.

Fire regime and vegetation

There was substantial variation in fire frequency (% times burned in 15 years) among vegetation types (see multivariate statistical model below), ranging from a mean of 2.9% across all sites in *Acacia* open woodlands ($n = 200$) to 43.2% ($n = 334$) in tropical EW and grasslands. Variation in fire frequency among vegetation types mostly echoed the relative position of vegetation types along the rainfall gradient (Table 2.2). There are no striking differences in fire seasonality among vegetation types, with the most obvious pattern a weaker tendency for variation in seasonality – compared with fire frequency – to reflect distribution along the rainfall gradient.

Fire regime and land use

Fire frequency (arcsine-transformed %) differed significantly among land use classes in the low rainfall zone ($F_{5,7766}$ = 3762, P <0.0001), being lowest in the few intensively used sites (1.24%) and highest in conservation parks and reserves (16.6%). All pair-wise comparisons were significantly different (P <0.05) except forestry with intensive grazing (Tukey's HSD P = 0.58), parks and reserves with Indigenous use (P = 0.99), and resource management with parks and reserves (P = 0.33). There were some associations of fire seasonality (FFY) with land use ($F_{5,5522}$ = 16.6, P <0.0001). Fires were latest (mean FM = 10.2) at points on more intensively used land and under native forestry (10.2). On Indigenous lands (8.6) and on parks and reserves (8.5) fires were significantly earlier than on extensive grazing (9.1) and resource protection lands (9.2) (Tukey's HSD, P <0.05).

Fire regime, vegetation and land use

Land use classification and vegetation types show strong associations. For example, in our random sample, lands under Indigenous ownership and use are dominated by hummock (chiefly spinifex) grasslands (57.3%), as are lands used for resource protection (67.3%). Hummock grasslands make up 15.6% of lands used for extensive grazing. In contrast, 92.9% of tussock grasslands included in our random sample were managed for extensive grazing.

Fire frequency in vegetation types varied somewhat among land uses. For example, under Indigenous management, EW were most often burned (mean of 34.3%). Under extensive pastoralism, tropical EW and grasslands were most often burned (33.5%) and tussock grasslands rarely burned (3.4%). Hummock grasslands under Indigenous and extensive pastoral management burned at approximately equivalent frequencies (15.2 and 12.4%, respectively). These apparent trends for fire incidence to vary with land use were confounded by inter-correlation among explanatory variables. In statistical models for fire frequency as response, land use was not included in the model with the best fit to the data (below).

Multivariate statistical model of association of fire regime with biophysical attributes

Our principal interest is to identify an appropriate boundary for application of a Carbon Farming Initiative (CFI) methodology for savannas with less than 1000 mm of rainfall. Ideally that boundary would be based entirely on a single rainfall variable to simplify application and understanding. However, given the lack of obvious discontinuities in rainfall patterns (evenness of gradients), we have also examined other biophysical variables (vegetation pattern and land use) that may affect interpretation of the influence of rainfall.

To support choice of a useful combination of variables, we examined a set of statistical models relating fire frequency to interactions of those biophysical variables with rainfall variables in various combinations. To reduce the number of candidate models to manageable levels, we confined analysis to those long-term rainfall variables that showed the strongest direct individual relationships with either fire frequency or fire seasonality.

Fire frequency

The explanatory variables considered for each point were MAR, NVIS vegetation type (VEG), land use, WI, RWQ, RDQ, and the ratio RWQ:RDQ. We omitted monthly equivalents (RWM and RDM) despite their relatively good individual fit because rainfall figures for single months in the dry season were considered insufficiently robust (e.g. before 2007 small negative values were extracted from the Bureau of Meteorology surface for some sites). Generalised linear modelling with binomial errors was used for all models of fire frequency. The sample included all relevant points in the lower rainfall (≤1000 mm) zone.

Candidates included models with natural logarithms of MAR, RWQ, RDQ and RWQ:RDQ. We did not consider three-way or higher interactions. With the exception of any combination of RDQ or RWQ with each other or with their ratio (RWQ:RDQ), the candidate set ($n = 82$) covered all potential combinations of the selected variables including two-way interactions.

The best model included VEG, logMAR and logRDQ and their two-way interactions (AIC weight = 0.9999; ΔAICc = 40.3 to next best model). The next best model included VEG, land use, and logRWQ and their two-way interactions. We have presented a simplified version of the best model (Table 2.4) omitting the interaction between the continuous variables because it is more readily interpreted and presented. Within the same vegetation types, fire tended to occur more frequently in the higher rainfall end of the distribution and frequency was higher still in regions of strong seasonality (low RDQ) (Table 2.4).

However, it is worth noting here the implications of a statistical interaction between MAR and RDQ, namely that shifts in fire frequency with a change in mean rainfall are not uniform across the range of RDQ (and *vice versa*). Fire frequency varies more acutely with changes in

Table 2.4. Parameters for a model relating frequency of fire (as binary variable years burned versus years unburned) over 15 years (1997–2011) to rainfall variables at randomly selected sites ($n = 7796$) in the savannas in the lower-rainfall (≤1000 mm) zone

Null deviance was 19 056 and residual deviance 10 464 (explained deviance 45.1%). Given the significance of VEG by logMAR and VEG by logDRQ interaction terms, separate intercepts and slopes relating these variables to fire frequency are presented for each vegetation type. The best model included an interaction term for logMAR*logDRQ, which added slightly to explained variance.

		Slope (s.e.)	
Vegetation type	**Intercept (s.e.)**	**logMAR**	**logRDQ**
ACSH	–11.86 (0.70)	2.64 (0.18)	–0.56 (0.09)
AFW	–3.97 (0.87)	0.63 (0.21)	–1.40 (0.11)
AOW	–0.89* (2.59)	–0.42* (0.73)	–1.26 (0.32)
CFW	–4.75* (2.45)	0.93* (0.69)	–0.96 (0.48)
EOF	–16.82 (4.83)	3.51 (1.06)	–0.94 (0.16)
EOW	–5.58 (0.41)	0.95 (0.10)	–0.87 (0.05)
EW	–2.42 (0.69)	2.50 (0.16)	–0.72 (0.03)
HG	–9.44 (0.30)	2.02 (0.08)	–0.22 (0.05)
MFW	–8.74 (1.22)	1.70 (0.28)	–0.11* (0.10)
OTH	–9.51 (0.91)	1.92 (0.22)	–0.84 (0.10)
TEWG	–10.98 (2.68)	2.31 (0.60)	0.29* (0.19)
TG	–16.39 (0.87)	3.33 (0.21)	–0.71 (0.10)

* $P > 0.05$ that the parameter is not different from zero.

MAR at higher RDQ (i.e. in less strongly seasonal sites), albeit from a lower base. But the proportion of our sample (surrogate for area) and hence total fire activity declines markedly at RDQ <20 mm. Choice of the lower annual rainfall (MAR) bound for a methodology will need to take account of seasonality (RDQ) to achieve the optimal mix of fire frequency (higher levels offering greater scope for efficient projects) and area of coverage (greater area offering scope for more projects and wider improvements in fire management).

Fire seasonality

Statistically weaker relationships were found in multivariate models of seasonality (FFY) as response variable on rainfall variables in the lower (≤1000 mm) rainfall zone. The best model ($\Delta AICc = -29.3$; $r^2 = 0.272$) included VEG, land use, and log(RWQ:RDQ), and all two-way interactions among these variables. Parameters of the best model (not presented here) are consistent with related univariate analyses presented elsewhere in indicating no striking variation in timing of fires with vegetation types and land uses compared with the relationships with individual rainfall quanta and their seasonality (e.g. Fig. 2.9).

Intra- and inter-annual dynamics of fire incidence and timing

Monthly time series of the number of points recording fire for both high and low rainfall zones are at Fig. 2.11. Seasonal variation is similar in both zones in most years, with the exception of somewhat more late-season fire in the low rainfall zone. Examination of time series decomposed into seasonal effects, trend and residual (using routine 'decompose' in R) confirmed this observation. In both zones the peak positive seasonal factor is for

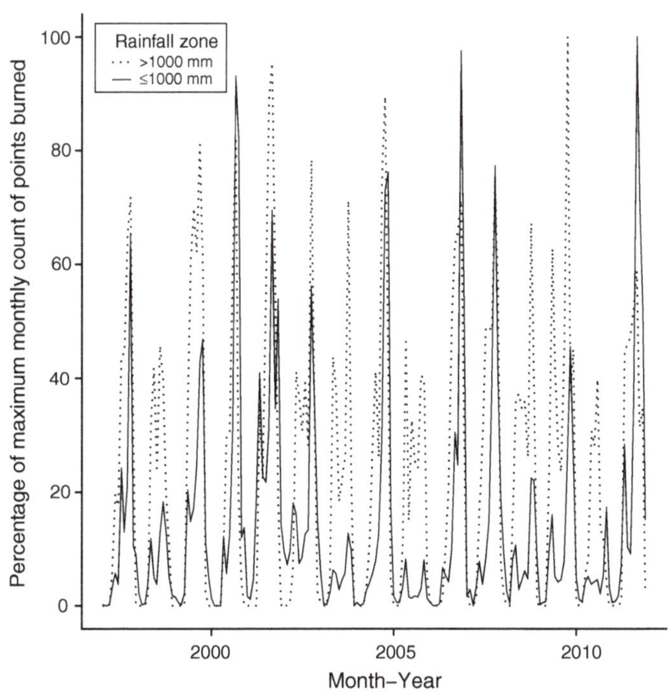

Figure 2.11. Monthly time series of number of fires recorded at points in high (>1000 mm) and low (≤1000) rainfall regions, shown as a percentage of the maximum number of fires recorded in a month at any time in the series.

October and nadir in February, but the next highest is November in the low rainfall zone and September in the high rainfall zone. Decomposition also confirmed absence of consistent long-term trends in fire incidence over the study period. In some years, the total incidence of fire was very much reduced in the lower rainfall zone (e.g. 2003, 2005, 2010) but not in the higher rainfall zone. Acute year-to-year variation is more obvious in a plot of annual fires (Fig. 2.4).

At this aggregated scale, significant autocorrelation (of total fire incidence across years) occurred only in the high rainfall zone, with fire incidence being reduced somewhat in the next calendar year following a site being burned (1-year lag: $r = -0.595$, $P = 0.025$).

Year-to-year fire incidence at individual sites – influence of antecedent rainfall

Statistical models relating year-to-year variation in incidence of fire at individual sites to prior rainfalls at the same site were mostly weak. Reduction in deviance (from a 'null' model using mean as fixed effect and point ID as random effect) was at best 5.7% for a model of RAIy for the immediately preceding wet season (ΔAIC = 144). Strength of relationships diminished quickly with longer lags between fire and rainfall events, with 'explained' deviance falling to 0.9% for RAIy lagged 1 year (i.e. two wet seasons preceding the year of fire) and lower still with longer lags. Models using aggregations of RAIy scores for up to 5 years were all weaker than the best single year model. There was significant interaction between long-term mean annual rainfall (logMAR) and RAIy (P <0.0001), with frequency of burning increasing more rapidly with RAIy in wetter regions (Fig. 2.12).

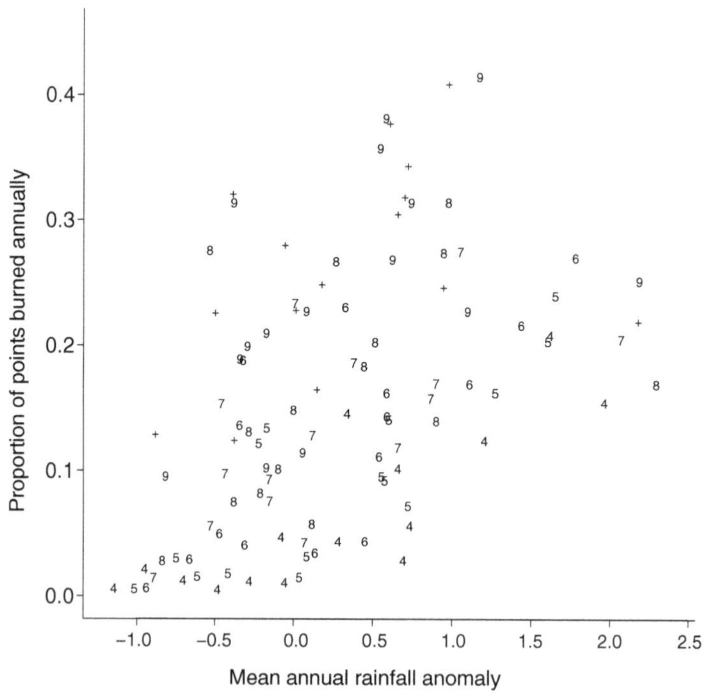

Figure 2.12. Relationship between the proportion of points recorded as burned in each year of the study and the rainfall anomaly (number of standard deviation departures from the long-term mean) for the wet season finishing in that year. Symbols are rainfall class in 100 mm increments from 4 (<400 mm) to 9 (800–900 mm) and + (900–1000 mm).

Relationships were also relatively weak (reduction in residual deviance 4.5%) using RDQ as explanatory variable in equivalent analyses. Fire incidence was reduced somewhat in years when more rain fell in the otherwise driest part of the year (the 'fire season').

Fire incidence – influence of prior fire

Consistent with the aggregated time series presented earlier, which pooled observations across sites, statistical models of associations between prior fire and subsequent fire at the same individual points within the lower rainfall zone (≤1000 mm) were relatively weak. Comparing lags of different lengths between fires individually for sites in the low rainfall (≤1000 mm) zone, the best model (ΔAIC = 278 to next best fit) for a 10-year dataset (2002–2011: to take account of loss of observations to lagging by up to 5 years) was for incidence of fire in the preceding year. Points burned in the preceding year burned slightly less often than those that had not been burned (11.3 and 11.9%, respectively). Adding interactions with long-term annual rainfall (in classes of 100 mm increments) to this analysis indicated that reductions were strongest in the 400–500 ($P = 0.018$ for rainfall class and burn interaction) and 500–600 mm ($P < 0.0001$) annual rainfall bands and non-significant in other bands ($P > 0.18$).

Models aggregating fire incidence over periods from 2 to 5 preceding years (without regard to sequence) within those periods were also weak, with the best aggregating over the all of the preceding 5 years (1.4% reduction in deviance over the null model; ΔAIC = 331 to next best). The revealed pattern was different from the 1-year lag. As the number of years of fire in the preceding 5 years increased, the probability of fire in the year immediately following that 5-year span also increased progressively. Of points that had not burned at all in the preceding 5 years, 6.8% burned in the succeeding year, while 34.4% of the small number of points that had burned in all of the preceding 5 years also burned in the succeeding year. Variances around these means increased greatly at higher frequencies of prior fire.

We interpret these weak patterns to indicate that while fire in the immediately preceding year may slightly reduce fire extent, perhaps due to reduced fuel loads, the probability of ignitions occurring in particular places is a more influential factor on fire incidence. Sites that in the past have been subject to regular (and given the time of year of most fire, principally anthropogenic) ignitions are likely to continue to be burned more frequently than the regional average. This is consistent with previous analyses at very large (national) scales that showed that places burned in the past are more likely to burn again (Russell-Smith *et al.* 2007).

Emission index for random polygons

The average annual emissions intensity over 15 years for 40 randomly placed 30 000 km^2 hexagons that fell within the 600–1000 mm annual rainfall region ranged from 3947 to 92 555 tonnes CO_2-e p.a. Precision of estimates ranged from 8.8 to 65.4% (SE/mean). Nine of the 40 sites showed coefficients of variation lower (more precise) than the WALFA baseline of 13.3% when calculated over 15 years. Over 10 years, only 1 of 40 sites had a coefficient of variation lower than the WALFA figure.

Discussion

Several projects to improve management of savanna fire have been formally declared under the Carbon Farming Initiative (DCCEE 2013). As a component of environmental offsets flowing from construction of its processing plant in Darwin, a major international natural gas explorer and producer, Inpex Corporation, has committed to support new savanna fire projects (http://gateway.icn.org.au/project/451/ichthys-lng-project, accessed 16 June 2014), potentially including regions with annual rainfall as low as 600 mm, despite the existing methodology covering only the area of Australian savannas north of the 1000 mm rainfall isohyet (Fig. 2.1). Reports of the very large areas subject to frequent fire in other parts of the savannas (Russell-Smith *et al.* 2007) have stimulated government, industry and landholder interest in extending the geographic reach of carbon farming incentives.

Essential features of the existing methodology that are also likely to be required of any new lower rainfall methodology include:

1. an active management regime that uses strategic EDS prescribed fires to reduce the risk and extent of LDS wildfires
2. planned and purposeful application of strategic burning rather than dependence on incidental control by processes such as heavy grazing
3. measurable reduction in methane and nitrous oxide emissions released by fire, taking account of area burned, fuel types consumed and the effects of timing or severity of fire on emissions
4. bounding the areas to which the methodology applies by reference to unambiguous mapping and specification of 'eligible' vegetation types for which sufficient information on fuels and emission factors are available to permit reliable accounting
5. a reliable baseline of pre-project fire patterns (area burned and timing of burning) with which to compare project emissions.

This chapter addresses particularly items (4) and (5). We interpret the task to require:

- description of the recent historical extent of fire and its timing to determine the scope for substantial reductions in emissions through active management
- analysis of variation in fire pattern through time to illustrate the potential to define a reasonable baseline of fire extent and timing (and hence emissions)
- analysis of biophysical correlates of fire patterns and their utility for defining the areas to which a new methodology might reasonably be applied
- the need, if any, for variations in the details of a new methodology from the existing one, additional to the area of application.

Scope for establishing savanna burning emissions abatement projects

Projects delivering significant abatement will require baselines of eligible areas large enough, burned frequently enough and late enough to generate substantial emissions under 'business as usual' conditions. We do not attempt to specify the levels of emissions needed to constitute a viable project because this will vary with management context. However, several features work against viable projects, particularly if present together, namely:

- uniformly low fire frequency
- strong negative temporal autocorrelation of fire events over substantial (multi-year) periods, indicating dependence of future fire on past fire
- history of extensive early season burning
- variance in potential biophysical drivers of fire extent and frequency high enough to compromise determination of a baseline and effectiveness of management interventions.

Fire frequency

Average fire frequencies are lower in the low rainfall zone than in the area covered by the existing methodology, declining substantially down the rainfall gradient, to annual averages straddling 10% at 700 mm or lower. However, there are very substantial areas with fire frequencies exceeding 1 year in 4 (Fig. 2.1), with most fire occurring late. Under these conditions, major ecological change, including general suppression of woody vegetation, and loss and potential local extinction of fire-sensitive plants and fauna, are likely (e.g. Murphy *et al.* 2009; Woinarski *et al.* 2009; Russell-Smith *et al.* 2010). Many sites offer substantial scope to reduce fire frequency and modify timing for emissions reductions.

Below 500 mm, only 4% of sites burn at least 1 year in 4. Above 500 mm, in all of the incremental 100 mm rainfall bands at least 22% and up to 52% of sites have fire frequencies exceeding this level. Above 500 mm, fire frequency mapping indicates at least some spatial clumping of higher frequencies (Fig. 2.1), making for more efficient projects. We acknowledge that subdivision of rainfall into 100 mm increments is arbitrary. Nonetheless we can confidently identify regions below 500 mm annual rainfall (MAR) as generally burning too infrequently and having too few 'pockets' of higher frequencies (Fig. 2.1) to favour management interventions for emissions abatement (see also Heckbert *et al.* 2012).

Regions with higher fire frequencies also tend to show acute seasonality of rainfall. Frequencies fall away rapidly when RDQ is much above 15–20 mm (Fig. 2.10). Because RDQ is relatively weakly correlated with MAR across north Australia as a whole ($r = -0.09$) with the relationship varying regionally, RDQ offers a useful second filter for setting boundaries.

Autocorrelation of fire events

One of the most important features of the high rainfall savanna burning methodology is that because the biophysical conditions for fire are reset each year, management action must also be taken in each and every year. If all or most available fuel was consumed and/or future production suppressed so that substantially more than one intervening wet season was required for sufficient accumulation of new fuels for fire over previously burned areas, uncertainties could emerge about the need for intervention and optimal timing, complicating determination of a baseline period demonstrably capturing one or more full fire cycles.

Aggregating over large scales, we found no evidence of temporal negative autocorrelation of fire events in the low (≤1000 mm) zone. Indeed, we confirmed previous reports (Russell-Smith *et al.* 2007) that sites burned more frequently over a period of years are likely to burn again in the future. At the scale of individual points we did find weak evidence for reduction in probability of fire (11.9–11.3%) in the year immediately following burning, but found this effect concentrated in <600 mm rainfall classes. There is no indication of widespread, beyond-annual cycles deriving from long-lasting suppression of fuel availability. To illustrate, in the low rainfall zone as a whole, 'high' (above average) fire years were consecutive in at least two separate periods (1999–2002 and 2006–2007: Fig. 2.4).

We conclude that at the spatial scales and levels of resolution relevant to selection of a boundary for a savanna fire methodology, at least above 600 mm annual rainfall, the savanna fuel–fire system is effectively reset annually. Effects of prior fire are small compared with the influence of rainfall. At the scale of our analyses burning can and frequently does recur in consecutive years at the same sites. The probability of fire recurring in successive dry seasons appears to be more strongly influenced by factors other than the effects of previous fire on fuel availability.

Levels of early burning

On average, fires occurred slightly later in the lower rainfall zone (Figs 2.7 and 2.11). There was no evidence of a strong skew to early fire in any part of the rainfall gradient. While somewhat more fires were mapped in the lower rainfall zone very early (in the period January to April: Fig. 2.5), the total proportion of EDS fires overall was very much lower. Fire is on average latest in the 600–700 mm band (Fig. 2.9), providing scope for management-directed shifts in seasonality which would be expected to contribute significantly to emissions abatement.

The existing high rainfall methodology uses time of fire as a surrogate for severity. Work is underway to offer more direct measures of fire severity (Edwards *et al.* 2013). Whether timing or more direct measures of intensity are used in future in savanna fire methodologies, it does not appear that implementation of fire management strategies based on early, cooler, patchier and hence lower severity preventive burns will be substantially inhibited by unusually high levels of EDS burning already implemented in the lower rainfall zone.

Variance in physical drivers of fire regimes

Long-term MAR and measures of intensity of rainfall seasonality are most strongly associated with fire frequency. Associations of rainfall with timing of fire are less pronounced. Year-to-year variability in both of these attributes (as a proportion of their mean) was greater in lower rainfall areas, probably due to the greater east to west ecological and social diversity of widely separated sites covered by drier bands (Figs 2.1, 2.13).

Annual rainfall

We quantified year-to-year variation in rainfall patterns by identifying positive and negative rainfall anomalies where rainfall departed by more than 1.5 standard deviations from the long-term mean for a point. If extent of fire and hence likelihood of being

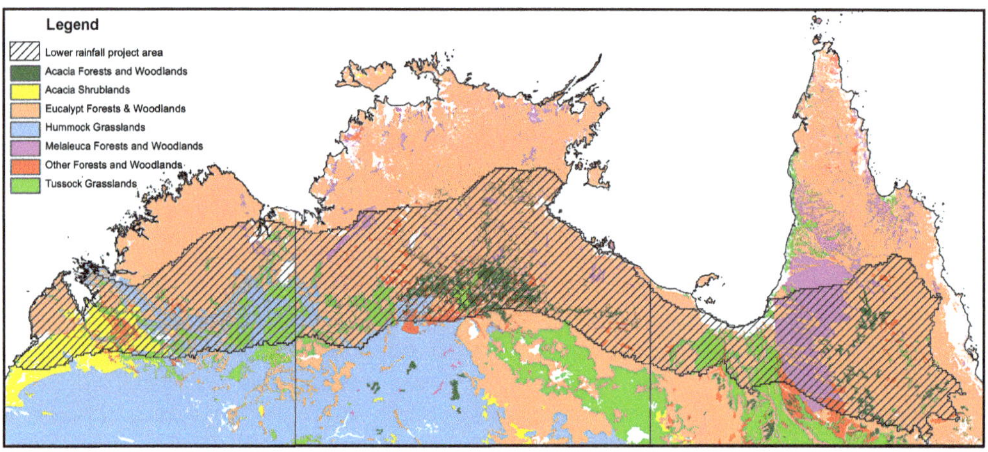

Figure 2.13. Map of proposed coverage for a savanna burning methodology for regions receiving between 600 and 1000 mm of rainfall annually and with the intense seasonality (RDQ ≤15 mm) that underpins the theory and practice of savanna burning for emissions abatement. The map also shows simplified vegetation classes, illustrating distribution of wooded savannas to which an abatement methodology is particularly relevant, relative to grassland systems where application is potentially problematic due to differences in fuel characteristics and fire behaviour.

fire-affected is influenced by fuel availability in landscapes with few other natural or anthropogenic barriers to fire spread, then we would expect greater proportions of these to be burned in fire seasons immediately following anomalously large rainfall and fewer to be burned after low anomalies. Although trends are weak, this is indeed the pattern observed (Fig. 2.12) and reflected in statistical analysis.

An important feature of this pattern is that while it is seen in all parts of the rainfall gradient, it is somewhat more pronounced in the lower rainfall bands (e.g. points labelled 4–6 in Fig. 2.12). This observation is consistent with the aggregated time series data for fire comparing the 1000+ mm zone with the ≤1000 mm zone (Fig. 2.4). This shows fire incidence falling to very low levels (especially in 2005 but also in 1998, 2003, and 2010) in ways that are not seen in the >1000 mm zone. Low fire incidence in 2005 coincides with the most extreme 'trough' of negative anomalies for rainfall observed during the period of fire mapping. In regard to positive anomalies (very high rainfall), the peaks straddling 2000 and in 2011 were also followed by peaks in fire, especially in the ≤1000 mm zone. These observations indicate that years in which fire incidence is substantially influenced by rainfall variation can be expected to recur often enough to be relevant to setting of baselines (see below).

Rainfall in driest quarter

Long-term RDQ was the measure of rainfall seasonality most strongly correlated with summary measures of fire regime [both fire frequency and average timing (month) over the whole of the lower rainfall zone]. Fire frequency was consistently higher in regions with <15 mm RDQ. Setting boundaries to exclude regions with long-term mean RDQ >15 mm, when combined with annual rainfall exceeding 600 mm, would capture most of the region that includes observations of higher fire frequency while excluding 92% of sites in the lower rainfall zone that experienced no fire during the 15 years of mapping.

Wet season timing and duration

We suggest that the relatively weak correlation of measures of the long-term average timing and length of the wet season with aggregated measures of fire incidence (frequency) or seasonality is most likely related to coarse temporal resolution (month) and arbitrary thresholds of rainfall to set beginnings and ends. Values fell into a small range dominated by a single month (e.g. 76.1% in April for wet season end in the lower rainfall zone, using first month with <50 mm rainfall). Finer descriptors using daily rainfall and more sophisticated assessments of rainfall continuity will be needed to better understand the shorter-term drivers of fire incidence and timing and their interaction with fire management operations.

Vegetation types

We excluded some vegetation types from our analyses on the grounds that patterns of burning and emission factors are likely to differ markedly from dominant systems, especially in wetter areas. However, these make up a relatively small proportion of our sample and the total area likely to fall within the bounds of a lower rainfall methodology. Choice of a 600 mm lower bound will incorporate relatively small proportions of hummock grasslands (Table 2.2 and Fig. 2.13), which present some specific challenges for fuel accumulation and emissions calculations, and which may therefore be excluded from early versions of a new methodology. Eligible vegetation types and their combustion and emission characteristics are dealt with in more detail in other publications (Yates *et al.*, Chapter 6).

Land use

All of the major land users deploy fire as a management tool but for different purposes in different ways on different parts of their (often large) land holdings. Perhaps the most notable observations are the high levels of fire on parks and reserves, highlighted elsewhere as a problem on government-managed conservation lands (Russell-Smith *et al.* 2009a, 2010). We found no compelling reasons to exclude any land use categories carrying largely natural vegetation from a methodology. Exclusions are most likely to arise more directly in urban and peri-urban areas and other highly modified agricultural landscapes because they lack substantial areas of eligible vegetation types or fire has already been effectively managed.

Setting baselines

For the existing methodology, emissions baselines are set as the annual mean estimated over 10 years, with an additional 5 years of fire mapping required to establish fuel loads at the start of the emissions baseline. For two declared projects for which data are available (Fish River and WALFA), the precision of baselines (standard error/mean) determined in this way is 6.0 and 13.3%, respectively (C Yates, unpublished data). With greater variability in biophysical influences on fire regimes and observed levels of fire frequency within the lower rainfall zone, baselines determined in the same way would – in randomly selected sites – be less precise. Achieving equivalent precision of empirical baselines in the lower rainfall zone could be addressed in at least two ways: (1) by careful choice of project sites to pick up areas in which fire has been relatively less variable, and (2) by extending baselines to include more annual observations to reduce error in estimation of the mean. The former if used exclusively would have the unwelcome effect of reducing applicability of a lower rainfall methodology to a narrow range of sites with unusually 'favourable' (for emissions estimation) fire histories and, potentially, use of highly artificial boundaries that compromise management effectiveness. The latter confronts difficulties in accessing fire mapping of the required standards over very long periods. We suggest that a new methodology should allow either approach: by requiring a specified level of precision in estimates to be met from a minimum run of 10 years of data or such longer period as might be necessary to meet a precision criterion.

Our crude estimates of variability in annual emissions intensity in randomly selected hexagons comparable in size to the WALFA project area were consistent with the expectation that lower rainfall sites would show less precise baselines when calculated over the same period, but when calculated over 15 years, a significant proportion (22.5%) were at least as precise as the WALFA estimates. This comparison should not be over-interpreted because site boundaries will never be determined randomly, but it does indicate that estimates of baselines at least equivalent in precision to the area on which the existing methodology was based are frequently achievable within the 600–1000 mm region, albeit more readily at the upper (higher rainfall) end of that range.

Interpretation and limitations

Australian savanna landscapes are highly combustible for much of each year, over huge areas uninterrupted by infrastructure or natural barriers, and support land uses that deploy fire as a management tool (Dyer *et al.* 2001). Savanna fire is mostly anthropogenic in origin (Russell-Smith *et al.* 2007) and its occurrence across the landscape is a product of complex interactions among patterns of human settlement and movement, dominant local land use, social history, cultural obligations, understanding (or its absence) of fire behaviour, and the control that landowners and managers are able to assert over fire use by residents and visitors, as well as biophysical attributes. The CFI and its methodologies are based fundamentally on providing incentives for changed human behaviours in land management. In the instance of savanna burning, present practice is not always well aligned with biophysical characteristics. No simply specified biophysical or land use envelope will neatly capture uniform management behaviours. It is therefore important to acknowledge that any proposed domain for a lower rainfall savanna burning methodology, as well as incorporating favourable sites, will also contain large areas in which there is little or no prospect of a viable savanna burning project now or in the mid-term future.

Defining the lower rainfall savanna burning envelope

We propose that an Australian lower rainfall savanna burning methodology apply north of 20°S, bounded by long-term annual rainfall of 600 mm. Variation in seasonality (<15 mm RDQ) should be used to guide adjustments at the eastern end of the proposed area, to confine application to regions that are clearly subject to intense rainfall seasonality and conform generally with the theory and practice of the existing savanna burning methodology (Fig. 2.13). The principal claim made for the proposed boundary – taking account of adjustments that may be made by identifying eligible and ineligible vegetation types (Yates *et al.*, Chapter 6) – is that it captures a large part of the total area with fire histories amenable to recurrent fire management, and thereby offers substantial opportunities for landholders presently challenged by clearly adverse fire regimes to access incentives to reduce greenhouse gas emissions through enhanced, active fire management.

We anticipate that the federal carbon farming regulator (presently the Clean Energy Regulator) will set a geographically explicit and precise boundary, as for the >1000 mm methodology, so that there will be no ambiguity about eligibility. Prospects for projects capable of substantially reducing emissions will, however, be variable within that boundary, increasing at annual rainfalls above the 600 mm minimum and with rainfalls in the driest quarter close to zero. Above all, owners and managers of potentially eligible sites will need to examine their actual long-term fire histories, with sites of higher than average fire frequencies and fires predominantly in the late dry season being more likely to offer opportunity.

Acknowledgements

Colleagues from CSIRO, Bushfires NT and Charles Darwin University, particularly Garry Cook and Stefan Meier, offered suggestions on approaches. The Australian Government provided financial support through the Department of Climate Change and Energy Efficiency. Comments from two anonymous referees helped improve the chapter.

References

Australian Government (2014) 'Emissions Reduction Fund White Paper.' Commonwealth of Australia: Canberra. www.environment.gov.au/resource/emissions-reduction-fund-white-paper (accessed April 2014).

Bayon R, Hawn A, Hamilton K (2006) *Voluntary Carbon Markets: An International Business Guide, What They Are and How They Work*. Earthscan: London.

Bureau of Resource Sciences (2010) 'The Australian Land Use and Management (ALUM) Classification Version 7, May 2010.' Bureau of Resource Sciences: Canberra. http://adl.brs.gov.au/landuse/docs/ALUM_Classification_V7_May_2010_detailed.pdf Accessed April 2013.

Burgess CP, Johnston FJ, Bowman DMJS, Whitehead PJ (2005) Healthy land: healthy people? Exploring the health benefits of Indigenous Land Management. *Australian and New Zealand Journal of Public Health* **29**, 117–122. doi:10.1111/j.1467-842X.2005.tb00060.x

Burnham KP, Anderson DR (2002) *Model Selection and Multimodel Inference: A Practical Information Theoretic Approach*. 2nd edn. Springer-Verlag: New York.

Cook GD, Liedloff AC, Eager RW, Chen X, Williams RJ, O'Grady AP, Hutley LB (2005) The estimation of carbon budgets of frequently burnt tree stands in savannas of northern Australia, using allometric and isotopic discrimination. *Australian Journal of Botany* **53**, 621–630. doi:10.1071/BT04150

Craig AB (1997) A review of information on the effects of fire in relation to the management of rangelands in the Kimberley high-rainfall zone. *Tropical Grasslands* **31**, 161–187.

Craig AB (1999) Fire management of rangelands in the Kimberley low-rainfall zone: a review. *The Rangeland Journal* **21**, 39–70. doi:10.1071/RJ9990039

Crawley MJ (2007) *The R Book*. John Wiley and Sons: Chichester, UK.

CSIRO, Bureau of Meteorology (2007) 'Climate change in Australia: technical report 2007.' CSIRO: Melbourne. www.climatechangeinaustralia.gov.au Accessed January 2013.

DCCEE (Australian Government Department of Climate Change & Energy Efficiency) (2010) 'National Greenhouse Gas Inventory: Accounting for the Kyoto Target.' May 2010. Australian Government Department of Climate Change & Energy Efficiency: Canberra. www.climatechange.gov.au/en/climate-change/emissions.aspx Accessed January 2013.

DCCEE (Australian Government Department of Climate Change & Energy Efficiency) (2013) Carbon Farming (Reduction of Greenhouse Gas Emissions through Early Dry Season Savanna Burning 1.1) Methodology Determination 2013. www.comlaw.gov.au/Details/F2012L01499 Accessed January 2013.

Dyer R, Jacklyn P, Partridge I, Russell-Smith J, Williams RJ (Eds) (2001) *Savanna Burning: Understanding and Using Fire in Northern Australia*. Tropical Savannas Cooperative Research Centre: Darwin.

Edwards AC, Russell-Smith J (2009) Ecological thresholds and the status of fire-sensitive vegetation in western Arnhem Land, northern Australia: implications for management. *International Journal of Wildland Fire* **18**, 127–146. doi:10.1071/WF08008

Edwards AC, Maier SW, Hutley LB, Williams RJ, Russell-Smith J (2013) Spectral analysis of fire severity in north Australian tropical savannas. *Remote Sensing of Environment* **136**, 56–65. doi:10.1016/j.rse.2013.04.013

Ellis S, Kanowski P, Whelan R (2004) 'National Inquiry on Bushfire Mitigation and Management.' Commonwealth of Australia: Canberra. www.coag.gov.au Accessed January 2013.

ESCAVI (Executive Committee for Australian Vegetation Information) (2003) 'Australian Vegetation Attribute Manual: National Vegetation Information System, Version 6.' Executive Committee for Australian Vegetation Information. Australian Government Department of the Environment and Heritage: Canberra.

Felderhof L, Gillieson D (2006) Comparison of fire patterns and fire frequency in two tropical savanna biomes. *Austral Ecology* **31**, 736–746. doi:10.1111/j.1442-9993.2006.01645.x

Fisher R, Vigilante T, Yates CP, Russell-Smith J (2003) Patterns of landscape fire and predicted vegetation response in the North Kimberley region of Western Australia. *International Journal of Wildland Fire* **12**, 369–379. doi:10.1071/WF03021

Grice AC, Slatter SM (Eds) (1997) 'Fire for the Management of Northern Australian Pastoral Lands.' Meat Research Corporation: Sydney.

Heckbert S, Russell-Smith J, Reeson A, Davies J, James G, Meyer C (2012) Spatially explicit benefit–cost analysis of fire management for greenhouse gas abatement. *Austral Ecology* **37**, 724–732. doi:10.1111/j.1442-9993.2012.02408.x

Jones DA, Weymouth G (1997) 'An Australian monthly rainfall data set.' Technical Report No. 70. Bureau of Meteorology: Melbourne.

Jones DA, Wang W, Fawcett R, Grant I (2006) 'The generation and delivery of Level-1 historical climate data sets.' Australian Water Availability Project Milestone Report. Bureau of Meteorology: Melbourne.

Meyer CP (2004) 'Establishing a consistent time-series of greenhouse gas emission estimates from savanna burning in Australia.' Report to the Australian Greenhouse Office, Canberra. CSIRO: Melbourne.

Murphy BP, Russell-Smith J, Watt F, Cook GD (2009) Fire management and woody biomass carbon stocks in mesic savannas. In *Culture, Ecology and Economy of Fire Management in North Australian Savannas: Rekindling the* Wurrk *Tradition.* (Eds J Russell-Smith, PJ Whitehead, P Cooke) pp. 361–394. CSIRO Publishing: Melbourne.

Murphy BP, Russell-Smith J, Prior LD (2010) Frequent fires reduce tree growth in north Australian savannas: implications for tree demography and carbon sequestration. *Global Change Biology* **16**, 331–343. doi:10.1111/j.1365-2486.2009.01933.x

Oliver JE (1980) Monthly precipitation distribution: a comparative index. *The Professional Geographer* **32**, 300–309. doi:10.1111/j.0033-0124.1980.00300.x

Preece N (2007) Traditional and ecological fires and effects of bushfire laws in north Australian savannas. *International Journal of Wildland Fire* **16**, 378–389. doi:10.1071/WF05079

R Development Core Team (2012) 'R: a Language and Environment for Statistical Computing.' R Foundation for Statistical Computing: Vienna, Austria. www.r-project.org Accessed 16 June 2014.

Richards AE, Brackin RD, Lindsay DAE, Schmidt S (2012) Effect of fire and tree-grass patches on soil nitrogen in Australian tropical savannas. *Austral Ecology* **37**, 668–677. doi:10.1111/j.1442-9993.2011.02335.x

Rossiter NA, Setterfield SA, Douglas MM, Hutley LB (2003) Testing the grass-fire cycle: alien grass invasion in the tropical savannas of northern Australia. *Diversity & Distributions* **9**, 169–176. doi:10.1046/j.1472-4642.2003.00020.x

Russell-Smith J, Yates CP, Edwards A, Allan GE, Cook GD, Cooke P, Craig R, Heath B, Smith R (2003a) Contemporary fire regimes of northern Australia: change since Aboriginal occupancy, challenges for sustainable management. *International Journal of Wildland Fire* **12**, 283–297. doi:10.1071/WF03015

Russell-Smith J, Whitehead PJ, Cook GD, Hoare JL (2003b) Response of *Eucalyptus*-dominated savanna to frequent fires: lessons from Munmarlary, 1973–1996. *Ecological Monographs* **73**, 349–375. doi:10.1890/01-4021

Russell-Smith J, Yates CP, Lynch B (2006) Fire regimes and soil erosion in north Australian hilly savannas. *International Journal of Wildland Fire* **15**, 551–556. doi:10.1071/WF05112

Russell-Smith J, Yates CP, Whitehead PJ, Smith R, Craig R, Allan G, Thackway R, Frakes I, Cridland S, Meyer CP, Gill AM (2007) Bushfires 'down under': patterns and implications of Australian landscape burning. *International Journal of Wildland Fire* **16**, 361–377. doi:10.1071/WF07018

Russell-Smith J, Edwards AC, Woinarski JCZ, McCartney J, Kerin S, Winderlich S, Murphy BP, Watt FA (2009a) Fire and biodiversity monitoring for conservation managers: a 10-year assessment of the 'Three Parks' (Kakadu, Litchfield and Nitmiluk) program. In *Culture, Ecology and Economy of Fire Management in North Australian Savannas: Rekindling the*

Wurrk *Tradition*. (Eds J Russell-Smith, PJ Whitehead, P Cooke) pp. 257–285. CSIRO Publishing: Melbourne.

Russell-Smith J, Murphy BP, Meyer CP, Cook GD, Maier S, Edwards AC, Schatz J, Brocklehurst P (2009b) Improving estimates of savanna burning emissions for greenhouse accounting in northern Australia: limitations, challenges, applications. *International Journal of Wildland Fire* **18**, 1–18. doi:10.1071/WF08009

Russell-Smith J, Whitehead PJ, Cooke PM (Eds) (2009c) *Culture, Ecology and Economy of Fire Management in North Australian Savannas: Rekindling the* Wurrk *Tradition*. CSIRO Publishing: Melbourne.

Russell-Smith J, Whitehead PJ, Cooke PM, Yates CP (2009d) Challenges and opportunities for fire management in fire-prone northern Australia. In *Culture, Ecology and Economy of Fire Management in North Australian Savannas: Rekindling the* Wurrk *Tradition*. (Eds J Russell-Smith, PJ Whitehead, P Cooke) pp. 1–22. CSIRO Publishing: Melbourne.

Russell-Smith J, Yates CP, Brock C, Westcott V (2010) Fire regimes and interval-sensitive vegetation in semi-arid Gregory National Park, Northern Territory. *Australian Journal of Botany* **58**, 300–317.

Russell-Smith J, Cook GD, Cooke PM, Edwards AC, Lendrum M, Meyer CP, Whitehead PJ (2013) Managing fire regimes in north Australian savannas: applying customary Aboriginal approaches to contemporary global problems. *Frontiers in Ecology and the Environment* **11**(SI), e55–e63. doi:10.1890/120251

Townsend SA, Douglas MM (2000) The effect of three fire regimes on stream water quality, water yield and export co-efficients in a tropical savanna (northern Australia). *Journal of Hydrology* **229**, 118–137. doi:10.1016/S0022-1694(00)00165-7

UNFCCC (1998) 'Kyoto Protocol to the United Nations Framework Convention on Climate Change.' United Nations, Bonn, Germany. http://unfccc.int/resource/docs/convkp/kpeng.pdf Accessed January 2013.

Walsh PD, Lawler DM (1981) Rainfall seasonality: description, spatial structure and change through time. *Weather* **36**, 201–208. doi:10.1002/j.1477-8696.1981.tb05400.x

Werner PA (2012) Growth of juvenile and sapling trees differs with both fire season and understorey type: Trade-offs and transitions out of the fire trap in an Australian savanna. *Austral Ecology* **37**, 644–657. doi:10.1111/j.1442-9993.2011.02333.x

Whitehead PJ, Bowman DMJS, Preece N, Fraser F, Cooke P (2003) Customary use of fire by Indigenous peoples in northern Australia: its contemporary role in savanna management. *International Journal of Wildland Fire* **12**, 415–425. doi:10.1071/WF03027

Whitehead PJ, Russell-Smith J, Cooke PM (2009) Fire management futures: new northern directions for environmental and socioeconomic benefit. In *Culture, Ecology and Economy of Fire Management in North Australian Savannas: Rekindling the* Wurrk *Tradition*. (Eds J Russell-Smith, PJ Whitehead, P Cooke) pp. 379–394. CSIRO Publishing: Melbourne.

Williams RJ, Griffiths AD, Allan GE (2002) Fire regimes and biodiversity in the wet-dry tropical landscapes of northern Australia. In *Flammable Australia: The Fire Regimes and Biodiversity of a Continent*. (Eds RA Bradstock, JE Williams, AM Gill) pp. 281–304. Cambridge University Press: Cambridge, UK.

Williams RJ, Woinarski JCZ, Andersen AN (2003) Fire experiments in northern Australia: contributions to ecological understanding and biodiversity conservation in tropical savannas. *International Journal of Wildland Fire* **12**, 391–402. doi:10.1071/WF03025

Williams RJ, Barrett D, Cook GD, Gill AM, Hutley L, Liedloff A, Myers B, Woinarski JCZ (2009) Landscape-scale fire research in northern Australia: delivering multiple benefits in a changing world. In *Culture, Ecology and Economy of Fire Management in North Australian*

Savannas: Rekindling the Wurrk *Tradition*. (Eds J Russell-Smith, PJ Whitehead, P Cooke) pp. 181–199. CSIRO Publishing: Melbourne.

Woinarski JCZ, Williams RJ, Price O, Rankmore B (2005) Landscapes without boundaries: wildlife and their environments in northern. *Australian Wildlife Research* **32**, 377–388. doi:10.1071/WR03008

Woinarski JCZ, Mackey B, Nix H, Traill B (2007) *The Nature of Northern Australia: Natural Values, Ecological Processes and Future Prospects*. Australian National University Press: Canberra.

Woinarski JCZ, Russell-Smith J, Andersen AN, Brennan K (2009) Fire management and biodiversity of the western Arnhem Land Plateau. In *Culture, Ecology and Economy of Fire Management in North Australian Savannas: Rekindling the* Wurrk *Tradition*. (Eds J Russell-Smith, PJ Whitehead, P Cooke) pp. 201–227. CSIRO Publishing: Melbourne.

Yates CP, Edwards AC, Russell-Smith J (2008) Big fires and their ecological impacts in Australian savannas: size and frequency matters. *International Journal of Wildland Fire* **17**, 768–781. doi:10.1071/WF07150

3

Fire extent mapping: procedures, validation and website application

Rohan Fisher and Andrew C Edwards

Summary

Our current understanding of global fire regimes is largely informed through satellite-derived fire mapping. The primary sources of satellite imagery are the continental-scale Moderate Resolution Imaging Spectroradiometer (MODIS) and Advanced Very High Resolution Radiometer (AVHRR) sensors from which are derived active fire, 'hot spot', and post-fire, burnt area, mapping. The mapping is particularly important for management in the highly fire-prone savanna landscapes of northern Australia, and also provides insights into the environmental, social and economic impacts of fire. Since 2003, fire mapping for northern Australia has been produced from MODIS data using a semi-automated object-based classification technique. Importantly, these data are updated regularly and made available through the North Australian Fire Information (NAFI) web portal for fire management planning and carbon and greenhouse gas emissions accounting. The robust nature of this dataset and its high public profile through NAFI have led to its becoming the baseline mapping legislated for use in the savanna burning greenhouse gas emissions abatement scheme, already approved for the higher rainfall regions of northern Australia.

In this chapter we describe the processes for producing the MODIS-derived burnt area mapping served through NAFI. We then undertake a thorough validation using an independent and extensive point dataset and compare it to a series of automated global fire mapping datasets through regression analysis of Landsat-derived burnt area mapping across five scenes in two seasons of 2 years.

The results demonstrate a high level of confidence in the product, with an overall accuracy of 88% when compared to the field data. When compared to other fire datasets the NAFI data was consistently more accurate. The accuracy of the global datasets decreased by nearly half compared to the NAFI and DLI AVHRR data when mapping smaller early dry season burnt areas. This is significant as these areas often have important implications for operational fire management. This suggests that although providing useful statistics (e.g. percentage area burnt, seasonality and frequency of fire), automated global and coarser resolution fire products are of limited use in operational land management. We suggest that for an operational product to be adopted by fire managers, a human operator is required in the mapping process to react to user requirements, and that an object-oriented classificatory approach, which more closely mimics human interpretation, provides a higher level of visual sophistication and accuracy to burnt area mapping.

Introduction

The AVHRR satellite sensor, launched in 1978, was the first to provide continental-scale mapping of thermal radiation and optical reflectance, thereby allowing for the development of large-scale active fire, 'hot spot', and burnt area datasets. The subsequent launch of the MODIS sensors in 1999 and 2000 provided a significant improvement in the spatial resolution of burnt area mapping from 1 km (AVHRR) to 250 m (MODIS) (Maier and Russell-Smith 2012). Key to the utility of these sensors is that they operate on multiple orbiting satellites, allowing for multiple image captures of the same place on the earth's surface each day. This high revisit frequency allows hotspot data to be updated every few hours, providing an important tool for active fire management. In addition, the high revisit frequency also reduces burnt area mapping omission errors due to cloud cover, and allows for the detection of cool or low fuel load fires that leave short-lived burn signatures.

Several global burnt area datasets have been produced from AVHRR and MODIS data through the application of automated algorithms. These algorithms track the reflectance response of an image pixel over time, flagging it as burnt when changes meet criteria that characterise the effects of fire. In this way fire may be effectively differentiated from other transient landscape changes due to such things as cloud cover, cloud shadow, soil moisture changes, tidal movement or flooding. These land cover types are hard to separate, which is why automated methods have higher commission errors, hence the requirement for a human operator. While automated mapping efficiently produces small-scale, consistent datasets, the results often vary between mapping algorithms. Furthermore, they all use a pixel-based classificatory approach which can produce a 'salt and pepper' effect whereby individual pixels meeting the burnt criteria are then erroneously classified as burnt. While only slightly affecting the overall accuracy or total burnt area, this speckled effect produces an end product that can be difficult for the end user (land manager) to interpret.

For regional fire data, semi-automated mapping methods are more commonly used. Semi-automated mapping relies on an experienced operator assessing the shape, scale and, importantly, context of fire with respect to landscape features, as well as the application of ancillary data (i.e. ignition point data provided by land managers and active fire map data) to identify burnt areas. While significantly more resource intensive, this mapping method produces 'cleaner', more user-friendly datasets free of the mapping anomalies common in automated data products. Thus, while automated mapping is an efficient way to map large areas and may produce statistically consistent data for overall analyses of fire regimes, semi-automated mapping can produce a product with more immediate utility to land managers.

One such service providing fire mapping to land managers is the North Australian Fire Information website (NAFI – www.firenorth.org.au). NAFI is used across north Australia to monitor active fires through hot spot data and to inform long-term fire management

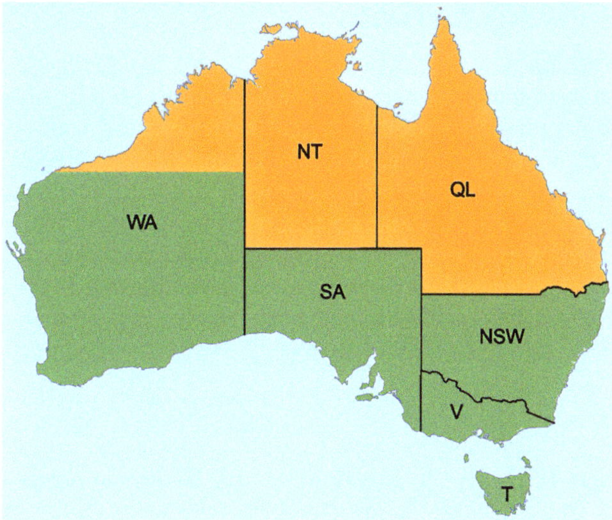

Figure 3.1. NAFI burnt area mapping coverage indicated in orange.

strategies with burnt area mapping. NAFI also hosts MODIS-derived burnt area mapping from 2000 to the present. This mapping has been produced using a consistent method by operators in two locales, one mapping all of the Northern Territory and Western Australia above 21°S, and the other mapping the state of Queensland (Fig. 3.1). Due to the extent, quality and accessibility of this mapping it has been selected as the data source for emissions and abatement calculations for Australia's first Carbon Farming Initiative: 'savanna burning', in the higher rainfall region of north Australia, and is presented here as it will be adopted in the methodology for the lower rainfall region.

A systematic accuracy assessment of the NAFI burnt area mapping is a requirement for its inclusion in the savanna burning methodologies. This assessment has been conducted over the past few years (2011–2013) by comparing the mapping to thousands of field 'ground truth' points attributed by burn status collected annually across northern Australia. This validation dataset is collected independently and does not inform any NAFI mapping activities. While resource intensive, this method produces a robust field dataset for validating the burnt area mapping.

For assessing global datasets the use of field data is logistically impractical. A more common approach is to use fire mapping derived from higher resolution satellite data, such as Landsat, for the comparison. Standard protocols have been developed for the assessment of MODIS burnt area mapping in this manner (Boschetti *et al.* 2006; Roy *et al.* 2005a).

This chapter describes the NAFI mapping and the field point data validation methods. In addition the NAFI mapping, along with several alternative burnt area datasets, both national and global, have been assessed for accuracy using Landsat derived mapping of selected scenes across northern Australia as the ground truth data. The purpose of these analyses is (1) to present the accuracy of the NAFI mapping using both field data and the internationally more common Landsat based approach, and (2) to contextualise the validity of NAFI data through comparison with other readily available fire datasets.

Burnt area mapping methods

Data pre-processing

An automated process acquires, pre-processes and archives MODIS imagery prepared for manual burnt area mapping. Raw MODIS image data is acquired from two satellites (Terra, Aqua) up to four times every day. These raw data (sourced from NASA: http://oceandata.sci.gsfc.nasa.gov) are processed (L1b processing with SeaDAS: http://seadas.gsfc.nasa.gov) to produce geo-located, at-sensor radiance calibrated imagery. Normalised top-of-atmosphere radiances are converted to top-of-atmosphere reflectances (i.e. no atmosphere correction is undertaken for these data), and an ocean mask is applied (Maier 2010). Imagery from the two 250 m MODIS bands are clipped to the NAFI mapping regions (WA, NT, QLD) and served in tiff file format via the internet for subsequent processing and classification.

Classification

The fire-affected area (FAA) mapping served through NAFI is processed every 1–2 weeks depending on fire frequency, cloud cover and operator availability. For each mapping period MODIS imagery with the least total cloud cover is selected or, where extensive cloud cover is persistent, the least cloud over areas of active fire as identified through active fire (hot spot) data.

The manual classification method uses an Object Based Image Analysis (OBIA) approach to segment the image into regular regions of similar spectral statistical classes. Image segmentation is the process of grouping multiple adjacent pixels into 'Objects' based on similar colour, shape and contextual characteristics. This differs from standard image analysis methods that classify each pixel in isolation of the contextual environment. The OBIA approach generally produces a cleaner map product, more meaningful to human users. The commercial eCognition® image analysis software is used to conduct the segmentation and initial classification. Pixels are grouped into image objects based primarily on the pre- versus post-fire change in the spectral response of the MODIS 250 m near-infrared (NIR, 841–876 nm) band. The occurrence and density of photosynthetic vegetation produces a strong spectral response in the NIR wavelengths, while the reflectance of dry grass is high over the entire spectral range. When burning occurs, the major reflectance change occurs in the NIR, particularly under dense tree cover, while the burning of dry grass causes an overall decrease in reflectance; thus the difference-image highlights fire-affected vegetation (Pereira 2003). These 'burnt' vector polygon objects are then separated into a set of fire classes based on threshold values of the mean reflectance change (Fig. 3.2).

Using a background image of pre- versus post-fire NIR change to guide selection, these classes are used to select FAAs manually.

To further assist the identification of burnt areas a range of ancillary data are overlayed on the imagery during the mapping process, including:

- active-fire data (hot spots). Derived from both AVHRR and MODIS imagery, they are used to help identify small fires and assist the mapping in cloud affected areas
- prescribed burning lines. Most prescribed burning of large areas in northern Australia is conducted using helicopter-delivered aerial incendiaries. Most of these burning activities are GPS logged and many land managers send this location data to NAFI to help guide mapping for these managed burns. This assures that their

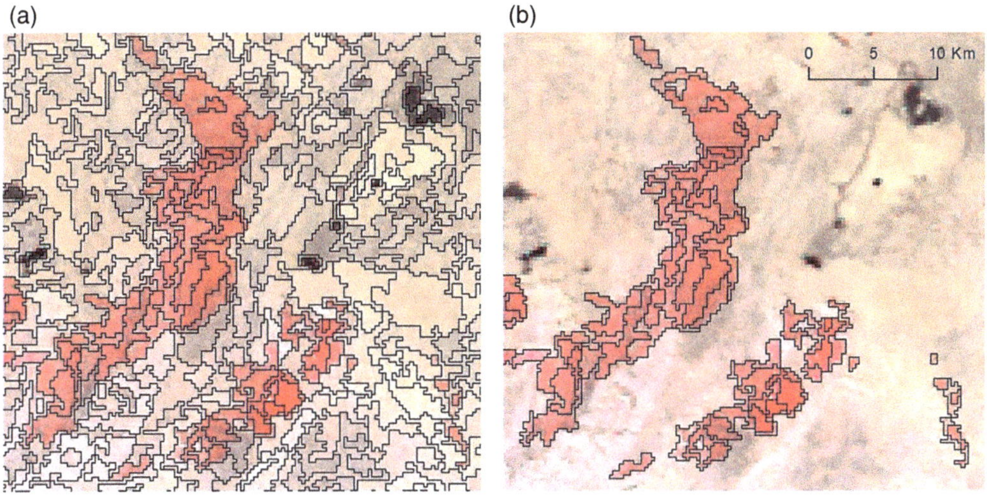

Figure 3.2. (a) Initial segmentation into image objects and (b) with threshold classification identifying burnt areas.

burns, as far as possible, are mapped correctly and quickly, thus increasing NAFI's utility as a planning tool while improving mapping accuracy
- hydrographic data. As rivers, floodplains and other bodies of water dry out through the course of the dry season, the change can appear similar to the effect of fire in the NIR. When these areas do burn, fire activity is confirmed through the use of active fire hot-spot data.

Another important component of the fire mapping process is user input. Land managers are encouraged to provide feedback on the mapping and identify areas that are incorrectly mapped. Responding to community input during the mapping not only increases the overall accuracy of the mapping, it also increases ownership of and confidence in the mapping product. The fact that the NAFI mapping is conducted in collaboration with land managers, not as an inflexible and uneditable automated process, is one reason for the widespread community engagement with this product.

The mapping is generally conducted at three approximate time periods, weekly, monthly and annually. The weekly mapping is produced to provide the most up-to-date burnt area mapping. Cloud cover often obscures parts of the mapping region during any weekly period. To improve coverage, additional mapping is conducted at the end of each month, using a selection of the best cloud-free imagery to capture regions previously covered by cloud. Finally, at the end of each year a retrospective assessment of the year's mapping is conducted using the best cloud-free imagery available for the whole year to once again capture any burnt areas not detected.

Data products

Once burnt area mapping is completed for a mapping period, each group of burnt area features is allocated a fire scar ID (FSID). The FSID is listed in an external Excel look-up table. Each FSID is attributed with information about the mapping period, the mapping region (NT, WA, QLD North, QLD South), the MODIS sensor and image dates used for that period of mapping, the month of the mapping period, and RGB values related to the mapping month to colour a raster file (.tif), exported in Arcmap (ESRI 2011). The final mapped data (the vector shape file and raster tif) are then uploaded to the NAFI site together with the FSID table. This allows users to download the data and associate it with the FSID data table for further analysis. In addition to this, base mapping a 'year since last burnt' raster layer is also produced. This data layer is derived by attributing all mapping for a given year with the value of that year (e.g. 2013, 2012, etc.) and overlaying the layers from the most recent back through time, illustrating the year a pixel was last affected by fire. In addition to these regular updates, longer term fire histories (i.e. Fire Frequency, Late Dry Season Fire Frequency) are compiled annually. These datasets, along with the archive covering earlier years, form the basis for much fire planning.

NAFI MODIS validation methods

Field point data accuracy assessment

The accuracy assessment method for MODIS-derived fire mapping has been improved and standardised in this region over previous decades (Edwards *et al.* 2001; Russell-Smith *et al.* 2003a,b; Edwards 2004; Edwards and Russell-Smith 2009) using a method initially developed for ground truth assessment for Landsat derived burnt area mapping in Kakadu National Park (Russell-Smith *et al.* 1997).

Ground truth data for the assessment are collected using a point-line intercept along randomly stratified transects in a helicopter travelling between 65 and 80 kn at an average height above ground of 500 feet (~150 m); the height varies based on the variability of the topography but is not usually more than 1200 feet (~365 m). Waypoints are collected using a GPS-enabled device and attributed as 'burnt', 'unburnt' or 'unknown' at regular (10 s) intervals. The 'unknown' points are discarded from the accuracy assessment. They generally occur where possible regrowth and ash dispersal have occurred due to the time since fire. In the tropical savannas this is most likely to occur 8 or more weeks after a low to moderate severity early dry season fire (Edwards *et al.* 2001; Russell-Smith *et al.* 2003a). We account for this when intersecting the point data and the burnt area mapping by removing all points that intersect fire scars detected in our burnt area maps that are older than 8 weeks from the date of the aerial transect. Transects are randomly stratified with two criteria: (1) transects cannot be more than 2.5 h flying time from the base (due to the lack of helicopter fuel in nearly all remote locations in northern Australia); and (2) transects should attempt to traverse as much burnt area as possible, although its availability is usually limited.

The remaining burnt/unburnt waypoints are labelled uniquely and buffered 100 m to compensate for spatial inaccuracies in the GPS-derived waypoint information. These buffered areas are intersected with the burnt area mapping and the result output to a spreadsheet. A new field is created; where the whole or any part of the buffered waypoint labelled as 'burnt' intersects a burnt area the result is 1, and similarly with a buffered waypoint labelled as 'unburnt' intersecting any part of an unburnt area. If the 'burnt' buffered waypoint wholly intersects an unburnt area or an 'unburnt' buffered waypoint wholly intersects a burnt area, the result is 0. A pivot table of the new field and the mapping is produced comparing the remotely sensed mapping to the field-collected waypoints. This is essentially the raw error matrix. Percentage errors of overall accuracy, omission and commission and the $\hat{K}$statistic are calculated, as given in Table 3.1. The kappa coefficient measures how well the classification performed compared to the probability of assigning points to their correct categories (Congalton 1991).

Table 3.1. The error matrix for a binary classification, used in the accuracy assessment of burnt area mapping

	Burnt area mapping			
Validation waypoints		Burnt	Unburnt	Total
	Burnt	A	B	A + B
	Unburnt	C	D	C + D
	Total	A + C	B + D	A + B + C + D
Overall accuracy	= (A + D) / (A + B + C + D)			
Omission error	Burnt = A / (A + C)			Unburnt = D / (B + D)
Commission error	Burnt = A / (A + B)			Unburnt = D / (C + D)
$\hat{K}$ statistic	$= (((A+D)*(A+B+C+D))-(((A+B)+((C+D)*(B+D))))/((A+B+C+D)^2-(((A+B)*(A+C))+((C+D)*(B+D))))$			

This method has been applied to assess the accuracy of NAFI fire mapping for three consecutive years, 2011–2013, amounting to over 10 000 km of aerial and ground transects and ~20 000 ground data points, Fig. 3.3.

Landsat-based accuracy assessment

NAFI data and three alternative burnt area map products, two global (MODIS) and one continental scale (AVHRR), were used for the Landsat-based comparative assessment. A regression approach, adapted from similar studies conducted in Africa (Eva and Lambin 1998) and North Australia (Yates and Russell-Smith 2002), was used to assess the accuracy of these data in comparison to Landsat-derived mapping. This method intersects the burnt area datasets, Landsat and MODIS/AVHRR, with a common grid layer dividing each dataset into several analysis cells, Fig. 3.4. For this analysis a 0.1° grid cell (~10 km) matrix

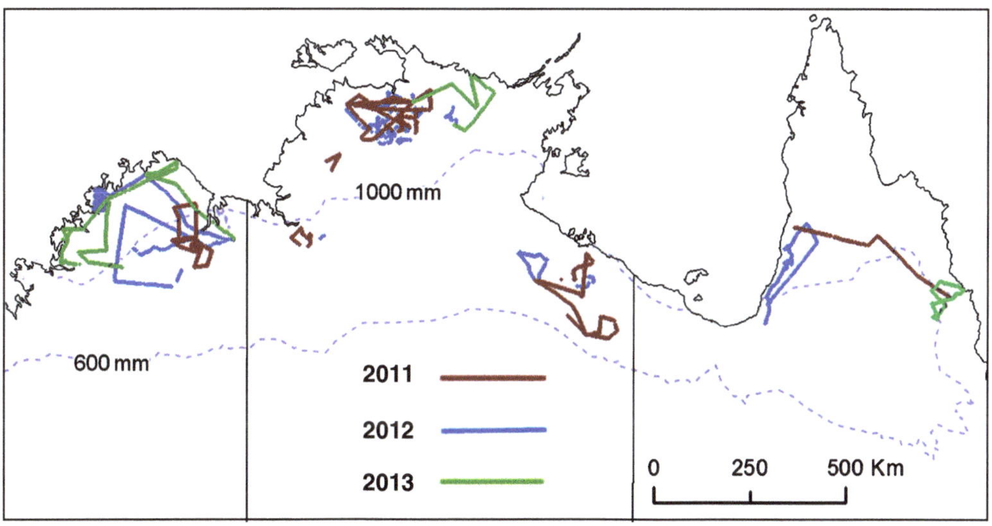

Figure 3.3. Map of north Australia illustrating the coverage by aerial and ground validation transects, 2011–13. (Grey dashed lines represent the 600 and 1000 mm annual rainfall isohyets).

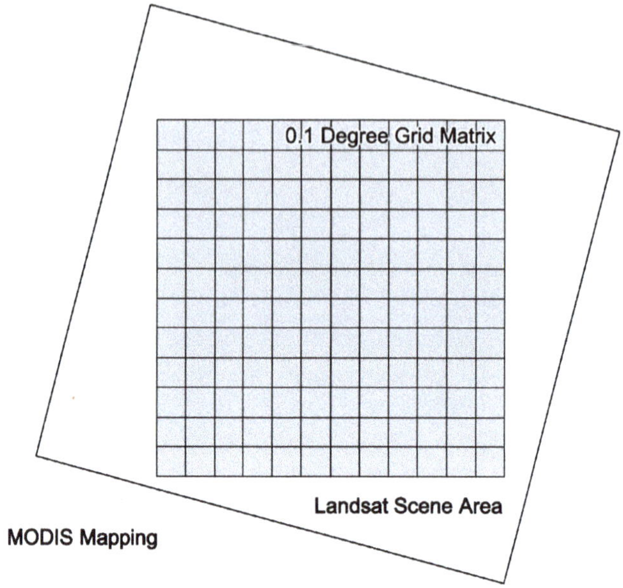

Figure 3.4. A schematic of the regression assessment method showing the intersection of a reference grid cell matrix with the Landsat and MODIS/AVHRR burnt area datasets.

was created that covered the Landsat-mapped area for each of the mapped regions. The burnt area for each grid cell is then mapped in a regression model.

This regression method allows for a robust comparison of the areal datasets, as it reduces the relative scale issues associated with comparing the finer resolution data with coarse resolution satellite imagery. In addition it is an assessment of not only cell by cell accuracy through the Pearson coefficient but also the slope of the line also characterises the overall mapping accuracy.

Five Landsat scenes across Northern Australia were selected to provide the 'ground truth' data. The scenes are in: the Kimberley (108–71); the Victoria River District (106–71); Arnhem Land (104–69/70); the NT Gulf (101–72), and Cape York (98–71), Fig. 3.5. For each

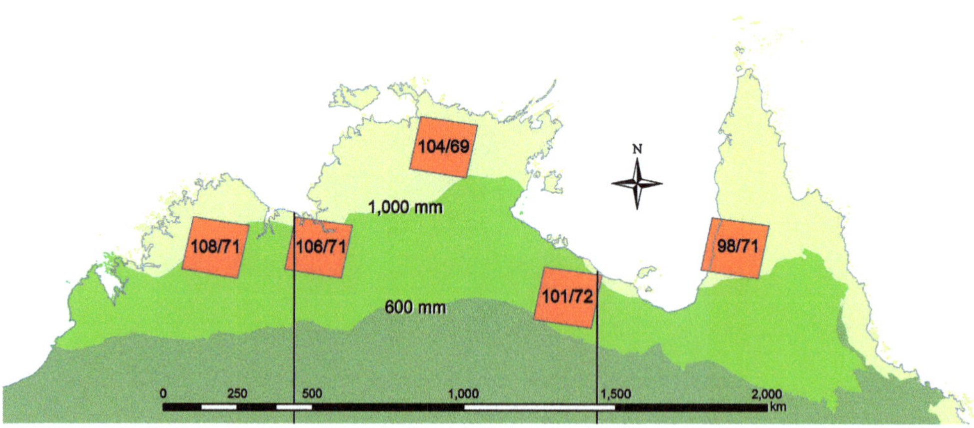

Figure 3.5. The five Landsat scenes across northern Australia used for the MODIS accuracy assessment. The high and low rainfall regions are represented in light and bright green respectively.

scene, burnt area mapping was produced for four mapping periods: the early and the late dry seasons of 2004 and 2011.

The Landsat burnt area mapping was conducted following the protocol outlined by Roy *et al.* (2005a). This approach requires the use of image pairs, and mapping fires between image dates. This allows for a temporal comparison between mapped fires from the MODIS imagery. In the tropics burned areas are a temporary land cover change, so it is therefore necessary to use image pairs in a time interval shorter than it takes for them to become undetectable. The amount of time that a burned area is visible depends on the changes in the post-burn surface. This can be due to vegetation regrowth, and/or dissipation of ash and charcoal by wind and rain. This is in stark contrast to the spinifex grasslands in central Australia or forest fires of southern Australia, where the fire effect may be visible for years. For this study image pairs were chosen with a period less than 2 months, as outlined in Edwards and Russell-Smith (2009), Table 3.2.

For the Arnhem Land scene, pre-existing mapping created for the West Arnhem Land Fire Abatement project (WALFA) was used (Edwards and Russell-Smith 2009). Also applying semi-automated object-oriented classificatory methods, it includes components of two scenes (104/69 and 70) to cover the Arnhem Plateau, a region of high conservation significance (Russell-Smith *et al.* 2012, 2009). All mapping was produced using a standard difference image using Landsat near-infrared and thermal bands to highlight change between image dates. This approach has been used extensively across northern Australia to map fire-affected areas. The accuracy of this Landsat mapping, routinely assessed using field data, has been shown to be consistently high (>80%) (Edwards *et al.* 2001; Fisher *et al.* 2003; Russell-Smith *et al.* 2003a; Yates and Russell-Smith 2003), providing confidence in the application of Landsat-derived burnt area mapping data as 'ground truth' for accuracy assessment in these landscapes. For this study, Ecognition image segmentation software (Definiens Imaging 2004) was then used to classify areas of significant spectral change due to fire. The classified dataset was then cleaned manually with reference to the difference image to produce the final burnt area map. This mapping was compared to the NAFI

Table 3.2. Landsat image dates used for the base line accuracy assessment data

	04 Early	04 Late	11 Early	11 Late
Kimberley	26/05/04	29/07/04	28/04/11	3/09/11
	27/06/04	30/08/04	15/06/11	2/08/11
VRD	26/04/04	07/31/04	19/04/13	19/07/11
	29/06/04	1/09/04	5/05/13	1/09/11
Gulf	09/05/04	29/08/04	27/04/11	17/08/11
	26/06/04	01/11/04	14/06/11	18/09/11
Arnhem Land	15/06/04	*	03/06/11	06/08/11
	08/08/04	*	05/07/11	23/09/11
Cape York	2/04/04	25/09/04	8/05/11	28/08/11
	5/06/04	27/10/04	25/06/11	31/10/11

* For the Arnhem Land scene no late dry season 2004 Landsat imagery was available for a time period that matched the NAFI MODIS mapping. Therefore to avoid errors due to a temporal mismatch of the mapping this period was excluded from the accuracy assessment.

Table 3.3. Attributes of burnt area products

Comparison datasets	Sensor	Spatial resolution	Spatial extent	Temporal extent	Update frequency	Latency	Map method
NAFI	MODIS	250 m	WA, NT,QL	2000–Present	Weekly	<week	Semi-auto
MCD45A1	MODIS	500 m	Global	2000–Present	Monthly	3 months	Auto
MCD64A1	MODIS	500 m	Global	2000–Present	Monthly	3 months	Auto
AVHRR DLI	AVHRR	1.1 km	Australia	Jul 89–Present	Yearly	1 year	Semi-auto

mapping, two global MODIS derived datasets, and one continental scale AVHRR dataset (Table 3.3).

The global MODIS products have been developed though collaboration between the University of Maryland and NASA. They have been used for this assessment as they are the most commonly used and referenced global fire datasets. They produce monthly gridded 500 m products, with each cell attributed by day of burn. For this study the data were accessed via the AusCover Remote Sensing Data facility (www.auscover.org.au/data/product-list).

MCD45A1. This product maps burnt areas using a multitemporal surface reflectance data approach based on methods developed by Roy *et al.* (2005b; Roy *et al.* 2002). This dataset has been developed by the Time Series Remote Sensing Team, CSIRO Marine and Atmospheric Research in Australia. The original data were supplied by the Land Processes Distributed Active Archive Center (LPDAAC), located at the USA Geological Survey (USGS) Earth Resources Observation and Science Center (EROS) (http://lpdaac.usgs.gov).

MCD64A1. This product maps burnt areas using a multitemporal surface reflectance data approach based on methods described by Giglio *et al.* (2009). The original data were developed by the MODIS Active Fire and Burned Area Products Team (http://modis-fire.umd.edu) and the Department of Geographical Sciences, University of Maryland (http://geog.umd.edu).

The **AVHRR dataset**, developed by the Western Australian Department of Administration (now Department of Land Information – DLI), is the longest fire history record for Australia. Burnt areas are mapped using a semi-automated change detection procedure (Craig *et al.* 2002) from NOAA-AVHRR daytime imagery using visible, near-infrared and thermal bands (Bands 1, 2, 5), every nine days of the repeat NOAA cycle. The AVHRR fire data is important because it was the only continental scale burnt area data available until MODIS burnt area mapping became available in 2003. The AVHRR data continues to be used to provide baseline fire history data back to 1990 for the Australian continent, and is used to estimate fuel loads for carbon accounting calculations before 2000. These data are publicly available for viewing through the DLI Landgate – Fire Watch website (http://firewatch.dli.wa.gov.au).

Analyses were conducted to show total accuracy, by location and by season (early and late dry season). Cells for which no fire was mapped in any of the burnt area products are

Table 3.4. Accuracy assessments using aerial validation data collected across north Australia throughout the fire season of the given year

Year	2011	2012	2013
Number of waypoints (n)	9761	7324	6527
Overall accuracy	**0.86**	**0.90**	**0.85**
Burnt omission error	0.14	0.04	0.09
Burnt commission error	0.09	0.20	0.10
Unburnt omission error	0.06	0.14	0.12
Unburnt commission error	0.17	0.02	0.06
Kappa statistic	0.71	0.79	0.70

not included. An additional analysis was conducted in cells less than 15% (~15 km^2) burnt in total to compare the ability of the various map products to map small burnt areas.

Point data accuracy assessment

The results, Table 3.4, demonstrate a high level of confidence in the NAFI burnt area mapping, with high overall accuracy in all years. The result also indicates consistently low commission and omission errors in burnt and unburnt areas. A good proportion of the error occurs on the burnt area edges, due to image resolution and the algorithm threshold.

Landsat data regression assessment

The NAFI mapping showed overall better results when compared to the other broad-scale datasets, particularly when mapping smaller burnt areas (Table 3.5). In addition it can be seen that the other datasets consistently under-mapped (m values < 1) fires when compared to the Landsat data. This is particularly clear in the AVHRR DLI-derived product, which showed a comparatively very low accuracy when mapping smaller burnt areas. While the AVHRR DLI dataset showed a high level of scatter, the estimations of total areas burnt (m) were consistently good. Despite this it is useful to note that the MCD64A1 map product produced a high level of overall accuracy although it also significantly under-mapped smaller fires.

Regression assessment by location

The five location areas were assessed individually, Fig. 3.6. The Cape and VRD mapping showed consistently good results across the three MODIS-derived mapping products

Table 3.5. Pearson's correlation coefficient and slope of line of best fit (constrained to zero) from linear regression analyses of Landsat vs. Global/National burnt area mapping

This table shows the results for all data and for cells where the percentage burnt was <15%.

	All data *n* = 1551		<15% cell burnt *n* = 997	
	R^2	m	R^2	m
NAFI	0.90	1.00	0.47	1.1
MCD64A1	0.86	0.86	0.42	0.50
MCD45A1	0.76	0.71	0.31	0.45
AVHRR DLI	0.69	0.95	0.20	1.0

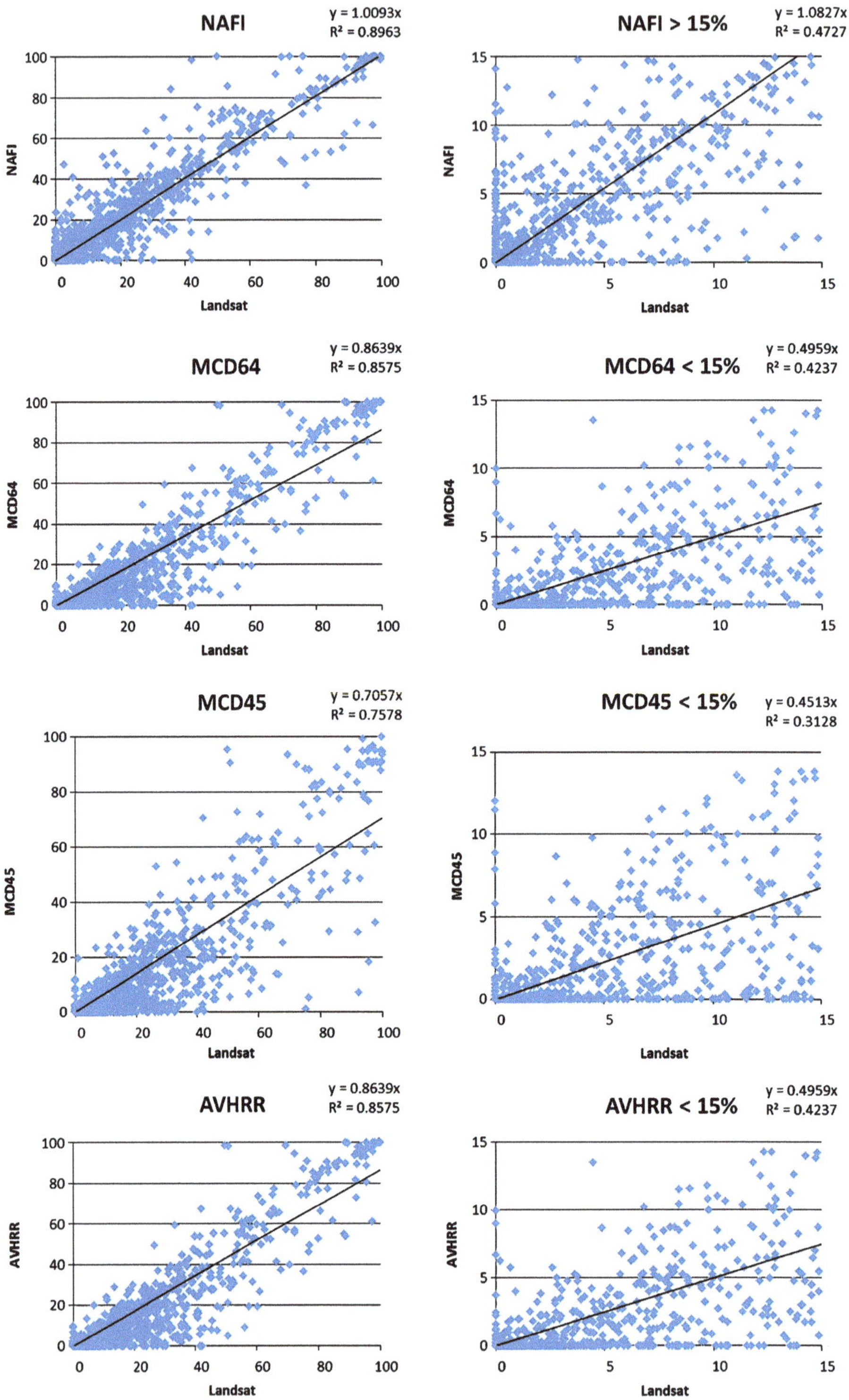

Figure 3.6. Landsat vs. National/Global Burnt Area Mapping product: regression analyses. All values on axes indicate the percentage of a cell burnt.

Table 3.6. Results from regression analyses (Landsat vs. AVHRR/MODIS-derived burnt area mapping) comparing the five locations across north Australia

		NAFI	MCD64	MCD54	AVHRR
Cape	R^2	0.95	0.94	0.81	0.69
n = 305	m	1.05	0.98	0.82	1
Gulf	R^2	0.9	0.87	0.73	0.79
n = 295	m	0.91	0.88	0.62	1.1
Kimberley	R^2	0.9	0.86	0.72	0.79
n = 282	m	0.91	0.87	0.62	1.07
VRD	R^2	0.87	0.87	0.82	0.6
n = 284	m	0.81	0.78	0.67	0.8
WALFA	R^2	0.77	0.65	0.59	0.55
n = 507	m	1.1	0.52	0.42	0.71

(R^2 >0.82), Table 3.6. The least accurate results were from mapping in the rugged dissected sandstone plateau country of the Western Arnhem Land Fire Abatement (WALFA) area. In WALFA all the map products other than NAFI significantly underestimate burnt area, Table 3.6. The mapping accuracies for the two scenes mostly below the 1000 mm isohyet (VRD, GULF) were high for the NAFI product and not markedly different from the above 1000 mm scenes across all the mapping products.

Analysis by season

All map products produced consistently poorer results in mapping early compared to late dry season fires, Table 3.7. Late dry season fires are more often larger, much less patchy and more severe, producing large and distinct fire scars that more often affect the upper canopy and are therefore more readily discernible on satellite imagery. Significantly, NAFI produced substantially better results (m = 1) mapping early dry season fires than the other products. The MCD64 product, which performed well overall, showed markedly poorer results when mapping early fires. The two MODIS automated products both substantially under-mapped fires in the early dry season.

Discussion

The results demonstrate a high level of confidence in the NAFI burnt area mapping, with high overall accuracy, 86%, 90% and 85% in 2011, 2012 and 2013 respectively, when assessed with the aerial survey data. Furthermore, when compared to other fire datasets

Table 3.7. Accuracy results from regression analyses (Landsat vs. AVHRR/MODIS) derived from early (1 January–30 July) and late (1 August–31 December) dry season mapping

		NAFI	MCD64	MCD45	AVHRR
Early	R^2	0.82	0.72	0.63	0.63
	m	1	0.68	0.51	0.89
Late	R^2	0.92	0.93	0.83	0.81
	m	0.98	0.94	0.79	0.92

the NAFI data was consistently more accurate and showed markedly better results mapping early dry season and smaller burnt areas. Some variation was shown in mapping accuracy between different sites across north Australia, particularly in the topographically complex sandstone landscapes of WALFA. It is also useful to note that the accuracy assessment in the lower rainfall areas (Gulf, VRD) showed minimal difference from that for the higher rainfall areas.

As presented in these analyses, mapping error was consistently higher in the early dry season when iterative fire management relies on accurate knowledge of the occurrence of fire, particularly with respect to the efficacy of prescribed burning activities. Early dry season fires form important strategic fire breaks and are often relatively small and patchy as ground fuel moisture is still high. The ability for NAFI mapping to identify these early dry season burnt areas more readily than other available burnt area mapping products is significant for operational planning purposes. Furthermore, with regard to greenhouse gas emissions and abatement calculations, the NAFI mapping was exceptionally accurate in the early dry season ($m = 1$). In addition to its higher accuracy the NAFI mapping process produces a product more easily interpretable by land managers. While the overall accuracy of the automated products was generally good, a visual assessment shows commission errors often appearing as scattered pixels, whereas commission errors in the NAFI mapping are more often spatially associated with correctly mapped burnt areas. This is due to the application of object-based image analysis for classifying burnt areas and a post-classification 'cleaning' conducted by a human operator using shape, pattern and contextual information. This method results in a product free of the mapping anomalies often present in pixel-based automated products. While providing useful statistics (e.g. % area burnt), it is clear from this assessment that automated global and coarser resolution fire products are of limited use in operational land management.

In addition to its 'usability', the ability for land managers to participate in the NAFI mapping process, through providing ancillary data such as burn lines and/or direct field-based feedback, creates a sense of collaboration with the service. This aspect of NAFI has helped to build user confidence, resulting in its becoming the default fire management web service across northern Australia. The way NAFI is used by land managers, with local knowledge and other datasets, for developing and implementing strategic fire management plans is described in detail by Jacklyn *et al.* in Chapter 5.

NAFI is providing a useful land management service, however the accuracy and utility of the mapping product could be further improved. The manual mapping process is, for example, more likely to miss burnt areas during cloudy periods than auto-mapping processes that use every available overpass. One way to reduce these cloud-induced omission errors is to use auto-mapping products, such as described in this chapter, for post-mapping quality assessment to identify potential omissions. The MCD64 dataset, for example, could assist understanding of the timing and extent of fires. Of interest here is the ongoing development of a 250 m burnt area product for Australia by CDU (Maier 2010). This could provide an opportunity to incorporate high temporal resolution auto-mapping data into the NAFI process at the same spatial resolution. In addition, national auto-products, under development at Landsat scale, will potentially provide complementary data to refine the spatial accuracy of fire information for northern Australian land managers. Another issue affecting NAFI is the potential for inconsistency between mapping staff, and the low temporal resolution, i.e. a week or more between fire map updates, during critical controlled burning periods. These issues are addressed in Chapter 5.

Acknowledgements

We would like to especially thank NASA for all the data they generously donate to the earth observation sciences all around the world; without these data none of this science would exist in such an advanced form. Thanks to Dr Stefan Maier at CDU for the excellent work he does in creating the daily image archive of MODIS imagery, and other derived datasets. We would also like to thank the West Arnhem Land fire managers for the Landsat mapping over that area.

References

Boschetti L, Brivio PA, Eva HD, Gallego J, Baraldi A, Gregoire JM (2006) A sampling method for the retrospective validation of global burned area products. *IEEE – Transactions on Geoscience and Remote Sensing* **44**, 1765–1773.

Congalton RG (1991) A review of assessing the accuracy of classifications of remotely sensed data. *Remote Sensing of Environment* **37**(1), 35–46.

Craig R, Heath B, Raisbeck-Brown N, Steber M, Marsden J, Smith R (2002) The distribution, extent and seasonality of large fires in Australia, April 1998–March 2000, as mapped from NOAA-AVHRR imagery. In *Australian Fire Regimes: Contemporary Patterns (April 1998 – March 2000) and Changes Since European Settlement* (Eds J Russell-Smith, R Craig, AM Gill, R Smith, J Williams). Australia State of the Environment Second Technical Paper Series (Biodiversity). Australian Government Department of the Environment and Heritage: Canberra. www.environment.gov.au/system/files/pages/7965c869-6e54-40e1-b7d4-a9b119a1e98e/files/part1.pdf

Definiens IMAGING (2004) *eCognition User's Guide 4*. Definiens Imaging: München, Germany.

Edwards A (2004) The validation of Landsat, MODIS and NOAA AVHRR derived fire products in tropical northern Australia. The 12th Australasian Remote Sensing & Photogrammetry Conference, Perth, WA.

Edwards A, Russell-Smith J (2009) Ecological thresholds and the status of fire-sensitive vegetation in western Arnhem Land, northern Australia: implications for management. *International Journal of Wildland Fire* **18**, 127–146. doi:10.1071/WF08008

Edwards A, Hauser P, Anderson M, McCartney J, Armstrong M, Thackway R, Allan G, Hempel C, Russell-Smith J (2001) A tale of two parks: contemporary fire regimes of Litchfield and Nitmiluk National Parks, monsoonal northern Australia. *International Journal of Wildland Fire* **10**, 79–89. doi:10.1071/WF01002

ESRI (2011) ArcGIS Desktop: Release 10. Environmental Systems Research Institute: Redlands, CA.

Eva H, Lambin EF (1998) Remote sensing of biomass burning in tropical regions: Sampling issues and multisensor approach. *Remote Sensing of Environment* **64**, 292–315. doi:10.1016/S0034-4257(98)00006-6

Fisher R, Vigilante T, Yates C, Russell-Smith J (2003) Patterns of landscape fire and predicted vegetation response in the North Kimberley region of Western Australia. *International Journal of Wildland Fire* **12**, 369–379. doi:10.1071/WF03021

Giglio L, Loboda T, Roy DP, Quayle B, Justice CO (2009) An active-fire based burned area mapping algorithm for the MODIS sensor. *Remote Sensing of Environment* **113**, 408–420. doi:10.1016/j.rse.2008.10.006

Maier SW (2010) Changes in surface reflectance from wildfires on the Australian continent measured by MODIS. *International Journal of Remote Sensing* **31**(12), 3161–3176.

Maier SW, Russell-Smith J (2012) Measuring and monitoring of contemporary fire regimes in Australia using satellite remote sensing. In *Flammable Australia: Fire Regimes, Biodiversity and Ecosystems in a Changing World.* (Eds RA Bradstock, AM Gill, RJ Williams) pp. 79–95. CSIRO Publishing, Melbourne.

Pereira JMC (2003) Remote sensing of burned areas in tropical savannas. *International Journal of Wildland Fire* **12**, 259–270. doi:10.1071/WF03028

Roy D, Lewis P, Justice C (2002) Burned area mapping using multi-temporal moderate spatial resolution data – A bi-directional reflectance model-based expectation approach. *Remote Sensing of Environment* **83**, 263–286. doi:10.1016/S0034-4257(02)00077-9

Roy D, Frost P, Justice C, Landmann T, Le Roux J, Gumbo K, Makungwa S, Dunham K, Du Toit R, Mhwandagara K (2005a) The Southern Africa Fire Network (SAFNet) regional burned-area product-validation protocol. *International Journal of Remote Sensing* **26**, 4265–4292. doi:10.1080/01431160500113096

Roy D, Jin Y, Lewis P, Justice C (2005b) Prototyping a global algorithm for systematic fire-affected area mapping using MODIS time series data. *Remote Sensing of Environment* **97**, 137–162. doi:10.1016/j.rse.2005.04.007

Russell-Smith J, Ryan PG, Durieu R (1997) A LANDSAT MSS-derived fire history of Kakadu National Park, monsoonal northern Australia, 1980–94: seasonal extent, frequency and patchiness. *Journal of Applied Ecology* **34**, 748–766. doi:10.2307/2404920

Russell-Smith J, Edwards AC, Cook GD (2003a) Reliability of biomass burning estimates from savanna fires: biomass burning in northern Australia during the 1999 Biomass Burning and Lightning Experiment B field campaign. *Journal of Geophysical Research – Atmospheres* **108**(D3), 8405. doi:10.1029/2001JD000787

Russell-Smith J, Yates C, Edwards A, Allan GE, Cook GD, Cooke P, Craig R, Heath B, Smith R (2003b) Contemporary fire regimes of northern Australia, 1997–2001: change since Aboriginal occupancy, challenges for sustainable management. *International Journal of Wildland Fire* **12**, 283–297. doi:10.1071/WF03015

Russell-Smith J, Whitehead P, Cooke P (Eds) (2009) *Culture, Ecology and Economy of Fire Management in North Australian Savannas: Rekindling the* Wurrk *Tradition.* CSIRO Publishing: Melbourne.

Russell-Smith J, Edwards AC, Price OF (2012) Simplifying the savanna: the trajectory of fire-sensitive vegetation mosaics in northern Australia. *Journal of Biogeography* **39**, 1303–1317. doi:10.1111/j.1365-2699.2012.02679.x

Yates C, Russell-Smith J (2002) An assessment of the accuracy of DOLA's Northern Australia NOAA-AVHRR Fire Affected Area (FAA) map products. *Assessing Fire Patterns and their Environmental Impacts.* (Eds J Russell-Smith, R Craig, AM Gill, R Smith, J Williams). State of the Environment Unit, Environment Australia: Canberra. www.environment.gov.au/soe/2001/publications/technical/fire/pubs/appendix2.pdf Accessed 5 October 2012.

Yates C, Russell-Smith J (2003) Fire regimes and vegetation sensitivity analysis: an example from Bradshaw Station, monsoonal northern Australia. *International Journal of Wildland Fire* **12**, 349–358. doi:10.1071/WF03019

4

Vegetation fuel type classification for lower rainfall savanna burning abatement projects

Dominique Lynch, Nicholas Cuff and Jeremy Russell-Smith

Summary

A key requirement for registered fire abatement projects under Australia's Carbon Farming Initiative (CFI) is a map of the vegetation fuel types. Designated vegetation fuel types are functional groups describing a relatively homogenous broad-scale area of flammable (fire-dependent or fire-sensitive) vegetation characterised by a common suite of fuel components and which exhibit characteristic emission responses to fire. Each type is fundamentally a structural vegetation group with a distinctive ground cover. The chapter first describes broad fuel types unique to the regional envelope encompassing the proposed savanna burning lower rainfall (1000–600 mm) zone. At least five vegetation fuel types (woodlands over tussock grasses; woodlands over mixed tussock and hummock grasses; woodlands over hummock grasses; open woodlands over hummock and tussock grasses; shrublands over hummock grasses) meet requisite criteria for inclusion in the CFI. The latter part of the chapter addresses issues relating to appropriate mapping and validation requirements.

Introduction

Savanna burning fire abatement projects are designed to reduce the extent of wildfire and thereby reduce emissions of accountable greenhouse gases through improved fire management. A core requirement for fire management and carbon accounting in CFI projects is suitable vegetation fuel type mapping. For northern Australia, various state and territory vegetation mapping datasets are available, but generally these are regional products (with the exception of Queensland) and variable in scale and classification.

The current CFI methodology for estimating emissions in higher rainfall savannas (above the 1000 mm isohyet) describes four vegetation fuel types: eucalypt open forest, eucalypt woodland, sandstone woodland, sandstone heath (DCCEE 2013). Description of these fuel types was based on research undertaken in the western Arnhem Land region (Edwards and Russell-Smith 2009) and, for the purposes of that abatement methodology, was applied generally across the higher rainfall savannas.

Myers (2006) notes that there are three classes of ecosystems exhibiting different fire regime responses: (1) fire-dependent ecosystems – those that need a specific regime of fire and are characterised by plant species with adaptations that respond positively to fire; (2) fire-sensitive ecosystems – those that are negatively impacted by most fires and where plant species are generally intolerant of fire; and (3) fire-independent ecosystems – those where fire normally plays little or no role. Those classes that are identified as fire-independent ecosystem classes are not applicable in fire abatement projects and the CFI methodology. Other vegetation class exceptions currently in the higher rainfall CFI methodology that would also be ineligible in the lower rainfall zone are seasonally inundated swamps, grasslands and woodlands typically dominated by *Melaleuca*, and eucalypt open forests and woodlands with ground layers infested with large exotic perennial and annual grasses. Ecosystems such as fire-sensitive ecosystems that benefit from early dry season burning practices to reduce late season wildfires (Edwards and Russell-Smith 2009) and would as a consequence have reduced greenhouse gas emissions would be eligible as a fuel type class for CFI methodology.

There are two major assumptions from the higher rainfall CFI methodology that will apply to fuel classification and mapping in the lower rainfall zone. The first is that only relatively unmodified native vegetation (i.e. uncleared, lightly grazed vegetation) qualifies for inclusion in the lower rainfall savanna burning abatement methodology. A description of unmodified native vegetation is presented in Thackway and Lesslie (2008). The second is that mapping of all vegetation fuel types requires the same spatial resolution of a minimum pixel size of 250 m.

The purpose of this chapter is to present a classification of vegetation fuel types for the CFI carbon accounting methodology for the region occurring between the 1000–600 mm rainfall isohyets, referred to as the 'lower rainfall' zone. These types are characterised by their vegetation structure (characterising system biomass) and described by common key component species and typical land type upon which they may occur. The major vegetation structural types are first described to give a broad overview of the major fuel types that occur in the north Australian lower rainfall zone. Possible fuel types are considered, and a final five are proposed for inclusion in a CFI methodology. Lastly, vegetation fuel type mapping and validation methods and associated issues, particularly those relevant to CFI methodological requirements, are discussed.

Vegetation fuel type classification

We define a vegetation fuel type as a functional group with a relatively homogenous broad-scale area of flammable (fire-dependent or fire-sensitive) vegetation that is characterised by a common suite of fuel components and exhibits characteristic responses (patchiness, burning efficiencies, and emission factors) to fire. Fuel components include woody debris material (heavy and coarse fuels), leaf litter, graminoid and herbaceous vegetation (fine fuels), and shrubs (refer Yates *et al.* Chapter 6 for definitions). These fuel components are derived from vegetation overstorey and understorey floristic and structural characteristics (Sanders 1985). For practical greenhouse gas emissions calculation purposes, and for meeting the requirements of Australia's National Greenhouse Gas Inventory (NGGI) (DCCEE 2013), a minimum set of vegetation fuel types is required.

The vegetation fuel type classification emphasises vegetation structure rather than species associations, although some may be differentiated based on their generic species characteristics. The structural formation terminology broadly uses the National

Vegetation Information System (NVIS) (ESCAVI 2003) scheme. NVIS Level IV would be the minimum set of vegetation attributes within vegetation datasets that is needed to describe potential fuel types for the purposes of savanna burning abatement projects.

Fuel types are named to reflect the structural formation, and the description includes dominant growth form, canopy height, and foliage projective cover (FPC) of the stratum. It is assumed that the dominant stratum is that with the greater biomass and is the major delineating component of the fuel type as it affects the proportion and accumulation of ground fuel components. It is important to note here that the woody stratum which makes up vegetation fuel types is divided into trees and shrubs. A tree is defined as any woody stem with a diameter at breast height (DBH) ≥5 cm (Edwards and Russell-Smith 2009). All other woody vegetation with DBH <5 cm is denoted as a 'shrub', including single and multi-stemmed shrubs and young trees. The ground layer is a major component in the fuel type delineation and usually dominated by graminoids, which in turn provide significant inputs to the fine fuel component. This is particularly pertinent in northern Australian savannas where fires are primarily ground-borne (Williams *et al.* 2002).

Development of the fuel type classification for the lower rainfall project zone began with collating described vegetation communities from available land systems and vegetation mapping in the Northern Territory (www.lrm.nt.gov.au/nrmapsnt), and Western Australia (Schoknecht and Payne 2011), and Regional Ecosystem mapping in Queensland (www.ehp.qld.gov.au/ecosystems/biodiversity/regional-ecosystems/maps). Most of these datasets classify vegetation at the floristic association or vegetation community level (equivalent to the Sub-Formation Level IV in the NVIS hierarchy, Table 4.1). The community descriptions recognise dominant structural attributes, life-forms and species in each stratum. Vegetation fuel classes were created based on dominant woody stratum structural attributes (FPC, height), and dominant graminoid life-form in the ground stratum. Those classes considered different in terms of apparent burning efficiencies were separated. To assess the efficacy of this classification, we cross-checked these mapped classes with equivalent attributes collected as part of extensive field data (refer Yates *et al.* Chapter 6) for reference.

NVIS Level III is equivalent to the Major Vegetation Groups (MVGs) which have been used to create a national map. There are 33 MVGs characterised by consistent floristic, structural and growth form attributes. To illustrate the type and extent of vegetation that may be found within the lower rainfall zone, the MVGs are listed in Table 4.2. The three largest mapped MVGs in the lower rainfall zone comprise non-forest types (hummock grasslands, eucalypt woodlands, *Acacia* shrublands); tussock grasslands and low open woodlands are also extensive (Department of the Environment and Water Resources 2007).

Vegetation fuel type class

The vegetation fuel type classification proposed for the lower rainfall CFI methodology is given in Table 4.3. The potential fuel types are discussed below and five fuel types are considered applicable. Those vegetation fuel types that do not have application in the lower rainfall abatement methodology are also noted in Table 4.3. Non-eligible areas are those that don't burn easily, such as perennially wet areas, or for which the management does not fit within the scope of a strategic early dry season (EDS) fire management regime. As per the higher rainfall savanna burning CFI methodology, other non-eligible areas include cleared areas, intensively grazed areas, and areas dominated by exotic (non-native) species,

Table 4.1. National Vegetation Information System (NVIS) classification hierarchy (ESCAVI 2003), Example *Eucalyptus* open woodland

Level	Description	Species	Growth form	Cover	Height
I	CLASS	–	1 dominant growth form for the dominant stratum	–	–
	Example	*Tree*			
II	STRUCTURAL FORMATION	–	1 dominant growth form for the dominant stratum	1 cover class for the dominant stratum	1 height class for the dominant stratum
	Example	*Open woodland*			
III	BROAD FLORISTIC FORMATION	1 dominant genus name for the dominant stratum	1 dominant growth form for the dominant stratum	1 cover class for the dominant stratum	1 height class for the dominant stratum
	Example	*Eucalyptus open woodland*			
IV	SUB-FORMATION	1 dominant genus name for each stratum ((max 3 strata; i.e. for U, M, G where substantially present)	1 dominant growth form for each stratum (max 3 strata)	1 cover class for each stratum (max 3 strata)	1 height class for each stratum (max 3 strata)
	Example	*+ Eucalyptus open woodland/Acacia tall sparse shrubland/Aristida open tussock grassland*			

and seasonally inundated wetlands including grasslands and woodlands typically dominated by *Melaleuca*. These non-eligible areas are not discussed.

In fire-prone savannas there are two major grass types: hummock grasses derived from spinifex (*Triodia* and related genera), and tussock grasses, which describes the remainder. These two grass types differ in fuel load accumulation and continuity characteristics (Yates *et al.*, Chapter 6). Perennial hummock grasses are also characteristically highly flammable given that they accumulate a dense core of dead material, and typically are highly resinous (Burrows *et al.* 2006). Each fuel type is categorised by dominance or mix of hummock or tussock grass in the ground strata.

The major structural vegetative types are represented; Open forests, Woodlands, Open woodlands and isolated trees (Shrublands and grasslands). Potential fuel types that are derived within each structural type are described.

Open forests

The four fuel types described for the higher rainfall abatement methodology (Russell-Smith *et al.* 2009) also occur in the lower rainfall region, although eucalypt open forest (EOF) is of very limited geographic extent based on available vegetation mapping sources and extensive field experience. Fuel types such as EOF that occur infrequently and/or as small patches are difficult to map, increasing mapping error (Lioubimtseva 2003). For practical purposes we propose merging EOF with the far more extensive woodlands.

Table 4.2. Major Vegetation Groups as described in national mapping products, occurring in the lower rainfall zone

MVG NVIS national mapping Ver4.1*		Map of the vegetation of the Australian Tropical Savanna **	
Eucalypt woodlands	**38%**	**Woodlands + tussock grassland**	**6%**
Tropical eucalypt woodlands/ grasslands	**6%**	**Woodlands mixed grasslands**	**27%**
Other forests and woodlands	3%	**Woodlands + hummock grassland**	**1.3%**
Eucalypt open forest	1%		
Eucalypt open woodlands	19%	Eucalypt open woodlands	7%
		Open woodlands + hummock grass	20%
Tussock grasslands	8%	Tussock grasslands	22%
Hummock grasslands	5%		
Acacia forest and woodlands	4%	Tall acacia woodlands	8%
Acacia shrublands	**4%**	**Acacia shrublands**	**5%**
Acacia open woodlands	0.5%		
Other shrublands	0.4%		
Melaleuca forests and woodlands	7%		
Other			

* NVIS Version 4.1 (Department of Environment 2012).
** Vegetation of the Australian Tropical Savannas (Fox *et al.* 2001).

Woodlands

Woodlands make up a greater proportion of the lower rainfall zone (Table 4.2) and have an average canopy height generally 10 m or taller. Woodlands may be split into three fuel classes characterised by the mix of tussock and hummock grasses.

1. Woodlands with tussock grass (WTu) (Fig. 4.1). This fuel type is typically dominated by a continuous ground layer of mixed perennial tussock grass that contributes to a higher fine fuel load component than in other woodland classes (Yates *et al.*, Chapter 6). The tree genera include eucalypts (incorporating both *Eucalyptus* and *Corymbia*) such as *E. tectifica*, *E. jensenii*, *C. terminalis* and *E. chlorophylla*. The WTu fuel type occurs on plains associated with heavier-textured, more fertile soils found on basic volcanics and alluvium. This promotes strong seasonal growth of perennial tussock grasses which make up the bulk of the fine fuels that are the distinguishing characteristic of this class.
2. Woodlands with mixed grass (WMi) (Fig. 4.2). These woodlands typically comprise mixtures of perennial tussock grasses that may co-occur with hummock grass. Typical tree taxa include *E. tetrodonta* and *E. miniata*, *C. ferruginea*, *Callitris* and *Erythrophleum*. These woodlands extend into higher rainfall areas in the north where they comprise the eucalypt woodland fuel type in the higher rainfall savanna burning abatement methodology (Russell-Smith *et al.* 2009). They are found on a range of upland landforms from undulating plains to low hills. Soils range from shallow sands to deeper earthy soils which are predominantly well drained. The presence of the mid-stratum in these woodlands is variable; it is often made up of immature trees and a mix of shrubs, although generally of less prominence than in areas receiving >1000 mm rainfall (Egan and Williams 1996).

Table 4.3. Fuel types in the lower rainfall zone

Fuel type code	Fuel type	FPC (%)	Canopy height	Characteristic species: Upper stratum	Characteristic species: Mid stratum	Characteristic species: Lower stratum	Substrate	Comments
WTu	*Woodland with tussock grass*	10–30 (CC >20)	generally >10 m	various *Eucalyptus* (e.g. *E. tectifica*), *Corymbia* (e.g. *C. opaca*) often with other taxa (e.g. *Erythrophleum, Terminalia*)	Well developed shrub layer may or may not be present	Dominated by native perennial tussock grasses	Generally deep well-drained soils to those with impeded drainage, typically on flat to undulating landtypes with fertile volcanic-derived substrates	Woodlands characteristic of more fertile soils
WMi	*Open forest with tussock grass*	30–70 (CC >50)	generally 10–30 m	Eucalypts (e.g. *E. tetrodonta, E. miniata*)	Various Well-developed shrub layer may or may not be present	Dominated by native perennial and annual tussock grasses	Well-drained deep soils, often sandy loams	Very restricted areas of open forest were merged with woodland
	Woodland, with mixed grass	10–30 (CC >20)	generally >10 m	Various *Eucalyptus* (e.g. *E. tetrodonta*) and *Corymbia* spp., often with other taxa (e.g. *Erythrophleum, Terminalia, Callitris*)	Well-developed shrub layer may or may not be present	Dominated by native perennial and annual tussock grass; may be associated with hummock (*Triodia*) grasses	Various situations including undulating to hilly landtypes on imperfectly to well-drained soils	Very restricted areas of open forest with mixed tussock/ hummock grasses are included in this class
WHu	*Woodland, with hummock grass*	10–30 (CC >20)	generally >10 m	Various *Eucalyptus, Corymbia* often with other taxa (e.g. *Erythrophleum, Terminalia, Xanthostemon*)	Well-developed shrub layer may or may not be present. Where present, may include woody heath taxa as listed for SHH	Hummock (*Triodia*) grasses usually dominant; tussock grasses may also occur	Rocky shallow soils derived typically from sandstone (quartzite); also lateritic hills and plateaux	Sandstone woodlands
OWM	*Open woodland, with mixed grass*	<10 (CC <20)	Occasional stems, generally <10 m	Various *Eucalyptus, Corymbia*, including *C. dichromophloia, E. leucophloia, E. brevifolia, E. pruinosa, E. tectifica*	Well-developed shrub layer may or may not be present. Where present, may include woody heath taxa	Hummock (*Triodia*) grasses usually dominant, or codominant with tussock grasses	Shallow substrates on undulating stony rises and rocky hills	

Table 4.3. (Continued)

SHH	*Shrubland (heath) with hummock grass*	<30 (shrubs) (CC <80)	<5 m	Sparse	Conspicuous shrub cover, commonly *Acacia* species and various other taxa (e.g. *Calytrix, Grevillea, Hibbertia, Hibiscus, Jacksonia, Tephrosia, Verticordia*)	Hummock (*Triodia*) grass, and/or other perennial restios (*Lepyrodia, Dapsilanthus*), sedges (*Schoenus sparteus*) or graminoids (e.g. *Lomandra, Xanthorrhoea*)	Sand plains often over laterite, or rocky, shallow substrates derived from sandstone, metamorphosed sandstone (e.g. quartzite)	Sandstone heaths, sandplain shrubland. Includes mulga (pindan) low woodlands
		10–70 (grass)	<2 m	sparse	Sparse		As for shrublands	Hummock grasslands
		<30	<5 m	*Melaleuca* species, *M. viridiflora, Eucalyptus pruinosa*	Sparse		Undulating plains and stony rises	Low *Melaleuca* open woodlands must be dominated by hummock grass
					Excluded fuel types			
*OW Tu**	*Grassland–open woodland, with tussock grass*	<10 (CC <20)	Occasional stems, generally <10 m	Occasional *Eucalyptus, Corymbia*, or other taxa (e.g. *Erythrophleum, Terminalia, Lysiphyllum*)	Shrub layer often absent, and where present may represent 'woody thickening'	Native perennial and annual tussock grasses	Generally well-drained deep soils to those with impeded drainage, typically on flat to undulating landtypes with relatively fertile soils with high clay content	Grasslands with open woodlands on fertile clay-derived soils
*Tu**	*Tussock grassland*	30–70 (grass)	<5 m	Sparse trees	Sparse	Tussock grasses, *Astrebla, Chrysopogon, Dicanthium, Eragrostis, Heteropogon, Sehima*		
*LCW** *Acw**	*Acacia tall woodland*	10–30	generally <10 m	*Acacia* species (*A. shirleyi, A. eriopoda, A. cambagei*)	Generally sparse	Typically sparse low tussock and hummock (*Triodia*) grasses	Shallow gravelly soils on low hills, rises and scarps	Lancewood, pindan Acacia woodlands/ forests

* Vegetation types not included as proposed fuel types.

Figure 4.1. Woodland with tussock grassland.

3. Woodlands dominated by hummock grass in the understorey (WHu) (Fig. 4.3). These woodlands occur on deep sand plains to rocky hills and sandstone scarps. This fuel type includes the sandstone woodlands as described in the higher rainfall methodology. Common tree species include *E. miniata*, *E. tetrodonta*, *C. dichromophloia*, *E. phoenicea* and *C. ferruginea*. A sparse to mid-dense shrub layer may be present consisting of heathy shrub genera (typically *Acacia*, *Calytrix* and *Jacksonia*) and a hummock grass understorey.

Open woodlands

Open woodlands (OW) make up ~20% (Table 4.2) of the lower rainfall zone. They become increasingly dominant as tree height, cover and basal area decline down the rainfall gradient (Williams *et al.* 1996; Hutley *et al.* 2011). These open woodlands typically have an

Figure 4.2. Woodland with mixed grassland.

Figure 4.3. Woodland with hummock grassland.

Figure 4.4. Open woodland with tussock grassland.

Figure 4.5. Tussock grassland.

average canopy height often much less than 10 m. The sparse tree layer is characterised by *Eucalyptus* and *Corymbia*. There are two OW fuel types characterised by dominance of either tussock or hummock grasses, respectively.

1. Open woodlands dominated by tussock grasses (OWT) (Fig. 4.4) and tussock grasslands (Fig. 4.5), are found on heavy textured or textured contrast soils common on broad alluvial plains and rises. *E. microtheca*, *Excoecaria*, *Bauhinia* and *E. tectifica* are common tree dominants. These lightly wooded plains are not a feasible fuel type for fire abatement projects given that they typically are heavily utilised for pastoral purposes and as such are typically seldom burnt (Russell-Smith *et al.* 2014). Both tussock grasslands and OWT types are also prone to woody plant invasion (woody thickening), especially under heavily grazed and seldom burnt conditions (Dyer *et al.* 2001; Crowley *et al.* 2009). As such these areas would not be typically managed with an early burning regime usual in savanna burning projects.
2. Open woodland with mixed grasses (OWM) (Fig. 4.6) forms a large component of the low rainfall zone along with woodland types. OWM occurs on scarps and plateaux, steep hills, and stony undulating rises and plains. Soils are typically shallow, gravelly and sandy to red earths often with rock. Predominant tree species include *E. leucophloia*, *E. brevifolia*, *E. leucophylla*; other associated species include *E. pruinosa*, *C. dichromophloia* and *E. tectifica*. The presence of a midstratum is variable and includes a sparse shrub layer often comprising *Acacia*, *Carissa*, *Grevillea* and *Petalostigma* species.

Shrublands and grasslands

The shrublands and grasslands form unique habitats across the low rainfall zone. The tall woody shrublands such as the lancewood forests, and sparsely wooded tussock grasslands as discussed with OWT, are not included as a vegetation fuel type. The shrublands associated with a hummock grass understorey and hummock grasslands are included together and benefit from a managed early burning fire regime.

Figure 4.6. Open woodland with mixed grassland.

1. *Acacia* forests/woodlands >5 m in height comprise <5% of the low rainfall zone (Table 4.2). Major regional examples comprise: lancewood thickets (*A. shirleyi*) and gidgee (*A. cambagei* or *A. georginae*) in the Northern Territory (NT) and Queensland (QLD); and brigalow (*A. harpophylla*), mulga (*A. aneura* s. lat.), bendee (*A. catenulata*), black gidgee (*A. argyrodendron*) in Queensland. Included here is 'pindan' vegetation dominated by various *Acacia* species (e.g. *A. eriopoda*, *A. tumida*, *A. platycarpa*) characteristic of the Dampier Peninsula in the Kimberley region, Western Australia. Lancewood is a widely occurring fire-sensitive species and is vulnerable to wildfires, particularly on thicket margins (Woinarski and Fisher 1995; Russell-Smith *et al.* 2010). Litter fuels in mature lancewood thickets (Fig. 4.7) comprise only a scatter of leaves, and other grass fuels typically are excluded (Russell-Smith *et al.* 2010). For the purposes of the savanna burning lower rainfall abatement methodology, relatively close-canopied

Figure 4.7. Lancewood, *Acacia shirleyi*, low open forest.

Acacia-dominated forests/woodlands are excluded on the basis that they are essentially fire-independent assemblages given that their fuels typically do not carry fire.

2. *Acacia* shrublands (isolated trees) comprise <5% of the lower rainfall zone (Table 4.2). *Acacia* shrublands (SHH) with a predominantly hummock grass ground storey (Fig. 4.8) occur extensively on sandy and sandstone-derived substrates, and are broadly analogous with the sandstone heath fuel type as described by Edwards and Russell-Smith (2009) for the higher rainfall zone abatement methodology. The SHH fuel type typically includes *Acacia*, *Grevillea*, *Jacksonia* and *Calytrix* taxa, <5 m in height. The ground stratum generally includes hummock (*Triodia*) grass, and/or other perennial restios (*Lepyrodia*, *Dapsilanthus*), sedges (*Schoenus sparteus*) or graminoids (e.g. *Lomandra*, *Xanthorrhoea*). For present purposes we have included also several other low shrubby vegetation assemblages with similar burning characteristics, including

Figure 4.8. Acacia shrubland.

hummock grasslands and *Melaleuca* low open-woodlands with dominant hummock grass understorey (Table 4.3).

Mapping fuel types

Fuel type maps are an essential element in savanna fire abatement projects to determine project boundaries, extent of eligible fuel types and non-eligible areas, and fire extent and severity. According to the high rainfall CFI methodology, areas classified as non-eligible include cleared areas, intensively grazed areas, and areas dominated by exotic (non-native) species, as well as non-applicable structural vegetation (fuel types). The non-eligible areas are masked out in the mapping process, leaving only eligible fuel types as defined in the CFI methodology. The purpose of the following section is not so much to explain the mapping process, which is documented in numerous texts including McKenzie *et al.* (2008), but to discuss the issues associated with fuel type mapping for the purpose of carbon accounting.

Remote sensing data – aerial photos and satellite imagery

Aerial photography and satellite imagery are commonly used for mapping across all scales. The technique of aerial photo interpretation (API) has been used successfully across northern Australian, including Land Unit mapping in the Northern Territory (www.lrm.nt.gov.au/nrmapsnt), and Regional Ecosystem (RE) mapping in Queensland (www.ehp.qld.gov.au/ecosystems/biodiversity/regional-ecosystems/maps). The heuristic interpretation offered by experienced API operators incorporates both structural and floristic information, and inferred ancillary information (e.g. landform pattern, geology) into the application of a classification scheme to the area of interest and affords a high degree of line and attribute accuracy to the final map product. Although expensive to initiate, labour-intensive to produce, and reliant upon user experience, studies in northern Australia have demonstrated that API continues to represent a cost-effective means of map production to high levels of accuracy when compared with other remotely sensed methods (Lewis *et al.* 2012).

The scale of mapping is limited by the resolution of the base data used for derivation of the map and its intended application. For the purpose of fire abatement fuel type mapping, the current CFI savanna burning methodology 1.1 requires a minimum pixel size of 250 m. A range of remotely sensed imagery that meets these ground resolution requirements is available, including Landsat (30 m pixels) or MODIS (250 m pixels) satellite imagery, both of which are cost-effective (free) and contemporary. Landsat and MODIS imagery have been used extensively for mapping land unit, vegetation and fuel types in northern Australia (Wilson *et al.* 1990; Edwards and Russell-Smith 2009). MODIS is broad-scale imagery used for regional and national mapping projects such as recent fuel mapping across the northern Kimberley in Western Australia (WA) and ongoing automated fire mapping across northern Australia (www.firenorth.org.au).

Differentiation of fuel types

Fuel type maps are essentially structural vegetation maps where the classes include a defined mix of hummock and tussock grass. The fuel types are associated with but not defined by a characteristic cohort of species typically described at the genus level (e.g. *Eucalyptus*, *Acacia*). One criterion in developing a fuel type is that it must be able to be differentiated across the landscape using remote sensing mapping techniques. This can prove difficult given that some fuel types have structurally similar characteristics in the dominant upper stratum, such as WTu and WMi, or where transition between types is gradual. This can obscure the level of resolution (either visual or spectral) required to differentiate fuel types that are strongly influenced by understorey life-forms and floristics. Ancillary information may be used to help map and classify fuel types, including geology, soils or other reliable landscape mapping available such as geomorphology and land systems. This ancillary information is additional to ground-based observations, particularly those from detailed field sites.

Extrapolation across the landscape

There are numerous approaches for extrapolation of on-ground information for the purpose of classifying whole landscapes. Fuel types are purposefully very generic and broad-scale and effectively mapped using a qualitative approach. It is widely accepted that there is no requirement to incorporate intense field sampling densities (e.g. as recommended in Thackway *et al.* 2008 and Neldner *et al.* 2012), which are necessary for deriving fine-scale floristic map products. Rather, the approach includes using effective qualitative approaches which are appropriate in northern Australian where landscapes are largely intact and there has been minimal disturbance of native vegetation. This enables extrapolation of field samples and map classification across broad areas using observed relationships between vegetation, geomorphological and landscape patterns (Hewitt *et al.* 2008). Wherever practicable, on-ground field samples are used in classification and then available informative mapping products (e.g. vegetation, geology, soils, land systems) and associated data are applied in the mapping process. The method is augmented with aerial survey collection of digital photography and calibration data points.

On-ground field data collection across northern Australia is constrained by the high cost of time and travel to remote locations. Access is often restricted due to lack of roads, which necessitates the use of a helicopter. The cost is a major factor limiting the high density field sampling required for quantitative surveys. High resolution geo-located aerial digital photography taken at low altitudes provides an alternative for obtaining extra contemporary field data for the purposes of a fuel type mapping exercise. The high resolution photos, however, must be supported by representative ground-based data sites for verification.

Scale

The aim of fuel type mapping is to differentiate homogeneous areas of fuel components across the landscape, not areas of homogenous vegetation communities *per se*. For this reason mapping-scale should be seen as distinct from classification-scale, which sets the conceptual framework for which suites of attributes may be applied across the landscape (Neldner *et al.* 2012). To differentiate the broad fuel types accurately using remote sensing approaches, a finer scale of mapping may need to be conducted to account adequately for the distribution of fuel types within the landscape. These finer classes (e.g. vegetation associations) are then able to be combined to produce a fuel map based on known vegetation community characteristics, specifically structural, floristic and life-form attributes.

Potential for deriving fuel maps from existing vegetation maps

Derivation of fuel maps from existing vegetation maps is appropriate when the map product has been produced at a classification scale and spatial resolution that is either equal to, or preferably finer than, the fuel type product required. The map classification must be at least at the detail of NVIS Level IV in order to allow reclassification of vegetation into structural vegetation classes that can be used to derive fuel types accurately. The spatial resolution of vegetation mapping must be at least 1:250 000.

A limited number of appropriately scaled, contemporary vegetation mapping datasets are readily available for public use across northern Australia; coverage is relatively restricted in the Northern Territory and Western Australia (Fig. 4.9). Available useful map products include Queensland Regional Ecosystems, and vegetation surveys across central Arnhem Land, the Mary River, McArthur River and Daly River catchments, and some other small project areas (all Northern Territory). These datasets contain the required level of classification detail to enable the derivation of fuel types in accordance with CFI methodology requirements; that is, dominant structural attributes and characteristic vegetation floristics in each recognised stratum of the map unit. Additionally, numerous 1:100 000 to 1:250 000 scale land resource mapping datasets exist in the Northern Territory from which the required structural and floristic vegetation attributes may be extracted from the core vegetation attributes associated with each survey. In each of these cases a look-up table can be constructed to attribute the spatial data to a fuel type based on unique map codes (or combinations of map codes) in the data attributes. In the first instance this allows an assessment to be made of the extent of eligible and ineligible areas within a project area and a determination made regarding fire project feasibility. Further calibration would be required for a property or project where the detail becomes discernible due to the finer scale of work.

Two continental-scale vegetation mapping products have been derived from the NVIS dataset, the MVGs and Major Vegetation Sub-groups (MVS). The MVG mapping categories are too broad to derive a useable fuel map for the purposes of a fire abatement project, being classified at NVIS Level III. Similarly, MVS mapping does not have the spatial resolution required for derivation of fuel maps. For example, significant areas of sandstone heath are misclassified in the higher rainfall zone as a result of the source data being generated at 1:1 000 000 scale and not being able to adequately represent homogeneous occurrences of these small patches in the landscape. These sorts of scale issues are particularly

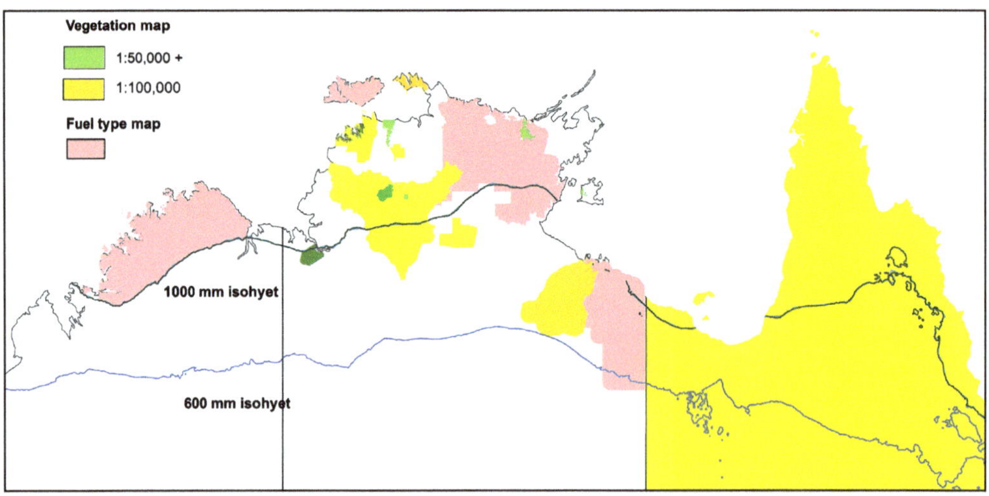

Figure 4.9. Currently available appropriately scaled vegetation mapping applicable for fuel type map derivation.

prevalent in the Northern Territory Top End, and Western Australian Kimberley, and are a result of the national NVIS product deriving seamless national vegetation map products using the best available scale and level of classification detail (NLWRA 2001; Thackway *et al.* 2001).

The resource-intensive task of producing and validating a fuel map, and lack of comprehensive coverage of suitable existing spatial data across northern Australia, have led to the investigation of other sources of mapping that could be utilised to support an abatement project. Over recent years several national initiatives have been undertaken to classify and monitor changes in landcover attributes using remote sensing techniques. These efforts have focused primarily on the MODIS (250 m) and Landsat (30 m) sensors and have included, but are not limited to, the following:

1. Australia's Forest Cover (ABARES 2014) – a multi-resolution, composite product of forest cover (trees >5 m tall and >30% foliage projective cover) in Australia.
2. Dynamic Land Cover (DLC) for Australia (Lymburner *et al.* 2010) – a MODIS 250 m scale Enhanced Vegetation Index (EVI) product to map ISO land cover standard (19144–2) categories across Australia.
3. Groundcover (Guerschman *et al.* 2012) – a MODIS 250 m scale analysis of photosynthetically active vegetation, senescent vegetation and bare soil fractions of landcover across Australia.
4. Persistent-Green (TERN/AusCover 2012). As discussed in detail in Chapter 7, the TERN AUSCover facility has produced calibrated and standardised estimates of continental-scale plant foliage projective cover (Persistent-Green) at the Landsat scale (30 m), providing a 2000–2010 time-series of persistently green (evergreen) FPC assumed to be the woody component of the spectral signal.

All of these products, despite offering significant potential at an appropriate spatial scale, are limited by the classification scheme associated with the data and/or the spectral sensitivity of the sensors. In their current forms, all of these products require reclassification or are unable to be reclassified into the complete range of vegetation cover classes on which the fuel type mapping is based. Similarly, these products are only able to give accurate indications of the cover characteristics of the dominant vegetation stratum (usually the canopy) and do not allow the user to derive fuel types which rely heavily on canopy floristics and ground stratum life-form for classification purposes.

Of these products the Persistent-Green data provides significant potential for the rapid resolution of cover attributes through raster reclassification in accordance with the NVIS cover categories at a scale and spectral sensitivity useful for the derivation of a fuel type map. However, case studies from Northern Territory savanna systems suggest consistent over-estimation of plant foliage projective cover when compared with field measurements (see Cuff and Brocklehurst, Chapter 7). This is potentially due to several characteristics of tropical Australian plant groups, including the leaf-architecture of overstorey taxa and the dominance of 'functionally evergreen' hummock grasses in the ground stratum. This over-estimation of cover results in potentially significant misclassification of pixels at the project scale, affecting overall map accuracy. It is unlikely that the required levels of fuel type map accuracy would be achieved utilising this data as the sole vegetation input to a mapping exercise, precluding its use as a baseline for fire abatement emissions accounting at the present time.

Fuel mapping

Recent land resource mapping methodology has successfully used object-based image analysis (OBIA) available with Ecognition software (www.eCognition.com) in a variety of land resource mapping applications (Hempel 2004; Blake 2005; Stefanacci 2006; Cuff *et al.* 2009; Edwards and Russell-Smith 2009; Cuff 2011; Tehrany *et al.* 2014). Two advantages with using OBIA are: first, the automatic segmentation of imagery into object features (groups of contiguous pixels) based on spectral homogeneity within defined parameters of shape and colour, and second, the ability to incorporate expert knowledge of landscape patterns in the setting of parameters when processing. Consequently field knowledge is essential to visually recognise suitable patterns and anomalous results. The segmentation and classification processes are iterative, with the mapper stopping at suitable scales of segmentation that are not too detailed, while avoiding under-segmentation with subsequent classification errors such as multiple class objects (Liu and Xia 2010).

The project area may be divided into meaningful regions such as stratification on land types to reduce area, simplify classification and subsequently reduce processing time. A supervised classification is carried out using a set of resultant mean spectral values or other suitable contextual feature data attributed to each object (polygon). The spectral data used in satellite imagery includes the visible bands (particularly red), near-infrared (NIR), short-wave infrared (SWIR), and the normalised difference vegetation index (NDVI). Additionally, digital elevation data (www.ga.gov.au/topographic-mapping/digital-elevation-data.html) and derived slope and roughness models are used to help stratify the landscape. Patterns that are easily and clearly distinguished use few conditions and are readily classified. These classes often include features such as water bodies, clearings, bare areas and distinct vegetation types such as floodplains and mangroves, which are subsequently masked or classified as not applicable. The project area is overlaid with geomorphology or other available and reliable ancillary mapping to help inform the classification. A 'nearest neighbour' (NN) classifier algorithm is used where classes are less easily defined, and uses a set of multiple conditions often including ancillary data and a representative set of samples based on field sites. The classified mapping is imported into GIS ArcMap software (ESRI 2012) for final analysis of any unclassified areas and validation.

Validation

Whatever the approach used to derive a fuel map, an independent validation will be required. Validation involves selection of a minimum set of geo-referenced waypoints per project area (refer DCCEE 2013), adequately sampling the variety of fuel types to provide a measure of accuracy and confidence in the map for the user and CFI Regulator.

Map validation is a well-defined process (e.g. Congalton and Green 1999; Stehman and Wickham 2011) which relies upon using independently collected data (reference data) to compare against a map product. Ideally, the spatial extent of reference data samples (e.g. a point, an area, a polygon) and assessment units (pixel, group of pixels, polygon) for comparison with the map product should be commensurate with the scale of the mapping to which they are being applied. This ensures a valid accuracy assessment is undertaken and minimises unnecessary error due to spatial inconsistencies. The independent reference data should also adequately represent the range of variation in fuel types within a project area.

In most northern savanna regions logistical and financial constraints result in validation assessments being conducted along transects from an aircraft in order to adequately sample across the area of a project and all vegetation fuel types. This is usually carried out using a helicopter, particularly given that observers can directly observe vegetation or fuel types beneath the aircraft. Substantial field experience with sampling of both calibration and validation waypoints suggests helicopter-based observations are readily obtained at 65 ± 5 kn and at ~400 feet above ground level. At these speeds and heights, a sampling 'point' typically represents a minimum sample area of ~50 m radius (Edwards and Russell-Smith 2009).

Recent project validation exercises have utilised Cybertracker software (Cybertracker Conservation 2009) on a Trimble Juno 3b hardware device to record waypoints at regular intervals (e.g. each 30 s) along a predetermined random-start flight-path (transect). Sampled points are buffered to a 50 m radius for validation purposes, which will ensure any positional error is also taken into account. These buffered areas are intersected with the fuel type map and a standard error matrix (contingency table) is derived, giving the resultant accuracy. The level of accuracy required for CFI projects is not less than 80% (DCCEE 2013).

References

ABARES (Bureau of Agricultural and Resource Economics and Sciences) (2014) 'Australia's State of the Forests Report 2013.' Australian Government Department of Agriculture: Canberra. www.agriculture.gov.au/abares/forestsaustralia/sofr/sofr-2013

Blake G (2005) Object oriented mapping of sandstone heath vegetation on the Arnhem Plateau. BSc (Hons) thesis. Charles Darwin University: Darwin.

Burrows ND, Ward B, Robinson AD, Behn G (2006) Fuel dynamics and fire behaviour in spinifex grasslands of the Western Desert. In *Life in a Fire-Prone Environment: Translating Science into Practice.* Bushfire Conference, Brisbane.

Congalton RG, Green K (1999) *Assessing the Accuracy of Remotely Sensed Data: Principles and Practices.* Lewis Publishers: Boca Raton, FL.

Crowley G, Garnett S, Shepherd S (2009) Impact of storm-burning on *Melaleuca viridiflora* invasion on grasslands and grassy woodlands on Cape York Peninsula, northern Australia. *Austral Ecology* **34**, 196–209. doi:10.1111/j.1442-9993.2008.01921.x

Cuff NJ (2011) 'Vegetation Mapping of the Daly River Catchment Version 1 – Spatial Data.' Northern Territory Government Department of Land Resource Management: Darwin.

Cuff NJ, Thompson T, Sparrow B, Brocklehurst PS (2009) 'Vegetation Survey and Mapping of the McArthur River Catchment, Northern Territory.' Technical Report No. 06/2009D. Northern Territory Government Department of Land Resource Management: Darwin.

Cybertracker Conservation (2009) CyberTracker – Software Tool that Allows Users to Collect Field Data with Handheld Computers or PDAs. South Africa. www.cybertracker.org

DCCEE (Department of Climate Change and Energy Efficiency) (2013) Reduction of Greenhouse Gas Emissions through Early Dry Season Savanna Burning – 1.1 Methodology Determination 2013. Australian Government. www.comlaw.gov.au/Details/F2012L01499

Department of Environment (2012) Australia – Present Major Vegetation Groups – NVIS Version 4.1. (Albers 100 m analysis product) www.environment.gov.au/metadataexplorer/full_metadata.jsp

Department of the Environment and Water Resources (2007) 'Australia's Native Vegetation: A summary of Australia's Major Vegetation Groups.' Australian Government: Canberra.

Dyer R, Jacklyn P, Partridge I, Russell-Smith J, Williams RJ (Eds) 2001 *Savanna Burning: Understanding and Using Fire in Northern Australia.* Tropical Savannas Cooperative Research Centre: Darwin.

Edwards AC, Russell-Smith J (2009) Ecological thresholds and the status of fire-sensitive vegetation in western Arnhem Land, northern Australia: implications for management. In *Culture, Ecology and Economy of Fire Management in North Australian Savannas: Rekindling the* Wurrk *Tradition*. (Eds J Russell-Smith, P Whitehead, P Cooke) pp. 229–255. CSIRO Publishing: Melbourne.

Egan JL, Williams RJ (1996) Life form distribution of woodland plant species along a moisture availability gradient in Australia's monsoonal tropics. *Australian Systematic Botany* **9**, 205–217. doi:10.1071/SB9960205

ESRI (2012) ArcGIS Desktop: Release 10.1. Environmental Systems Research Institute: Redlands, CA. www.arcgis.com/en/arcgisdesktop/10.0

Executive Steering Committee for Australian Vegetation Information (ESCAVI) (2003) National Vegetation Information System (NVIS). Version 6.0. Australian Government Department of the Environment and Heritage: Canberra. www.environment.gov.au/topics/science-and-research/databases-and-maps/national-vegetation-information-system

Fox ID, Neldner VJ, Wilson GW, Bannink PJ, Wilson BA, Brocklehurst PS, Clark MJ, Dickinson KJM, Beard JS, Hopkins AJM, Beeston GR, Harvey JM, Thompson EJ, Ryan TS, Thompson SL, Butler DW, Cartan H, Addicott EP, Bailey LP, Cumming RJ, Johnson DC, Schmeider M, Stephens KM, Bean AR (2001) The Vegetation of the Australian Tropical Savannas. (Digital coverage scale of 1:1,000,000) Queensland Herbarium, Environmental Protection Agency, Brisbane and the Cooperative Research Centre for the Sustainable Development of Tropical Savannas, Darwin. http://savanna.cdu.edu.au/research/vegetation_map.html

Guerschman JP, Oyarzabal M, Malthus TJ, McVicar TM, Byrne G, Randall LA, Stewart JB (2012) 'Validation of the MODIS-based vegetation fractional cover product.' CSIRO Land and Water Science Report: Canberra.

Hempel C (2004) 'Land Cover Mapping for Nitmiluk National Park.' Unpublished Technical Report, Department of Land Resource Management, Northern Territory Government.

Hewitt AE, McKenzie NJ, Grundy MJ, Slater BK (2008) Qualitative survey. In *Guidelines for Surveying Soil and Land Resources*. 2nd edn. (Eds NJ McKenzie, MJ Grundy, R Webster, AJ Ringrose-Voase) pp. 285–306. CSIRO Publishing: Melbourne.

Hutley LB, Beringer J, Isaac PR, Hacker JM, Lucas A (2011) A sub-continental scale living laboratory: spatial patterns of savanna vegetation over a rainfall gradient in northern Australia. *Agricultural and Forest Meteorology* **151**, 1417–1428. doi:10.1016/j.agrformet.2011.03.002

Lewis D, Phinn S, Pfitzner K (2012) Pixel-based image classification to map vegetation communities using SPOT5 and Landsat5 Thematic Mapper data in a tropical savanna, northern Australia *Canadian Journal of Remote Sensing* **38**(5), 570–585. doi:10.5589/m12-047

Lioubimtseva E (2003) An evaluation of VEGETATION-1 imagery for broad-scale landscape mapping of Russia: effects of resolution on landscape pattern. *Landscape and Urban Planning* **65**(4), 187–200.

Liu D, Xia F (2010) Assessing object-based classification: advantages and limitations. *Remote Sensing Letters* **1**(4), 187–194. doi:10.1080/01431161003743173

Lymburner L, Tan P, Mueller N, Thackway R, Lewis A, Thankappan M, Randall L, Islam A, Senarath U (2010) '250 metre Dynamic Land Cover Dataset of Australia.' 1st edn. Geoscience Australia: Canberra.

McKenzie NJ, Grundy MJ, Webster R, Ringrose-Voase AJ (Eds) (2008) *Guidelines for Surveying Soil and Land Resources*. 2nd edn. CSIRO Publishing: Melbourne.

Myers RL (2006) Living with Fire – Sustaining Ecosystems & Livelihoods Through Integrated Fire Management. The Nature Conservancy, Global Fire Initiative.

National Land and Water Resources Audit (2001) Australian native vegetation assessment 2001. National Land and Water Resources Audit: Canberra. p. 177. http://npsi.gov.au/files/products/national-land-and-water-resources-audit/pr010286/pr010286.pdf

Neldner VJ, Wilson BA, Thompson EJ, Dillewaard HA (2012) 'Methodology for Survey and Mapping of Regional Ecosystems and Vegetation Communities in Queensland.' Version 3.2. Queensland Herbarium, Queensland Department of Science, Information Technology, Innovation and the Arts: Brisbane.

Russell-Smith J, Murphy BP, Meyer CP, Cook GD, Maier S, Edwards AC, Schatz J, Brocklehurst P (2009) Improving estimates of savanna burning emissions for greenhouse accounting in northern Australia: limitations, challenges, applications. *International Journal of Wildland Fire* **18**(1), 1–18. doi:10.1071/WF08009

Russell-Smith J, Yates CP, Brock C, Westcott VC (2010) Fire regimes and interval-sensitive vegetation in semiarid Gregory National Park, northern Australia. *Australian Journal of Botany* **58**(4), 300–317.

Russell-Smith J, Yates CP, Evans J, DeSailly M (2014) Developing a savanna burning emissions abatement methodology for tussock grasslands in high rainfall regions of northern Australia. *Tropical Grasslands – Forrajes Tropicales* **2**(2), 175–187.

Sanders KD (1985) Rangeland fire effects: a symposium: proceedings of a symposium sponsored by Bureau of Land Management and University of Idaho at Boise, Idaho, State Office, USDI-Bureau of Land Management, United States.

Schoknecht N, Payne A (2011) 'A. Land Systems of the Kimberley Region, Western Australia.' Technical Bulletin No.98. Western Australian Government Department of Agriculture & Food: Perth.

Stefanacci J (2006) Automated stand delineation and fire fuels mapping. *ASPRS 2006 Annual Conference Proceedings*, Nevada, USA.

Stehman SV, Wickham JD (2011) Pixels, blocks of pixels, and polygons: choosing a spatial unit for thematic accuracy assessment. *Remote Sensing of Environment* **115**(12), 3044–3055. doi:10.1016/j.rse.2011.06.007

Tehrany MS, Pradhan B, Jebu MN (2014) A comparative assessment between object and pixel-based classification approaches for land use/land cover mapping using SPOT 5 imagery *Geocarto International* **29**(4), 351–369.

TERN/AusCover (2012) Persistent Green-Vegetation Fraction and Wooded Mask – Landsat, Australia coverage. www.auscover.org.au/xwiki/bin/view/Product+pages/Persistent+Green-Vegetation+Fraction

Thackway R, Lesslie R (2008) Describing and mapping human-induced vegetation change in the Australian landscape. *Environmental Management* **42**, 572–590. www.ncbi.nlm.nih.gov/pubmed/18521659

Thackway R, Sonntag S, Donohue R (2001) 'Compilation of National Vegetation Information System (NVIS) Vegetation 2000 data set, final report BRR13.' National Land and Water Resources Audit: Canberra.

Thackway R, Neldner J, Bolton M (2008) Vegetation. In *Guidelines for Surveying Soil and Land Resources*. 2nd edn. (Eds NJ McKenzie, MJ Grundy, R Webster, AJ Ringrose-Voase) pp. 115–142. CSIRO Publishing: Melbourne.

Williams RJ, Duff DM, Bowman DMJS, Cook GD (1996) Variation in the composition and structure of tropical savannas as a function of rainfall and soil texture along a large-scale climatic gradient in the Northern Territory, Australia. *Journal of Biogeography* **23**, 747–756. doi:10.1111/j.1365-2699.1996.tb00036.x

Williams RJ, Griffiths AD, Allan GE (2002) Fire regimes and biodiversity in the wet-dry tropical landscapes of northern Australia. In *Flammable Australia: The Fire Regimes and Biodiversity of a Continent.* (Eds RA Bradstock, JE Williams, AM Gil) pp. 281–304. Cambridge University Press: Cambridge, UK.

Wilson BA, Brocklehurst PS, Clark MJ, Dickinson KJM (1990) 'Vegetation Survey of the Northern Territory, Australia.' Technical Report No.49. Northern Territory Government: Darwin.

Woinarski CZ, Fisher A (1995) Wildlife of lancewood (*Acacia shirleyi*) thickets and woodland in northern Australia: 2. Comparison with other environments of the region (Acacia woodlands, Eucalyptus savannah woodlands and monsoon rainforest). *Wildlife Research* **22**, 413–443. doi:10.1071/WR9950413

5

The operational role of satellite-based fire data in savanna burning methodologies

Peter Jacklyn, Shaun Ansell, Aaron Crosbie, Jarrad Holmes, Aidan Joseph, Sarah Legge, Ben Lewis, Anna Pickworth, Tom Vigilante and Jake Weigl

Summary

This chapter investigates the operational use of satellite-sourced fire data by projects implementing the savanna burning methodology, drawing on the knowledge of the fire planners involved in these projects. It shows that there is extensive use of satellite-sourced fire data in the projects, and that implementation of the methodology depends on these data. The use of these data extends throughout the fire season and involves a range of formats and tools. Almost all these data are sourced from, or used with, the North Australian Fire Information (NAFI) service. The importance of the NAFI service for the implementation of the savanna burning methodology underlines the need to effectively manage risk and ensure the service is reliable.

Introduction

The Australian Government's initiatives to encourage abatement of greenhouse gas emissions by the land sector include methodologies for savanna fire management (hereafter savanna burning methodologies or SBM) developed under the Carbon Farming Initiative (CFI), which rely on the integration of advanced technologies with the expertise and field duties of practitioners. CFI methodologies succeed when project activities or technologies can be reasonably applied by proponents (DCCEE 2011) and the tools required for such practical application should be planned, funded, and updated. Projects applying SBM, like most other Australian savanna fire managers, need to monitor fire activity and burnt areas across large tracts of remote country as part of their day-to-day burning operations (e.g. see Edwards and Cooke 2010). They must undertake these tasks with a relatively small number of fire managers, which would be difficult without ready access to satellite-sourced data on fire activity to help guide fire operations (Edwards and Cooke 2010; Ansell and Koenig 2011; Ens 2012).

As described in Chapter 3, it is apparent that projects applying SBM are using the Moderate Resolution Imaging Spectroradiometer (MODIS) burnt area mapping sourced from the North Australian Fire Information (NAFI) service in day-to-day fire operations – a usage distinct from the role these data have in emissions accounting in the CFI methodologies. Unlike the other technologies used in the application of the methodology, such as vehicles and incendiary devices, the satellites that produce these data and the websites the data are accessed through are recent innovations. For example, the NAFI website was developed in 2002 by the community of fire researchers and fire managers involved in the Tropical Savannas Cooperative Research Centre to allow the general community of northern fire managers access to regularly updated maps of active fires and burnt areas. Even over this short time, the internet-based technologies and IT systems used to access these satellite data have evolved significantly and this rate of change is likely to continue.

The SBM projects, however, are expected to continue operation over many years (e.g. Whitehead *et al.* 2008). If satellite mapping and web-based tools are widely used to implement these methodologies, then long-term planning should consider the support, funding and overall resourcing of these or similar services with appropriately updated functionality. In particular, likely changes in ITC technology need to be included in any planning.

To clarify the importance of resourcing and planning for future provision of satellite data, this chapter identifies the extent to which fire management operations that use the SBM depend on the application of MODIS burnt area mapping and other satellite-sourced data on fire activity. It also seeks to clarify the role that delivery platforms for these data have in the implementation of the methodology. The connection between the use of data tools and the application of CFI methodology in turn affects planning and resourcing of the satellite-based data and content delivery platforms.

It is apparent that the same satellite data and delivery platforms are used more broadly for fire monitoring across northern Australia (Ludwig *et al.* 2007; Manson 2010) and, while the detailed usage of satellite data by these groups is beyond the scope of this chapter, as with the SBM projects, if this broader range of fire managers intend to continue to use these delivery platforms they will need to be involved in their planning and resourcing.

Methods

The SBM projects rely on a key principle based on the timing of fires. By implementing strategically placed burns in the cooler, early dry season, the frequency and extent of the larger and more intense fires that occur later in the dry season will be reduced, as well as overall fire frequency. This brings various benefits including reduced greenhouse gas emissions. The broad steps involved in the process follow a typical seasonal pattern:

1. Plan strategic burning before the start of the fire season.
2. Implement strategic burning, much of it in the early dry season (EDS).
3. Assess the progress of strategic burning.
4. Monitor wildfires in the late dry season (LDS) and respond if needed to limit wildfires that are not checked by burnt areas.
5. Assess fire management at the end of the fire season (e.g. see Edwards and Cooke 2010).

The use of satellite-sourced fire data in each of these stages was addressed by the authors, who include fire planners working with projects implementing the SBM across northern Australia. Fire planners in these projects will often collate spatial data, create maps of fire history, property boundaries, topography etc. for delivery at fire planning workshops with staff or stakeholders. Projects involving Indigenous landholders typically hold fire planning workshops with traditional owners to decide on what fire strategy to pursue on their country and then work with the Indigenous rangers to implement the plan. Fire planners have a good overview of how the SBM projects use fire information, including satellite-sourced fire information, from their own use of this information, and from their knowledge of how rangers use fire information.

The authors have worked with a range of SBM projects across northern Australia. As shown in Fig. 5.1, these project locations extend from the Kimberley in Western Australia, across the NT Top End to the Gulf of Carpentaria and Cape York Peninsula in Queensland. The projects cover an area of ~218 million hectares – slightly less than the area of the state of Victoria. The areas included most of the projects currently registered for the high rainfall methodology, two of the largest areas, the West Arnhem Land Fire Abatement Project (WALFA) and the Central Arnhem Land Fire Abatement Project (CALFA),[1] that are preparing to register under the high rainfall methodology, and major project areas such as those operated by the Australian Wildlife Conservancy (AWC) looking to register under both the high and low rainfall methodologies.

For each stage of the fire season described above we addressed key aspects of the use of satellite-based fire information: its format, its source, and the way it was applied in the project.

Results

The results are shown in Table 5.1. The left-hand column shows the satellite-sourced fire data used in the different stages of the projects. A brief description of how these data are used appears below the data description. The usage is then described in more detail below. The right-hand columns show, for each group engaged in SBM projects, the data types used, the format used and the source of the data.

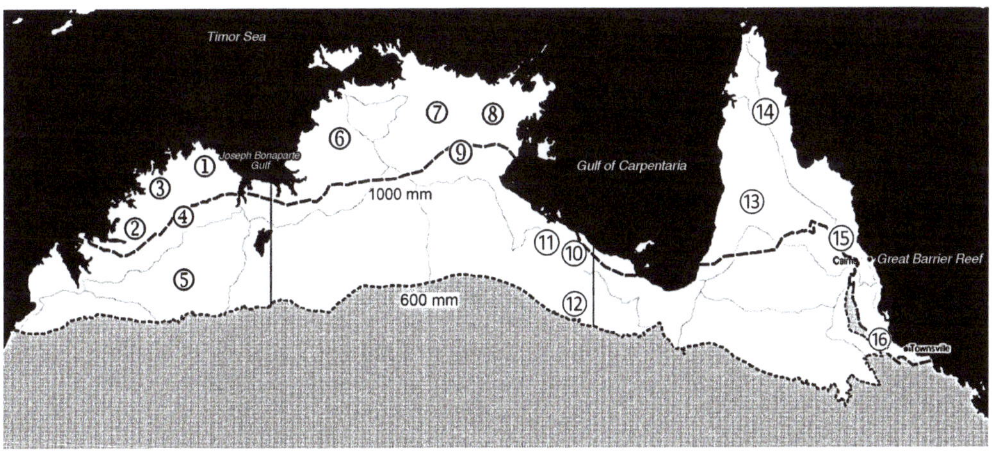

Figure 5.1. Location of areas covered by SBM projects in this study

1. North Kimberley – Balanggarra
2. North Kimberley – Wilinggin, Dambimangari
3. North Kimberley – Wunambal Gaambera
4. Ecofire area (north) – includes pastoral properties and AWC Artesian Range (near 2)
5. Ecofire area (south) – includes pastoral properties and AWC areas
6. Fish River Station
7. West Arnhem Land Fire Abatement (WALFA) project
8. Central Arnhem Land Fire Abatement (CALFA) project (extends south of the 1000 mm isohyet)
9. AWC Wongalara
10. AWC Pungalina–Seven Emu
11. Garawa Aboriginal Land Trust
12. Waanyi/Garawa Aboriginal Land Trust
13. Mitchell-Coleman project properties
14. AWC Piccaninny Plains
15. AWC Brooklyn
16. AWC Mt Zero–Taravale.

Table 5.1. Summary of satellite fire data usage in SBM projects

	Usage, formats and sources of satellite fire data for each SBM project							
Areas – the numbers refer to the areas in Fig. 5.1	**Nth Kimberley 1, 2**	**Nth Kimberley 3**	**AWC-EcoFire 4, 5, 9, 10, 14, 15, 16**	**Fish River 6**	**WALFA 7**	**CALFA 8**	**NT Gulf 11, 12**	**Mitchell – Coleman 13**
Planners and organisation involved in SBM operations in table	Anna Pickworth (KLC)	Tom Vigilante (WGAC)	Sarah Legge (AWC)	Shaun Ansell (ILC)	Jake Weigl (WLM)	Ben Lewis (NLC)	Jarrad Holmes (NLC)	Aaron Crosbie, Aidan Joseph (CYSF)
Satellite fire data used (*Primary use for the data*)								
	Planning for EDS burning							
MODIS long-term fire history *Habitat/fire proneness*	NAFI[1]	NAFI[1]	NAFI[1]	Not generally used	NAFI[2]	NAFI[1]	NAFI[1]	Not generally used
MODIS recent years' burnt area *Fuel available*	NAFI[1]	NAFI[1]	NAFI[1]	NAFI[1]	NAFI[1]	NAFI[1]	NAFI[1]	NAFI[1]
Last year's hotspots (MODIS, AVHRR) *Check on missed firescars*	Not generally used	Not generally used	Not generally used	Not generally used	Not generally used	NAFI[1]	NAFI[1]	Not generally used
	Implementing EDS burning							
MODIS long-term fire history *Habitat/fire proneness context for firelines*	NAFI[1]	Not generally used	NAFI[1]	Not generally used	NAFI[2]	NAFI[1]	Not generally used	Not generally used
MODIS recent years' burnt areas *Fuel context for firelines*	NAFI[1]	NAFI[1]	NAFI[1]	NAFI[1]	NAFI[1] NAFI[2]	NAFI[1]	NAFI[1]	NAFI[1]
MODIS current year burnt areas *Fuel context for firelines*	NAFI[1]	NAFI[1]	NAFI[1]	NAFI[1]	NAFI[1] NAFI[2]	NAFI[1]	NAFI[1]	NAFI[1]
Last year's hotspots (MODIS, AVHRR) *Check on missed fire scars*	Not generally used	Not generally used	NAFI[1]	Not generally used	Not generally used	NAFI[1]	Not generally used	Not generally used
	Assessing EDS burning							
MODIS current burnt areas *Fuel context for firelines*	NAFI[1] NAFI[2]	NAFI[1] NAFI[2]	NAFI[1] NAFI[2] NAFI[3]	NAFI[1] NAFI[2] NAFI[3]	NAFI[1] NAFI[2] NAFI[3]	NAFI[1] NAFI[2] NAFI[3]	NAFI[1] NAFI[2]	NAFI[1] NAFI[2]
Current hotspots (MODIS, AVHRR) *Effectiveness of burns*	NAFI[2]	NAFI[2]	NAFI[1] NAFI[2] NAFI[3]	NAFI[2] NAFI[3]	NAFI[2] NAFI[3]	NAFI[2] NAFI[3]	NAFI[2]	NAFI[2]

Table 5.1. (Continued)

MODIS raw imagery *Fuel/Effectiveness of burns*	Not generally used	Not generally used	Not generally used	GA[1]	Not generally used	Not generally used	Not generally used	Not generally used
Analysis of burnt area extent *Assessing progress to targets*	NAFI[4]	NAFI[4]	Not generally used	ILC (NAFI)[5] NAFI[4]	Not generally used	BFNT (NAFI)[5] NAFI[4]	BFNT (NAFI)[5] NAFI[4]	NAFI[4]
	Monitoring wildfire							
MODIS long-term fire history *Fuel & habitat in path of wildfire*	Not generally used	Not generally used	NAFI[1] NAFI[2] NAFI[3]	Not generally used	Not generally used	Not generally used	Not generally used	Not generally used
MODIS current year burnt areas *Fuel in path of wildfire*	NAFI[2]	NAFI[2]	NAFI[1] NAFI[2] NAFI[3]	NAFI[2] NAFI[3]	NAFI[2] NAFI[3]	NAFI[2] NAFI[3]	NAFI[2]	NAFI[2]
MODIS recent years' burnt area *Fuel in path of wildfire*	NAFI[2]	NAFI[2]	NAFI[1] NAFI[2] NAFI[3]	NAFI[2] NAFI[3]	NAFI[2] NAFI3	NAFI[2] NAFI[3]	NAFI[2]	NAFI[2]
Current hotspots (MODIS, AVHRR) *Progress of wildfire*	NAFI[2]	NAFI[2]	NAFI[1] NAFI[2] NAFI[3]	NAFI[2] NAFI[3]	NAFI[2] NAFI[3]	NAFI[2] NAFI[3]	NAFI[2]	NAFI[2]
	Assessing/reporting year's work							
MODIS current year burnt areas *Assessment of fire management*	NAFI[1]	NAFI[1]	NAFI[1]	NAFI[1]	NAFI[1]	NAFI[1]	NAFI[1]	NAFI[1]
Landsat raw imagery *Assessment of fire management*	Not generally used	Not generally used	USGS[1]	Not generally used	Not generally used	Not generally used	Not generally used	Not generally used

(*Continued*)

Table 5.1. (Continued)

MODIS recent years' burnt area *Fuel context and comparison for assessment*	NAFI[1]	Not generally used	NAFI[1]	NAFI[1]	NAFI[1]	NAFI[1]	Not generally used	NAFI[1]
Analysis of burnt area extent *Quantifying progress*	NAFI[4]	NAFI[4]	Not generally used	ILC (NAFI)[5] NAFI[4]	DCBR (NAFI)[5]	NAFI[4]	BFNT (NAFI)[5] NAFI[4]	CYSF (NAFI)[5] NAFI[4]
Year's hotspots (MODIS, AVHRR) *Used for animations*	Not generally used	WA DPAW (Landgate WA)[5]	NAFI[1]	Not generally used	Not generally used	Not generally used	BFNT (NAFI)[5]	NAFI[1]
	Estimating emissions							
Analysis of burnt area extent *To assist with emission calculation*	Not generally used	NAFI[4] DCBR (NAFI)[5]	AWC (NAFI)[5]	Not generally used	DCBR (NAFI)[5]	NAFI[4]	Not generally used	Not generally used
Full methodology process *Calculating emissions*	NAFI[4] SavBAT	SavBAT	SavBAT	NAFI[4] SavBAT	SavBAT	NAFI[4] SavBAT	NAFI[4] SavBAT	NAFI[4] SavBAT

1 Data is downloaded from the data source website indicated either as a shapefile, geo-referenced image or screen shot.
2 Data is viewed on the website map-viewer indicated.
3 Data from indicated data source is viewed on either Google Earth (mainly) or Virtual Earth map-viewers.
4 Graphed and tabulated data from indicated data source is downloaded as a report.
5 The organisation indicated has directly assisted in generating the data product using the data source indicated in brackets.

SavBAT – the Savanna Burning Abatement tool that uses NAFI-MODIS burnt area data to calculate emissions.

(AVHRR = Advanced Very High Resolution Radiometer, AWC = Australian Wildlife Conservancy, BFNT = Bushfires NT, CYSF = Cape York Sustainable Futures, DCBR = Darwin Centre for Bushfire Research, Charles Darwin University [ex BFNT], GA = Geoscience Australia-Satellite Imagery, ILC = Indigenous Land Corporation, KLC = Kimberley Land Council, MODIS = Moderate Resolution Imaging Spectroradiometer, NAFI = North Australian Fire Information, NLC = Northern Land Council, SavBAT = NAFI-hosted Savanna Burning Abatement Tool, USGS = United States Geological Survey, WA DPAW = WA Department of Parks and Wildlife, WGAC = Wunambal Gaambera Aboriginal Corporation, WLM = Warddeken Land Management Ltd.).

Note that the NAFI hotspot data are originally sourced from Landgate WA and Geoscience Australia.

Planning for EDS strategic burning (early in the year)

As shown in Table 5.1, all the SBM projects downloaded the recent year's MODIS burnt area fire history data from NAFI for planning, and some downloaded last year's hotspots as a proxy for recent fire history. In the far north, the reliable rainfall and minimal grazing pressure in the abatement areas means that the time since an area was last burnt is a good predictor of grassy fuel levels (Cook and Meyer 2009). As grassy fuels regrow almost fully in a few years, it is recent fire history that is a guide to grassy fuel levels.

The fire history data are often downloaded as shapefiles so they can be displayed using GIS software with other map layers displaying topography, tracks, property boundaries, significant features etc. A generic example is shown below (Fig. 5.2) where the last three years of fire history (year since last burnt) are displayed together with creek lines, roads and tracks and significant sites. Areas with different fuel levels can be identified, and this can then help to develop fire plans that utilise the decreased fire spread in areas with recent burns and reduced fuel levels.

The number of years of fire history mapped could be adjusted depending on the type of vegetation present: the Wunambal Gaambera group considered longer fire histories

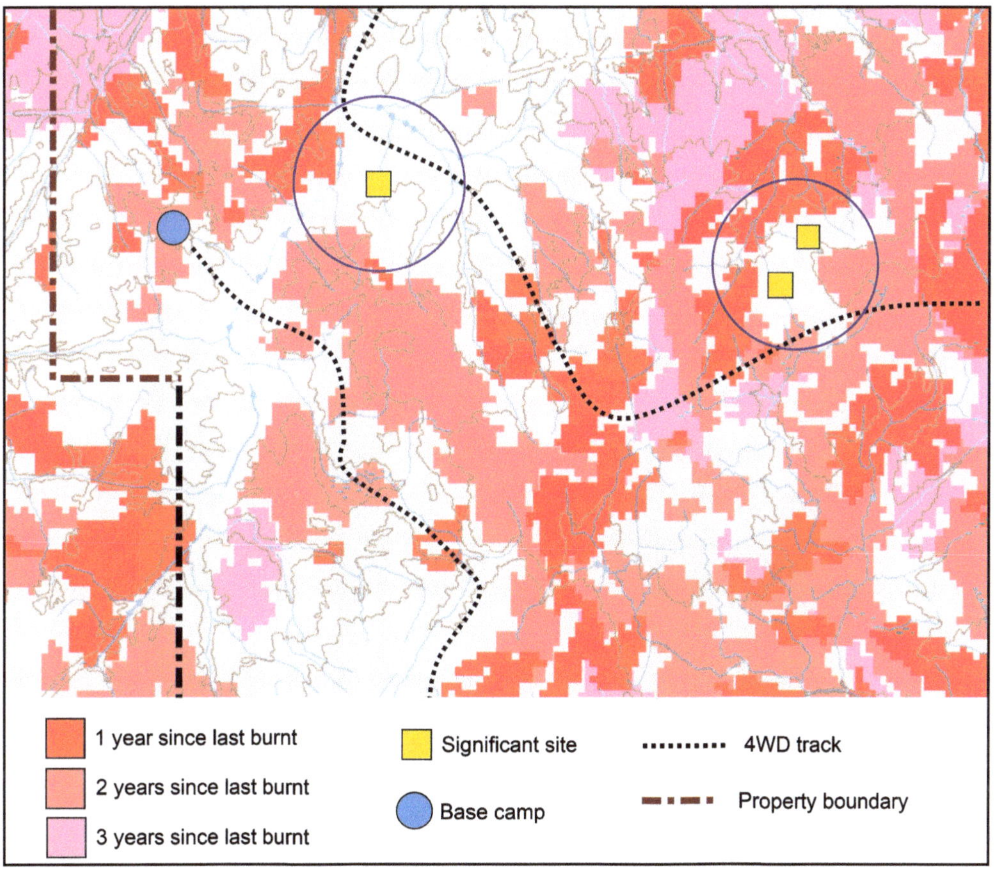

Figure 5.2. Generic example of a map created with a GIS and used in planning EDS burning operations. The circled areas may be targeted for protection with fire breaks as they have significant sites in areas unburnt for 3 years.

important for sandstone/spinifex country, but not needed for perennial tussock grass. In country with longer unburnt areas, such as the broken terrain in the Kimberley and western Arnhem Land, the longer term (over 10 years) fire history data were also downloaded and used for fire planning. For example, 13-year fire frequency maps were used by some groups to identify long unburnt habitat for fire-sensitive species, or to locate areas that were more or less prone to being burnt.

One point emphasised by fire planners was that fire planning is fundamentally based on a good knowledge of the country – the vegetation types, how they will burn, how these factors vary through the year, and how topography and local weather affect fire. Once you have this knowledge, the relatively coarse satellite maps of fire history can then be used effectively to help plan strategic burning.

Implementing EDS burning operations (April–July)

With a burning plan in place, implementation of the strategic EDS burns needs to be completed in the few months available after the country has dried out enough to carry fire and before it becomes too dry, hot and windy for fires to be controlled easily. The often large areas involved, both within the burns and between them, dictate that much of this work needs to be done from the air, by dropping incendiary capsules. Most of the fire planners reported that the downloadable NAFI-MODIS recent fire history data and some of the other spatial layers used in planning the burning operations were loaded onto a portable device, usually linked to a GPS, and carried in the aircraft during burning operations (Fig 5.3). This fire history was then used to help orient the aerial operations to the planned burning. Some of the groups used a simplified base map/fuel map background to streamline the locating of ignitions, monitoring and mapping for fire managers in the field.

Figure 5.3. Uunguu Ranger Neil Waina guides the implementation of an aerial EDS burning operation in the north Kimberley using a tablet with downloaded map information. Photo: Anna Pickworth.

The Indigenous rangers engaged in the burning operations in the Coleman-Mitchell area of southern Cape York Peninsula often cannot use aircraft to implement EDS burning themselves. Instead, they rely on aerial burning operations conducted by either the Queensland Parks and Wildlife Service or the Queensland Fire and Rescue Service, who use GPS points and local directions to conduct aerial burning operations. However the NAFI-MODIS fire history data will inform these local directions and have been used to help guide much of the on-ground burning.

Assessing EDS burning operations (April–July)

Fire managers in SBM projects need to know if a particular burning operation has been successful in creating the desired burnt area fuel breaks or patches in the right location. For the extended areas burned in aerial operations, they rely on the satellite information to assess burning operations. If a burn operation has been successful (i.e. not gone out) it will often be detected as active fires or 'hotspots', which for a given area are updated around four times a day by the satellites (Fig. 5.4). Because it is difficult to access this data in the field due to the lack of internet access, once burning operations are complete fire managers will view the NAFI website or view the NAFI hotspots[2] on Google Earth back at base to see if the burns have produced hotspots. This practice was reported by all groups. Jake Weigl of Warddeken said it involved 'purple star anxiety,' as the most recent hotspots (less than

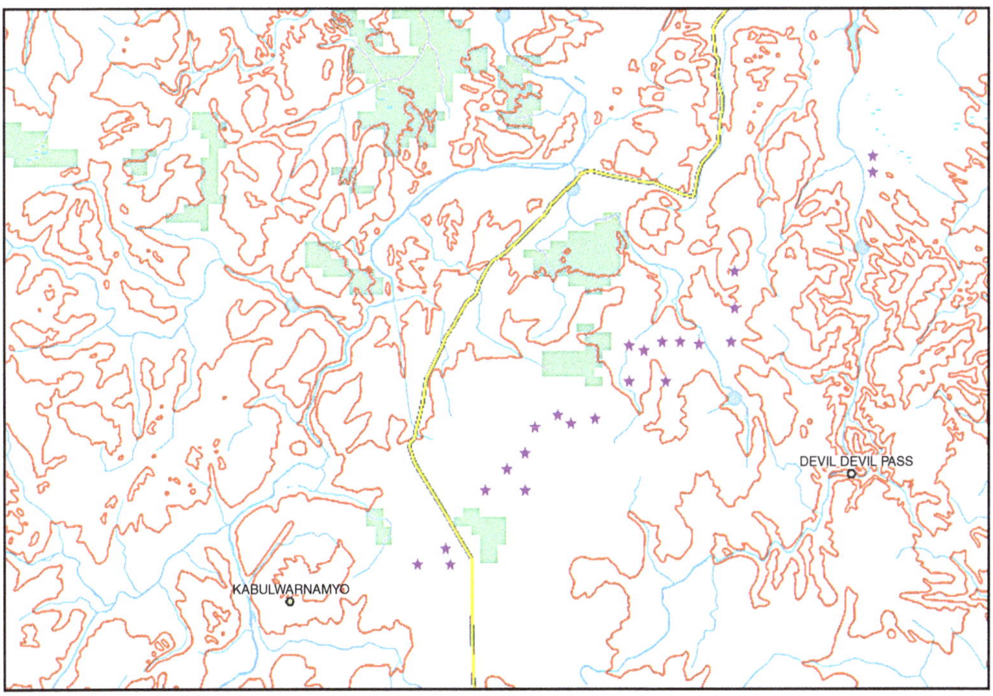

Figure 5.4. Purple star anxiety: a common practice reported was the viewing of the NAFI website immediately after an aerial burning operation to check to see if the burns had been successful (not gone out) by looking for recent hotspots indicated by purple stars on the site. This map indicates success – the purple stars appeared, signifying that the aerial ignitions continued to burn. The green patches indicate burnt areas from earlier aerial operations.

six hours old) will be coloured purple on the site and a successful burn is cause for some relief.

A more accurate indication of the extent of the burn is produced by the NAFI-MODIS burnt area mapping that is regularly updated every week or so during the fire season. To improve the accuracy of burnt area mapping in these areas, fire managers will send through shapefiles of their burn lines to the NAFI mappers, who can then pay particular attention to the fire scar mapping in the area of the recent fire operations (see Chapter 3). This regularly updated burnt area mapping from earlier burning operations acted as a crucial guide for later burning operations, as noted by one fire planner:

> *[A] really crucial thing for us in implementing burns across such a massive area is getting the updated fire scars every week. This helps us see where our fires have burnt from the previous weeks' burns and modify (or not) our burn plans for the coming weeks to implement strategic burns. This helps save a lot of money as it makes our operations a lot more efficient and effective.*
>
> *Anna Pickworth, Kimberley fire planner*

As the current 1 August cut-off date for the end of the EDS (under the high rainfall SBM) approaches, the project fire managers find it useful to estimate the area of EDS burnt area breaks or patches they have implemented, as this is an indication of progress made with respect to their fire management and greenhouse gas emission targets. Rather than use mapping software like ArcGIS, which can be time consuming, many fire planners now use a reporting tool on NAFI which generates a pdf report on burnt areas for each month within a pre-defined project area, as shown in Table 5.1.

Monitoring wildfire (August–December)

In the late dry season (LDS), from August on, the dry and hot conditions make uncontrollable wildfires more likely. For fire abatement projects, it can be important to check that these wildfires are not getting through the network of burnt areas that have been created earlier in the year. To do this satellite-based fire hotspots and recent burnt areas will be viewed on the NAFI website, or these NAFI data will be viewed on Google Earth.

Fig. 5.5 shows why these data are operationally useful. Because the time since the last fire is a reliable indicator of the levels of grassy fuel in the fire abatement areas, the maps of recently burnt areas and hotspots can be used to visualise the passage and direction of fires and forecast the fuel matrix and breaks in the landscape, like rivers and roads, that lie before a moving LDS fire – and do this on operationally useful timescales.

> *[I]n every fire I light I watch the hotspots for a few days on NAFI to see how far it goes and remember what it looked like when you flew over it – it's about learning about fuel: what it looks like from the air and then how it behaves by looking at the fires on NAFI – and the fire mapping obviously behind that – the feedback about how things went.*
>
> *Ben Lewis, Arnhem Land fire planner*

These maps do have their limitations: in many areas even recently burnt areas with low fuel can still carry a fire in the LDS, if they are well cured under extreme fire weather. In many projects, if the satellite maps indicate a wildfire is threatening to breach the network

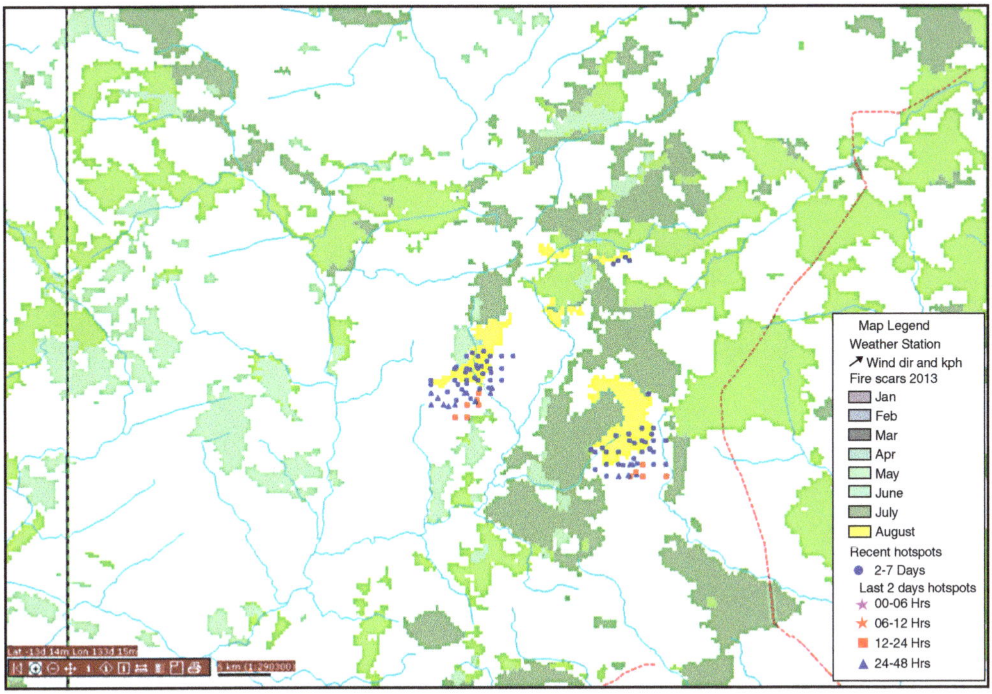

Figure 5.5. NAFI display of two LDS fires in western Arnhem Land: the active fire front is shown by the hotspots with the recently burnt ground indicated by the yellow (August) scars immediately to their north. Both fires are moving through country that has not been recently burnt. These fires are in fact controlled burns that are trickling into August (R. Geddes pers. comm.) but they illustrate the principles used to monitor wildfires.

of EDS burnt areas, then managers will have time to respond by putting in an additional fire break.

Assessing the year's work

Once the fire season is over the burnt area data for the year showing EDS burnt areas and any subsequent wildfires are downloaded from NAFI and printed out on large maps so people can sit around a table and work out how the fire management has gone and how it might be improved on for the next fire season. In some cases such meetings are part of the pre-season planning. For example, Fig. 5.6 shows the completed fire scar pattern for 2012 in south-east Arnhem Land – part of the Central Arnhem Land Fire Abatement (CALFA) Area.

The NAFI-MODIS fire history layers from recent years are also downloaded from NAFI and viewed against other relevant layers and printed out on maps to help estimate how levels of grassy fuel may have influenced the fire season.

Some groups will also download the hotspot data for their area from NAFI and the use GIS software to animate the hotspots so that the outbreak and spread of fires throughout the year can be visualised. In the case of the Mitchell-Coleman catchment project in Queensland, both the hotspots and the subsequent fire scars are animated and shown at the end-of-season and pre-season planning sessions.

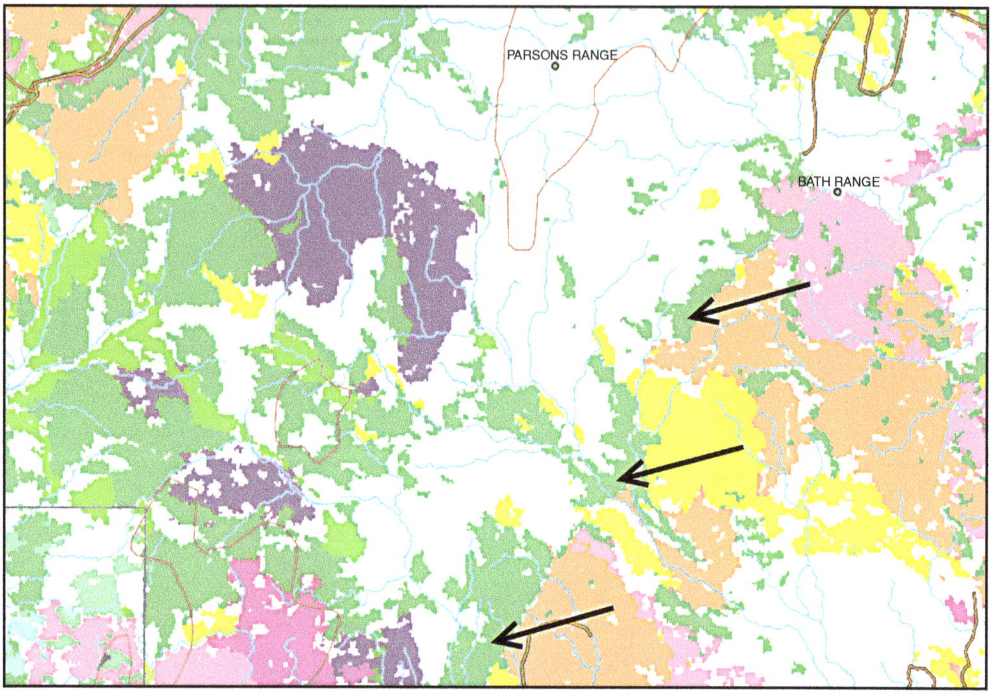

Figure 5.6. EDS burnt areas that worked. The olive green burnt areas (arrowed), implemented in July, have succeeded in pulling up late dry season fires (in yellow, orange and pink) at lower right driven by south-easterly winds, protecting the area to the north and west which has remained unburnt.

> *So the landholders have a look at these animations … together usually, so they all sit in a room … and they can identify exactly where these fires are coming from, and in the planning that we've done, the fires have been starting from the same property, so we could quickly identify where the problem was and what the risk was, and then try and mitigate that risk (in the next fire season). With everyone in the room – that was quite important.*
>
> *Aaron Crosbie, Mitchell-Coleman fire planner*

Again, many abatement projects would make use of the NAFI fire reporting tool at the end of the year to get a quick estimate of the areas burnt in each month. These area burnt figures were also used by land management groups associated with abatement projects as part of their reports to funding bodies.

Estimating emissions

All projects stated that if they were not doing so already, they would be using the Savanna Burning Abatement Tool (SavBAT) to calculate their emissions. This is an online tool hosted by the NAFI website that largely automates emissions calculations under the high rainfall SBM (Russell-Smith *et al.* 2009a). Proponents upload a map of their eligible fuel types, clipped to their project boundaries. The tool will then apply the tables in the SBM and the relevant NAFI-MODIS burnt area data to produce the emissions calculations for that project area. The tool produces a report with a unique checksum, corresponding to the input data and the code settings. An auditor need only re-upload the fuel map to SavBAT, make the emissions calculation and, if it generates the same checksum, be assured that the calculation is reliable. SavBAT is, however, an accounting and auditing tool rather than an operational tool. Several fire planners have requested a more flexible emissions calculator for operational tasks, such as progressive emissions calculations during the year and calculation of emissions for different fire management scenarios.

Discussion

Some broad points are worth noting. First, there is extensive use of satellite-sourced fire data in the abatement projects. The usage extends throughout the fire season and uses fire information in a range of formats and through a range of tools. Second, almost all these data are sourced from or used with the NAFI service. Of the 172 data access operations reported by the fire planners in Table 5.1, all but three involved the NAFI service. Third, all fire planners are of the opinion that the abatement projects would not be operationally feasible without the satellite-based fire information for planning and assessing fire operations throughout the year, given the large areas of country that needed to be burnt in a relatively short window of time in the EDS.

What are the implications for the resourcing of this usage? To help answer this question we examine the two main elements involved in the use of satellite fire data as indicated by Table 5.1: the manual MODIS mapping and the NAFI web service.

Manually mapped MODIS burnt areas

Table 5.1 shows that the MODIS manually mapped burnt area data is used operationally throughout the year in different formats and contexts: as fire histories for planning, guiding EDS implementation and for post-season assessment; as frequently updated current burnt area maps to guide EDS implementation and to assess the threat of wildfires; and as the basis for periodic estimation of areas burnt, usually using the NAFI reports.

Why is it so valued operationally? First, this mapping is the same dataset that is used to calculate the emissions under the current and proposed SBMs and hence the income a project generates, which gives it an operational value to abatement projects that should not be underestimated. However it is also valued for its *frequency* of updates and for its *accuracy*. The NAFI MODIS burnt area mapping is manually processed at weekly intervals throughout the northern fire season, and this updating was valued by all fire planners as it allowed them to guide burning operations in relation to the latest burn patterns. One additional advantage of this process is that it makes it easier for the accuracy of the mapping to be improved through feedback each week from fire managers (see Chapter 3 this volume). As outlined in Chapter 3 there are several automated MODIS burnt area mapping algorithms available that would require considerably less resources than that involved in the manual MODIS mapping, which requires two part-time mappers. However, these algorithms were generally less accurate than the NAFI manual mapping, particularly for the mapping of the smaller fires that are important to abatement operations in the EDS. Chapter 3 also notes that the object-based image analysis used in the manual process gave a cleaner mapping that was easier for fire managers to interpret than the pixel-based analysis used in the automated algorithms.

A detailed cost–benefit analysis of manual vs. automated algorithms is beyond the scope of this chapter, but given the investment in MODIS manual mapping in the abatement area is less than $150 000 each year and it effectively supports abatement projects across northern Australia that involve investment of millions of dollars annually and potential incomes of similar amounts, it would be a risky undertaking to switch to automated mapping in order to save money – until the feasibility of using automated burnt area mapping algorithms in the abatement projects is more fully investigated. Similarly the continued development of tools that add value to the burnt area data, such as SavBAT, should be encouraged as they have low costs and generate significant cost savings for the abatement projects.

The NAFI web service

The NAFI web service comprises servers, a website with a map viewer, maintenance and development services, and user engagement services that make the satellite fire data useful to fire managers. Like the MODIS mapping itself, it requires a reasonable investment.

How crucial is this web service to the implementation of the methodology and can it be made more efficient? While Table 5.1 shows the NAFI website is universally popular among fire planners, a significant portion of the usage involves accessing NAFI-MODIS burnt area data files from the NAFI server, and when the NAFI map viewer is used it is also used to view these data along with hotspots. Thus the value of the web service cannot be easily separated from that of the data it provides, as it is the only source of the manually processed MODIS burnt area data. There was, however, a benefit of the site, distinct from the data it provided, that was often mentioned by fire planners: its widespread usage.

> *[E]veryone talks NAFI, everyone talks about NAFI. Late last year I was talking to [cattle station manager], so here's a guy who's a third or fourth generation cattleman, stockman, buffalo catcher who … you know … is very, very old school – absolutely old school (claps hands) – but he was like 'yeah I was watching that fire on NAFI' … I mean everybody uses it.*
>
> *Shaun Ansell, Fish River fire planner*

Workable collaboration with neighbours is important for effective fire management in abatement projects, as it is for any savanna fire manager, and this can be difficult given the diverse mix of individuals, cultures and industry sectors that manage fire across the tropical savannas (Russell-Smith *et al.* 1999; Dyer *et al.* 2001). Fire planners indicated that this widespread use of NAFI allows neighbours to monitor and talk about fires of common interest using a *lingua franca* of background maps, coloured fire scars and hotspot symbols.

This common mapping language – the colours used for burnt areas and hotspots, the mix of burnt area histories developed, the background maps – indeed the focus on manual burnt area mapping itself – has been driven by feedback from fire managers. The same feedback has driven the development of the NAFI pdf report tools and the Google Earth viewing option. In many ways this feedback has been enabled by the NAFI web service and map viewer as it has allowed a varied cross-section of fire managers to provide feedback to the data providers. As described above, the NAFI map viewer also allows abatement project fire managers to help the MODIS burnt area mappers improve their accuracy in mapping EDS burns.

So while it is hard to separate out the value of the web service from the value of the MODIS burnt area data it provides, the reverse is also true as the value of the burnt area data to fire managers has been continually improved by the engagement with these managers through the web service. While the technology used to view these data will change, it is this engagement between different fire managers in the landscape, and between fire managers, data providers, researchers and technology specialists that has been enabled by the web service, that is an important part of effective fire operations in the Savanna Burning Methodologies – and this needs to be maintained.

Risk management

There is some risk involved in the current arrangements used to provide satellite-sourced fire data to abatement projects. Currently the NAFI service provides fire information through a range of centralised convenient tools, but uses systems that fire managers cannot control and cannot easily work around if those systems should fail – and failure in the delivery of satellite-based fire information would have a significant impact on the abatement projects as indicated above. The principal areas of concern are:

- Most of the fire information is based on imagery from MODIS sensors carried by two NASA satellites (Terra and Aqua) which have already exceeded their average six year design lifetimes. Hotspot data from a new Visible Infrared Imaging Radiometer Suite (VIIRS) sensor on the bridging Suomi-NPP satellite is now available (2014) but the two operational replacement satellites, the Joint Polar Satellites, are not due to be launched until 2017 and 2022 which could produce a potential data gap (US Government Accountability Office 2012). If this reliance on a small number of satellite-borne sensors persists and more than one was to degrade or fail before around 2018, then the fire information provided to abatement projects would be significantly degraded.
- The satellite-based fire data used by the NAFI service sit in databases hosted by Charles Darwin University in Darwin and these databases are needed to provide the continually updated fire data to the NAFI and Google Earth map viewers. Darwin and its network connections to the wider internet are subject to events like cyclones and flooding. If a catastrophic failure to the networks were to occur it may take some time for the NAFI map viewer to be back online.
- The regular supply of MODIS burnt area maps by NAFI during the northern fire season depends on the availability of two experienced mappers who have acquired familiarity with mapping northern burnt areas and who are in turn dependent on pre-processing of the satellite imagery. Unexpected loss of one or both mappers during the fire season might significantly degrade the regular supply of burnt area maps.
- The MODIS burnt area mapping and the NAFI service are currently funded through fixed term government-funded contracts and longer term funding is not secure.

Recommendations

It is recommended that the following actions are undertaken to ensure that satellite fire data continues to be useful to SBM project fire managers (including planners) and that there are options in place should some satellite data feeds fail.

1. Build capacity around satellite data mapping for SBM projects. This includes support for image pre-processing, automated mapping algorithms, and back-up fire mappers.
2. Maintain the current engagement between SBM project fire managers, data providers, researchers and technology specialists that allows the satellite-sourced fire data to be packaged and delivered in ways that meet the fire managers' evolving needs.

3. Build capacity to use fire data (not just satellite fire data) more generally in the SBM community to build resilience.
4. Work with relevant institutions to develop a strategy to support the actions above. This strategy should address:
   (a) security of data access from MODIS or other satellites, including back-up plans for data provision in the event that the data feed from the primary satellites should fail
   (b) security of data storage and processing, including a review of the viability of durable service provision by Charles Darwin University. One option here is to make the NAFI hotspots and burnt area data available through other well-supported databases that enable viewing on a range of map viewers and websites to reduce dependency on a single data centre
   (c) long-term funding options for the continued engagement with users and development of satellite fire data products. One option here is to investigate how support for the NAFI service can be built into arrangements under the carbon and other offsets markets. Any funding options need to take account of the fact that the satellite fire data products are used to achieve broader goals than emissions abatement, such as conservation of biodiversity and other economic, social and cultural goals, and that these data are used by many fire managers outside the SBM projects.

Acknowledgements

The authors would like to acknowledge Ron Steffens, Steve Sutton, Andrew Edwards and Rohan Fisher for their valuable comments on the draft and Richard Geddes for his information on Arnhem Land fires.

Notes

1. By early 2015 part of the Central Arnhem Land Fire Abatement Project area was registered as WALFA2 under the CFI.
2. These data are originally sourced from Landgate WA and Geoscience Australia but are stored locally and displayed with timings and symbology to suit northern fire management.

References

Ansell S, Koenig J (2011) CyberTracker: An integral management tool used by rangers in the Djelk Indigenous Protected Area, central Arnhem Land, Australia. *Ecological Management & Restoration* **12**(1), 13–25. doi:10.1111/j.1442-8903.2011.00575.x

Cook G, Meyer CP (2009) Fire, fuels and greenhouse gases. In *Culture, Ecology and Economy of Fire Management in North Australian Savannas: Rekindling the* Wurrk *Tradition*. (Eds J Russell-Smith, PJ Whitehead, P Cooke) pp. 313–337. CSIRO Publishing, Melbourne.

DCCEE (Australian Government Department of Climate Change and Energy Efficiency) (2011) Template for Submitting Methodologies. Carbon Farming Initiative: Canberra. www.climatechange.gov.au/reducing-carbon/carbon-farming-initiative/methodologies/guidelines-submitting-methodologies

Dyer R, Jacklyn P, Russell-Smith J, Williams R (2001) Introduction. In *Savanna Burning: Understanding and Using Fire in Northern Australia*. (Eds R Dyer, P Jacklyn, I Partridge, J

Russell-Smith, R Williams) pp. 1–4. Tropical Savannas Cooperative Research Centre: Darwin.

Edwards A, Cooke P (2010) 'West Arnhem Fire Management Agreement Annual Report 2009.' Unpublished report to Northern Territory Government.

Ens E (2012) Monitoring outcomes of environmental service provision in low socio-economic indigenous Australia using innovative CyberTracker Technology. *Conservation & Society* **10**(1), 42–52. doi:10.4103/0972-4923.92194

Ludwig JA, Bastin GN, Wallace JF, McVicar TR (2007) Assessing landscape health by scaling with remote sensing: when is it not enough? *Landscape Ecology* **22**, 163–169. doi:10.1007/s10980-006-9038-6

Manson G (2010) 'The Value of the Sentinel Hotspots System to Emergency Managers and the Australian Community.' Unpublished report for Geoscience Australia.

Russell-Smith J, Allan G, Thackway R, Rosling T, Smith R (1999) Fire management and savanna landscapes in northern Australia. In *ACIAR Proceedings: Fire and Sustainable Agricultural and Forestry Development in Eastern Indonesia and Northern Australia.* (Eds J Russell-Smith, G Hill, S Djoeroemana, B Myers) pp. 95–101. ACIAR: Canberra.

Russell-Smith J, Murphy BP, Meyer CP, Cook GD, Maier S, Edwards AC, Schatz J, Brocklehurst P (2009a) Improving estimates of savanna burning emissions for greenhouse accounting in northern Australia: limitations, challenges, applications. *International Journal of Wildland Fire* **18**, 1–18. doi:10.1071/WF08009

US Government Accountability Office (2012) 'Polar-Orbiting Environmental Satellites: Changing Requirements, Technical Issues, and Looming Data Gaps Require Focused Attention.' GAO-12-604. Washington, DC.

Whitehead P, Purdon P, Russell-Smith J, Cooke P, Sutton S (2008) The management of climate change through prescribed savanna burning: emerging contributions of indigenous people in northern Australia. *Public Administration and Development* **28**, 374–385. doi:10.1002/pad.512

6

Fuel accumulation, consumption and fire patchiness in the lower rainfall savanna region

Cameron P Yates, Jeremy Russell-Smith, Brett P Murphy, Mark Desailly, Jay Evans, Sarah Legge, Fay Lewis, Dominique Lynch and Andrew C Edwards

Summary

This chapter outlines field studies and subsequent analyses undertaken for the purposes of deriving robust parameter values for fuel load accumulation (FLA), fire patchiness (P), and burning efficiency factors (BEF) for incorporation into a savanna burning emissions abatement methodology for areas of lower rainfall (600–1000 mm p.a.). We apply a broadly similar methodology to that developed for the current higher rainfall (>1000 mm p.a.) savanna burning abatement methodology. FLA data were assembled from 648 transects sampling the full rainfall envelope over a wide variety of landscape conditions that support characteristic unmodified flammable savanna vegetation structural types. Patchiness data were assembled from 3325 observations utilising three assessment methods. Fuel combustion (pyrolysis) and BEF data were derived from 90 fire treatments under a range of fire severity conditions, and categorisation of fire severity under early dry season (EDS) and late dry season (LDS) conditions was derived from observations of 392 fires at 267 permanent plots over the period 2009–2013.

Mean total fuel loads in the early (5.66 t ha^{-1}) and late (6.15 t ha^{-1}) dry seasons, and those of constituent components (fine fuels, coarse woody fuels, heavy woody fuels, shrub fuels), were less than those sampled under higher rainfall conditions. Based on data presented in Chapter 7, most input of litter fuel (leaf, twigs) occurred in the LDS (73%) (after August). Accumulation of fine fuels was effectively described by Olson curve and regression models for each of six vegetation/fuel types (described in Chapter 4), but not for lancewood and tussock grasslands fuel type combinations. Fires were observed to be more patchy early (79% of area burnt) than late (97%) in the dry season. BEFs for fuel components were observed to vary little between EDS and LDS, and overall were lower for coarse (sticks) and heavy (logs) woody debris components, and shrub fuels. Fine fuels exhibited high BEFs given a much larger grass component and subsequent more complete pyrolysis although, overall, our data suggest that fire treatments were of lesser intensity than observed under higher rainfall conditions. This assessment indicates that, for implementing a savanna burning abatement project in lower rainfall areas, two parameters are of particular importance – accumulation of greater proportions of fuel loads with litter inputs later in the season, and patchier early season fires.

Introduction

This chapter reports a range of fuel and fire parameters, based on substantial fieldwork conducted from 2009, that are necessary for the development of a lower rainfall (1000–600 mm p.a.) savanna burning emissions abatement methodology. These parameters are fuel load accumulation (FLA), fire patchiness (P), and burning efficiency factors (BEF). The methods mostly follow procedures as described by Russell-Smith *et al.* (2009) for the development of the higher rainfall (>1000 mm p.a.) methodology, but draw also on results from other chapters describing fuel types (associated with dominant vegetation types: Lynch *et al.*, Chapter 4), seasonality of litter fall (Cuff and Brocklehurst, Chapter 7), and seasonal fire severity characteristics (Edwards and Russell-Smith, Chapter 8).

For ease of presentation, we address methods and results for FLA, P, and BEF parameters in turn. A final table summarises FLA, P, and BEF parameters for inclusion in a lower rainfall savanna burning abatement methodology, and as applied in a subsequent chapter (Russell-Smith *et al.*, Chapter 10) which provides worked examples of the proposed methodology in idealised regional project settings.

Field program

Assembled data were collected primarily by the Darwin Centre for Bushfire Research (DCBR), with the aid of Indigenous ranger groups in the Gulf region (both Queensland and the Northern Territory), Arnhem Land (Northern Territory), and the Kimberley (Western Australia); with additional data collected by the Australian Wildlife Conservancy (AWC) in the Kimberley.

Fuel load accumulation (FLA) was assessed by DCBR at 162 plot locations across the tropical savannas (69 in the Kimberley, 57 in the Gulf of Carpentaria, and 36 in central Arnhem Land). In most cases sampling was undertaken with at least three replicate transects at each location, giving a total of 413 transects. AWC assessed fuels at 82 plot locations in the southern Kimberley with a total of 235 transects. Fine fuels data was augmented with an additional 115 transects in Gregory National Park (Russell-Smith *et al.* 2010). As in the higher rainfall savanna burning abatement methodology (Russell-Smith *et al.* 2009), we considered replicate transects at a single site (400 m–1 km separation) as independent observations given substantial fire history heterogeneity within and between transects. Plots were geographically widely distributed over a rainfall envelope ranging from a long-term mean annual average of 582–1062 mm p.a. (Fig. 6.1).

Plots were stratified by vegetation / fuel type based on structural characteristics (e.g. foliage projective cover and canopy height) and the dominant tree, shrub and grass species composition (refer Lynch *et al.*, Chapter 4), and by time-since-fire (TSF), based on detailed fire history assessments using Landsat imagery for DCBR plots and MODIS imagery for AWC plots.

Sampling was undertaken predominantly in the LDS, from 2006 to 2012, following the methodology described by Russell-Smith *et al.* (2009). The bulk of the sampling conducted by DCBR was undertaken in 2009–2010, and by AWC over 2011–2012. Given that our aim was to sample ambient fuel loads under relatively unmodified savanna conditions, where possible we avoided heavily grazed, weedy, and cleared situations. At each transect we recorded apparent bovine (feral animal) grazing intensity using a semiquantitative scale of low (n = 301 transects), moderate (n = 250), and heavy (n = 21) apparent grazing impacts.

Fuel load accumulation

Field methods

Fuel load accumulation (FLA) data for four fuel components (fine fuel, coarse woody debris, heavy fuel, shrubs) was measured following the methodology outlined in Russell-Smith *et al.* (2009) for a total of 648 transects, comprising 227 in the Gulf, 110 in central Arnhem Land, and 311 in the Kimberley. Each transect was 100 m in length. Fuel components were collected as follows:

Fine fuels (<6 mm diameter): All fine fuel was collected from five 1 × 1 m quadrats at 20 m intervals along the 100 m transect. Fine fuels (<6.0 mm diameter) were collected separately for standing grass and litter components, and weighed in the field with digital scales. Moisture contents for the grass and litter components were estimated by taking a composite subsample and drying and weighing in the laboratory. Moisture content was used to convert fresh weight to dry weight.

Coarse fuels (woody debris ≥0.6 cm – <5 cm diameter): All coarse fuel was collected from ten 1 × 1 m samples at 10 m intervals along the 100 m transect, and weighed in the field. Again, fuel moisture content was used to convert fresh weight to dry weight.

Heavy fuels (woody debris ≥5 cm diameter): Sampling was undertaken in a 5 × 100 m swath, recording the length, approximate diameter and proportional hollowness of all heavy fuel lying on the ground within the swath. Assuming each piece of heavy fuel was cylindrical in shape, we estimated the total volume of heavy fuel, and total mass assuming a specific gravity of 0.995 t m^{-3} (approximating that of eucalypt wood; Eamus *et al.* 2000).

Shrub fuels (<5 cm diameter at breast height – DBH): Live and dead shrubs and juvenile trees were counted in five 1 × 20 m sections along the length of the 100 m transect. Shrubs were grouped into in four height classes: <0.5 m; 0.5–1.0 m; 1–2 m; and >2 m

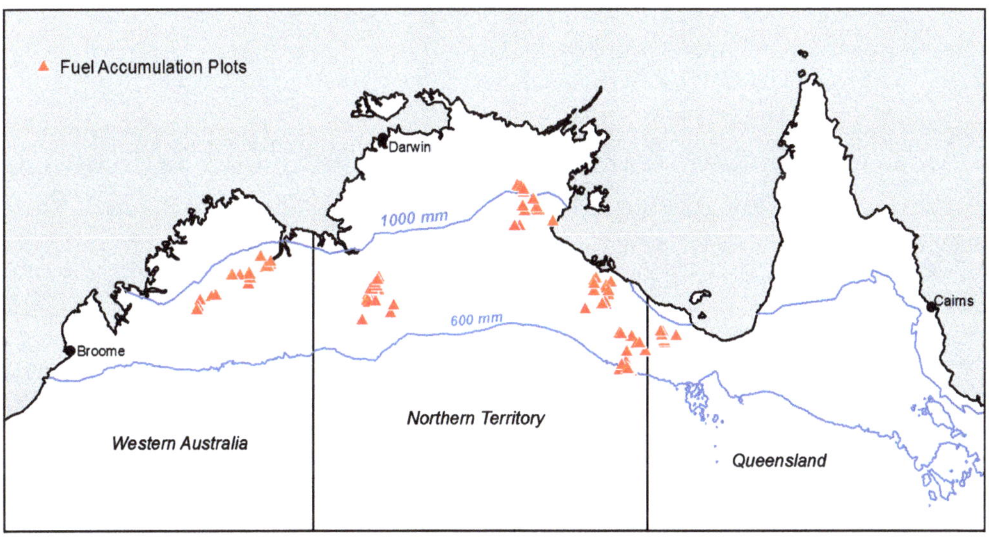

Figure 6.1. Location of fuel accumulation sampling plots with respect to 1000 and 600 mm rainfall isohyets. Rainfall data derived from mean annual wet season rainfall, 1969–2009 (Bureau of Meteorology).

but <5 cm DBH. Live shrubs were identified to species, but dead shrubs were not. To estimate shrub biomass, we harvested and weighed (corrected for moisture) the aboveground parts of at least three replicates for each size class of 105 common species for a total of 229 samples (some species sampled more than once). Species were allocated to one or more vegetation types in which they commonly occur and, for each vegetation type, mean shrub mass per height class was calculated. For each field count, shrub densities were converted to aboveground dry biomass by multiplying the mean shrub mass calculated for the appropriate vegetation type by the shrub counts per height class.

Analyses

Data from all transects were used in FLA analyses, with the exception that heavily grazed transects were excluded from analyses. Simple linear regression was used to examine the relationship between fuel components (fine, coarse, heavy, and shrub) and TSF for each of the fuel types. A natural log-transformation was applied both to fuel load response (given positive skews in most datasets) as well as TSF (given apparent non-linearity). In those cases where statistically significant relationships ($P < 0.05$) were not observed between fuel load and TSF, we follow the methodology set out in the higher rainfall region with fuel loads being described by a simple mean. We also report relationships based on use of the Olson curve (Olson 1963) given that FLA is described in Australia's National Greenhouse Gas Inventory (DoE 2014) using such relationships.

Separate analyses were conducted for groups of transects defined by fuel type (Lynch *et al.*, Chapter 4). For this chapter we report on the six vegetation/fuel types following, but, as discussed both in Chapter 4 and 10, the open woodland over tussock grasses class is excluded from the final methodology:

- woodland (10–30% foliage cover) over tussock grasses (WTu)
- woodland over mixed tussock and hummock (i.e. spinifex (*Triodia*)) grasses (WMi)
- woodland over hummock grasses (WHu)
- open woodland (<10% foliage cover) over tussock grasses (OWT)
- open woodland over mixed hummock and tussock grasses (OWM)
- shrubland heaths (foliage cover <10–30%) over hummock grasses (SHH).

Given limited assembled samples for the SHH class, we included an additional 115 fine fuel measures from Gregory National Park sampled in an analogous fashion (but incorporating ten 1 × 1 m quadrats) following the methodology as described in Russell-Smith *et al.* (2010).

Results and discussion

Fuels in the lower rainfall region predominantly comprise fine fuels. Excluding fine fuel data from Gregory National Park, see above, 55% of the total fuel loads consisted of grass (32%) and litter (24%), with lesser, more-or-less equal proportions of coarse, woody and shrub fuel components (Table 6.1). While shrub fuels contributed just 14% of total fuels (Table 6.1), most of this was derived from taller shrubs, especially woody stem components (Table 6.2).

Fine fuel loads were significantly related to TSF in all lower rainfall fuel types, generally with R^2 values >0.5 for linear models (Table 6.3), with lower R^2 values observed for Olson

Table 6.1. Mean fuel loads (t ha^{-1}, SE) and proportions (%), of fuel components in the lower rainfall region for respective fuel types, where n = number of observations used in respective analyses

Standard error for untransformed data shown.

	Grass			Litter			Total fine			Coarse			Heavy			Shrub			
	t ha^{-1}, (SE)	(%)	n	t ha^{-1}, (SE)	(%)	n	t ha^{-1}, (SE)	(%)	n	t ha^{-1}, (SE)	(%)	n	t ha^{-1}, (SE)	(%)	n	t h^{-1}, (SE)	(%)	n	t ha^{-1}
WHu	2.17, (0.28)	24	42	2.17, (0.24)	24	42	4.33, (0.39)	47	42	1.85, (0.34)	20	39	1.14, (0.21)	12	42	1.84, (0.62)	20	14	9.17
WMi	1.04, (0.08)	18	84	1.50, (0.09)	25	84	2.53, (0.12)	43	84	0.69, (0.09)	12	68	2.02, (0.37)	34	86	0.66, (0.12)	11	59	5.9
WTu	1.94, (0.12)	27	93	2.45, (0.20)	34	93	4.39, (0.27)	60	93	1.39, (0.14)	19	81	1.25, (0.34)	17	114	0.27, (0.06)	4	31	7.3
OWM	2.21, (0.12)	38	241	0.98, (0.06)	17	241	3.19, (0.15)	54	241	0.75, (0.10)	13	207	0.80, (0.09)	14	246	1.12, (0.10)	19	152	5.87
OWT	1.95, (0.11)	34	190	1.48, (0.11)	26	190	3.43, (0.16)	60	190	0.68, (0.08)	12	156	0.81, (0.09)	14	192	0.75, (0.09)	13	118	5.68
SHH	2.05, (0.16)	41	18	1.36, (0.17)	27	18	3.26, (0.21)	65	133	0.73, (0.34)	14	7	0.17, (0.06)	3	13	0.89, (0.24)	18	13	5.05
Total	1.94, (0.06)	32	668	1.48, (0.05)	24	668	3.39, (0.08)	55	783	0.90, (0.06)	15	558	0.97, (0.08)	16	779	0.89, (0.06)	14	387	6.15

Table 6.2. Proportion of shrub biomass in each height class, where n = number of samples

Height class	Proportion of total shrub biomass (%)	Proportion stem (%)	Proportion leaf (%)
<0.5 m (n = 56)	9.17	42.43	57.57
0.5–1.0 m (n = 65)	19.57	49.08	50.92
1.0–2.0 m (n = 62)	31.04	63.90	36.10
>2.0 m (n = 46)	40.23	79.16	20.84

Table 6.3. Linear model for fine fuel accumulation: (ln (fuel + 0.01) = a * ln(TSF) + b) where fuel is the fuel load (t ha^{-1}) and TSF is time since fire (years)

Significant ($P < 0.05$) models given in bold. Standard errors of the regression coefficients are shown.

Fuel type (n)	a	b	P	R^2
WHu (42)	**0.44 ± 0.03**	**0.96 ± 0.07**	**<0.0001**	**0.87**
WMi (84)	**0.25 ± 0.02**	**0.80 ± 0.04**	**<0.0001**	**0.64**
WTu (93)	**0.31 ± 0.03**	**1.19 ± 0.06**	**<0.0001**	**0.54**
OWM (241)	**0.30 ± 0.01**	**0.90 ± 0.04**	**<0.0001**	**0.71**
OWT (190)	**0.21 ± 0.02**	**0.99 ± 0.04**	**<0.0001**	**0.38**
SHH (133)	**0.47 ± 0.04**	**0.57 ± 0.06**	**<0.0001**	**0.55**

Table 6.4. Significant Olson curve model for fine fuel accumulation: Fuel = Xss*(1 – exq(–k*TSF)), where fuel is the fuel load (t ha^{-1}) and TSF in time since fire (years)

Significant (P <0.05) models given in bold. Standard errors of the regression coefficients are shown.

Fuel type	Xss	k	R^2
WHu	**6.61 ± 1.78**	**0.36 ± 0.23**	**0.47**
WMi	**3.26 ± 0.18**	**1.21 ± 0.23**	**0.44**
WTu	**7.30 ± 0.87**	**0.54 ± 0.14**	**0.42**
OWM	**4.20 ± 0.17**	**1.23 ± 0.27**	**0.39**
OWT	**4.33 ± 0.25**	**1.05 ± 0.19**	**0.25**
SHH	**10.08 ± 3.48**	**0.14 ± 0.07**	**0.45**

Table 6.5. Linear model for coarse woody debris accumulation: (ln (fuel + 0.01) = a * ln(TSF) + b) where fuel is the fuel load (t ha^{-1}) and TSF is time since fire (years)

Significant (P <0.05) models given in bold.

Fuel type (n)	a	b	P	R^2
WHu (39)	–0.20	0.16	0.6210	0.01
WMi (68)	0.37	–1.054	0.0590	0.05
WTu (81)	0.46	–0.55	0.0539	0.05
OWM (207)	**0.51**	**–1.75**	**0.0012**	**0.05**
OWT (156)	0.06	–1.34	0.7580	0.00
SHH (7)	4.01	–7.89	0.5370	0.08
OW	**0.31**	**–1.51**	**0.0089**	**0.02**
W	**0.36**	**–0.69**	**0.0111**	**0.03**

Table 6.6. Significant Olson curve model for coarse woody debris accumulation: Fuel = Xss*(1 – exq(–k*TSF)), where fuel is the fuel load (t ha^{-1}) and TSF is time since fire (years)

Significant (*P* <0.05) models given in bold.

Fuel type	Xss	k	R^2
WHu			
WMi	0.84	1.08	0.03
WTu	1.55	1.45	0.01
OWM	**0.92**	**0.64**	**0.01**
OWT			
SSH			
OW	**0.31**	**–1.51**	**0.02**
W	**0.36**	**–0.69**	**0.03**

Table 6.7. Linear model for heavy fuel accumulation: (ln (fuel + 0.01) = a * ln(TSF) + b) where fuel is the fuel load (t ha^{-1}) and TSF is time since fire (years)

Significant (*P* <0.05) models given in bold.

Fuel type (*n*)	a	b	*P*	R^2
WHu (42)	–0.43	–0.32	0.439	0.02
WMi (86)	–0.04	–0.49	0.7481	0
WTu (114)	0.18	–1.47	0.165	0.02
OWM (246)	**0.13**	**–1.42**	**0.0425**	**0.02**
OWT (192)	–0.07	–1.50	0.585	0
SHH (13)	–0.001	–2.45	0.999	0
OW	0.09	–1.50	0.1320	0.01
W	0.03	–0.99	0.683	0

Table 6.8. Significant Olson curve model for heavy fuel accumulation: Fuel = Xss*(1 –exq(–k*TSF)), where fuel is the fuel load (t ha^{-1}) and TSF is time since fire (years)

Significant (*P* <0.05) models given in bold.

Fuel type	Xss	k	R^2
WHu			
WMi			
WTu	1.95	0.57	0.02
OWM	**0.86**	**1.94**	**0.01**
OWT	0.90	1.57	0.003
SHH			

curve models (Table 6.4). Significant relationships (but with low R^2 values) were observed for coarse fuels in OWM and WTu classes (Table 6.5), for heavy fuels in the WTu class (Table 6.7), and for shrub fuels in WMi, OWM and OWT classes (Table 6.9) for linear models, and generally with even lower R^2 values observed for corresponding Olson curve models (Tables 6.6, 6.8, and 6.10). Significant relationships (but with low R^2 values) with

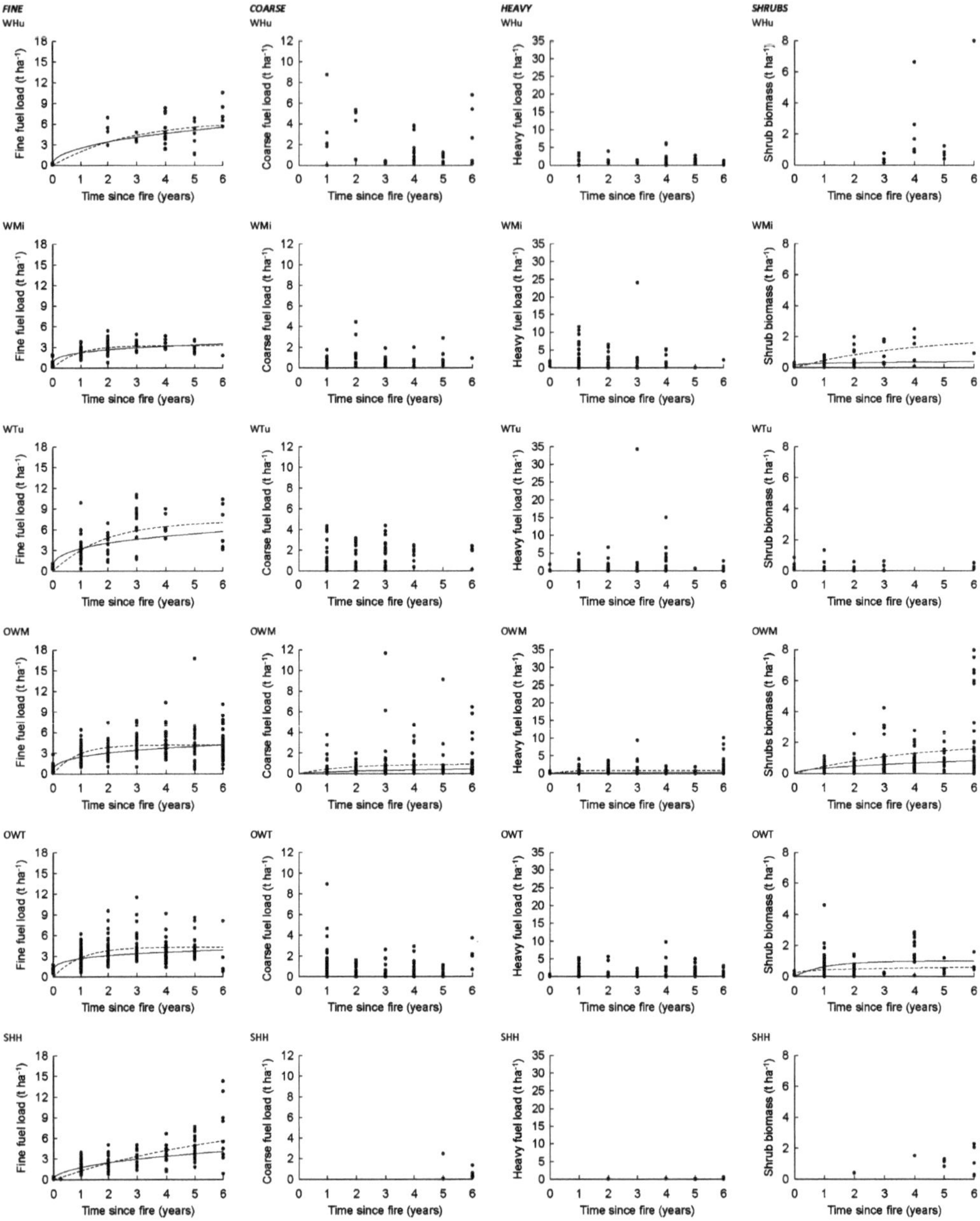

Figure 6.2. Relationship between accumulation of respective fuel components (fine, coarse, heavy, shrubs) and time-since-fire, in respective vegetation types: WHu; WMi; WTu; OWM; OWT; SHH. The solid regression lines refer to the results of linear regressions of fuel load (natural log-transformed) vs. time since fire (natural log-transformed). The dashed lines refer to fitted Olson curves.

TSF were observed for coarse, heavy, and shrub fuel components in analyses where observations were pooled into broad vegetation structural types (i.e. woodland, open woodland) rather than by respective fuel types (Tables 6.5, 6.7, 6.9 and Fig. 6.2). Corresponding Olson curve relationships are given in Tables 6.4, 6.6, 6.8, 6.10, and Fig. 6.2, respectively. Given

Table 6.9. Linear model for shrub fuel accumulation: (ln (fuel + 0.01) = a * ln(TSF) + b) where fuel is the fuel load (t ha^{-1}) and TSF is time since fire (years)

Significant (P <0.05) models given in bold.

Fuels class (*n*)	a	b	*P*	R^2
WHu (14)	2.30	–3.17	0.0839	0.23
WMi (59)	**0.16**	**–1.23**	**0.0254**	**0.08**
WTu (31)	–0.08	–1.77	0.2890	0.04
OWM (152)	**0.49**	**–1.07**	**0.0004**	**0.08**
OWT (118)	**0.20**	**–0.87**	**0.0189**	**0.05**
SHH (13)	0.47	–0.93	0.4200	0.06
OW	**0.29**	**–0.87**	**0.0000**	**0.07**
W	**0.13**	**–1.21**	**0.0212**	**0.05**

Table 6.10. Significant Olson curve model for shrub fuel accumulation: Fuel = Xss*(1 – exq(–k*TSF)), where fuel is the fuel load (t ha^{-1}) and TSF in time since fire (years)

Significant (P <0.05) models given in bold.

Fuel type	Xss	k	R^2
WHu			
WMi	**0.84**	**1.08**	**0.03**
WTu	**1.55**	**1.45**	**0.01**
OWM	**0.92**	**0.64**	**0.01**
OWT			
SSH			

that significant relationships for all fuel types were not obtained for coarse, heavy and shrub fuels, averages for these fuel components are used following the methodology described in Russell-Smith *et al.* (2009).

Fire patchiness

Fires do not consume all fuel within a fire-affected area – some fuel remains unburnt. Available optical satellite sensors lack the resolution to accurately map fire patchiness particularly within pixels (e.g. Price *et al.* 2003). To account for this, we measured fire patchiness using three different methods across the lower rainfall region.

Methods

Linear transects: Patchiness was measured along walked transects using ground surveys under differing habitat conditions and in different seasons (EDS, LDS) following the methodology outlined in Price *et al.* (2003) and Russell-Smith *et al.* (2009). Sampling was undertaken within the fire-affected perimeter using transects of differing lengths depending on fire size. Observations were recorded for each 100 × 1 m section, noting: patchiness (% burnt), average scorch height (m), bare earth (%), rockiness (%). Patchiness was sampled at a total of 549 fire events in the EDS and 61 in the LDS, equating to a total linear sample length of 61 km.

Treatment plots: Post-fire assessments were undertaken on established FLA transects that were deliberately burnt following fuel assessments. A total of 62 burnt transects were sampled for patchiness, comprising 41 in the EDS and 21 in the LDS. At each transect, patchiness was measured in five 1 × 20 m sections along the 100 m transect and then averaged.

Aerial photographic assessments: AWC estimated fire patchiness for different seasons, in different vegetation/fuel types, and for a range of fuel ages (TSF) using aerial photography. The aerial photography involved taking high resolution photographs of fire scars using a hand-held digital camera from a helicopter flying at 500–800 feet above ground level. At this height, patches too small to be sampled using medium-resolution satellite imagery such as MODIS (i.e. 1–250 m) were easily identified. The plane of the camera view was roughly parallel to the ground, with part of the helicopter skid included in the frame as a point of reference. Because we estimated patchiness as the proportion of a fire scar that was burnt, rather than by area, the exact height of the helicopter and the angle of the photograph was not material. Each photograph was matched to a GPS location record. Photographs were excluded that were blurry, over- or under-exposed, not perpendicular to the ground, or had some other ambiguity. The total sample comprised 2503 photographs for EDS fires taken over four consecutive years (2010–13), and 174 LDS photographs taken over two years (2011–12). While we sampled fewer LDS fire scars, we note that variation in patchiness was substantially less. Using Photoshop (Adobe 2002) we manually traced the boundaries of burnt and unburnt areas in each photograph, summed the pixels in the burnt and unburnt areas, and calculated the proportion of all pixels that were burnt. Photographs were scored by the same person. We checked for scoring repeatability by tracing the boundaries a second time for a subset of photographs; the results of the second sample were within 5%, and most were within 2%,

Table 6.11. Fire patchiness in early (EDS) and late (LDS) dry seasons, comparing the three different collection methods

SE = Standard error of the untransformed mean.

	EDS		LDS	
Source (*n* EDS, *n* LDS)	**Mean %**	**SE**	**Mean %**	**SE**
Walk (548, 38)	80	1.14	93	1.6
Post-fire (41, 21)	84	3.56	95	2.87
Photo (2503, 174)	78	1.74	99	0.11
Weighted mean (3092, 233)	79		97	

Table 6.12. Fire patchiness in early (EDS) and late (LDS) dry seasons, comparing the three dominant grass understorey components

SE = Standard error of the untransformed mean.

	Dominant understorey					
	Hummock mean %	**SE**	**Mixed mean %**	**SE**	**Tussock mean %**	**SE**
Early	74	0.84	81	0.6	80	0.85
Late	94	1.95	99	0.29	93	2.13

of the first tracing. As presented here, patchiness data were combined and grouped by season (EDS, LDS) for dominant vegetation/fuel types.

Results and discussion

Combining data from all three procedures, resultant patchiness was 79% in the EDS and 97% in the LDS (Tables 6.11, 6.12) for the five fuel types included in the methodology. These figures differ somewhat from that determined for the >1000 mm abatement methodology (cf. 71% EDS, 89% LDS: Russell-Smith *et al.* 2009), which is likely attributable to a more open tree canopy (Cuff and Brocklehurst, Chapter 7) that allows for a more continuous layer of grassy fuels, and hence less patchy fires in lower rainfall regions.

Burning efficiency factors (BEFs)

Fuel combustion data were collected at a subset of fire-treated transects established and measured for FLA then burnt and re-measured. The proportion of each fuel component remaining unburnt or consumed was measured (i.e. either volatilised or combusted to form charred material and white ash), following exposure to fires of varying severity. In contrast to terminology used describing combusted material (i.e. 'ash') in the earlier study of Russell-Smith *et al.* (2009), here we use the broader description, 'residual combusted material'. Burning efficiency is defined as the mass of fuel exposed to fire that is volatilised.

To convert these combustion data observed under three fire severity classes (see below) to BEFs, we applied the seasonal (EDS, LDS) distributions of fire severity classes as observed from 392 fires recorded from 267 long-term monitoring plots established in the lower rainfall abatement region for the purposes of deriving relationships between fire regime and woody biomass change. Fire severity was assessed from photos taken at annual site visits, and checked against available Landsat and MODIS imagery where season of burning required verification. Plot-based observations of seasonal fire severity were made over the period 2009–2013, and are discussed in more detail in Chapter 8.

Table 6.13. Combustion rates for each fuel component for low/moderate and high severity with SE on untransformed data, fires and estimates of BEF for EDS and LDS fires based on relative frequency of fire severity

	Fuel consumption (%)			
(a) Combustion				
Fire severity	**Fine, SE**	**Coarse, SE**	**Shrub**	**Heavy, SE**
Low	77.94, 1.88	8.67, 1.53	9.37, 1.15	5.18, 1.53
Moderate	81.76, 1.94	14.68, 2.63	10.50, 1.58	9.47, 2.51
High	84.75, 1.71	30.12, 4.42	11.81, 2.00	16.85, 3.5
(b) Fire severity/seasonality				
Season	**Low**	**Moderate**	**High**	
EDS % (*n*)	72.6 (199)	24.8 (68)	2.6 (7)	
LDS % (*n*)	13.6 (16)	38.1 (45)	48.3 (57)	
(c) BEF				
Season	**Fine**	**Coarse**	**Shrub**	**Heavy**
EDS	79.92	10.9	9.82	6.65
LDS	83.28	20.16	10.98	11.92

Table 6.14. Summary of parameters for calculating emissions for the lower rainfall region

Fuel accumulation (t ha^{-1})							
		TSB					
Fine fuels		1	2	3	4	5	> 5
Early	WHu	1.75	3.14	4.12	4.80	5.27	5.60
	WMI	2.00	2.69	2.89	2.95	2.97	2.98
	WTu	2.71	4.48	5.52	6.12	6.47	6.68
	OWM	2.79	3.66	3.91	3.98	4.01	4.01
	OWT	2.54	3.53	3.87	3.99	4.04	4.05
	SHH	1.00	2.11	3.08	3.93	4.68	5.32
Late	WHu	2.01	3.40	4.38	5.06	5.53	5.86
	WMI	2.28	2.97	3.17	3.23	3.25	3.26
	WTu	3.05	4.82	5.86	6.46	6.81	7.02
	OWM	2.97	3.84	4.09	4.16	4.19	4.19
	OWT	2.81	3.80	4.14	4.26	4.31	4.32
	SHH	1.28	2.39	3.36	4.21	4.96	5.60
Non-fine fuel averages							
		Coarse	*Heavy*	*Shrub*			
	WHu	1.85	1.15	1.84			
	WMi	0.69	2.02	0.66			
	WTu	1.39	1.25	0.27			
	OWM	0.76	0.8	1.13			
	SHH	0.73	0.17	0.87			
Patchiness (%)							
	Early	*Late*					
	79	97					
BEF (%)							
	Early	*Late*					
Fine	79.92	83.28					
Coarse	10.90	20.16					
Shrub	9.82	10.98					
Heavy	6.65	11.92					
Residual material remaining: 3.79%							

Methods

The severity of 90 imposed fire treatments was assessed on the basis of leaf-scorch height, following the methodology of Russell-Smith and Edwards (2006), where: low severity = leaf scorch <2 m, or fires patchy with >20% groundcover remaining unburnt; moderate severity = leaf scorch higher than 2 m but upper canopy not scorched; high severity = upper canopy scorched. Of the 90 applied fire treatments, 41 were of low severity, 30 were of moderate severity, and 19 of high severity. Post-fire transects were sampled as follows:

Fine fuels: Five 1 × 1 m quadrats were sampled at equal distances along the transect, but only in situations where fire had affected 100% of the sample quadrat. Separate measures of remaining grass, litter and ash were recorded with digital scales.

Coarse and heavy fuels: The proportion of coarse and heavy fuel components consumed was measured in five 1 × 20 m strips along the length of the 100 m transect. The consumption of individual sticks and logs was based on the relative length of the 'ash trail' left after the fire.

Shrubs: Consumption was measured in five 1 × 20 m samples along the length of the transect, recording the proportion of shrubs totally consumed, and the proportion where leaves only were consumed, relative to respective pre-fire counts for each height class.

Fuel combustion: To convert our fuel combustion rates and associated fire severity observations to equivalents for EDS and LDS fires separately, we applied the frequency of low/moderate and high severity fires in both seasons derived from numerous plot photographs, as reported by Edwards and Russell-Smith (Chapter 8).

Residual combusted material (char and ash): Post-fire abundance of combusted material was measured at 21 of the transects used for post-fire fine fuel assessments. Dry weight measurements of combusted material and unconsumed post-fire fuels were undertaken separately in five 25 × 25 cm quadrats per transect. From these measurements we calculated that 40% of the post-fire residue comprised combusted material.

BEFs: Given that mean total pre-fire fuel load at the 90 treatment transects was 6.12 t ha^{-1}, and the mean fuel residue remaining after fire treatments was 0.64 t ha^{-1} (9.6% of original fuel) of which 40% was combusted material, the total mean residue conversion rate was 3.79%. Given further that we could not reliably determine which fuel components contributed to the residue, we assumed that post-fire residues sampled systematically in post-fire fine fuel quadrats represented residue production (including redistribution of some entrained ash from the smoke plume) in proportion to respective rates of observed fuel combustion.

Fuel combustion rates for fuel components (Table 6.13) were adjusted for the average fraction of fuel remaining after combustion. In effect, this required adding the calculated residue from coarse, heavy and shrub fuels to respective combustion rates, and subtracting that combined sum from the fine fuels combustion rate. In the absence of more comprehensive data we applied this rate equally to fires of low/moderate and high severity.

The product of fuel combustion data adjusted for residual combusted material (Table 6.13a) and seasonal distribution of fire severity determined from long-term monitoring plots (see above; Table 6.13b), were used to derive BEFs (Table 6.13c).

Discussion and summary

Sampling to derive fuel accumulation, fire patchiness, and combustion parameters was conducted over the full 600–1000 mm rainfall gradient envelope over a wide variety of landscape conditions supporting characteristic unmodified flammable savanna vegetation mostly in the Northern Territory and Western Australia. Assembled datasets for deriving FLA, P and BEF parameters were substantially larger than those assembled for the higher rainfall abatement methodology. For example, FLA and BEF assessments for this lower rainfall abatement methodology were derived from 783 transects (for fine fuels) and

90 transects respectively, whereas data from 219 and 42 transects, respectively, were used for analogous higher rainfall abatement methodology assessments (Russell-Smith *et al.* 2009).

Fuel accumulation

Overall, available fuels were dominated by fine fuels, especially grasses, with lesser relatively equal contributions of coarse woody, heavy woody and shrub fuel components. Mean total fuel load, and mean fuel loads of constituent fuel components, were consistently lower than those sampled under higher rainfall conditions. For example, the mean fuel load reported here was 5.66 t ha^{-1} in the EDS and 6.15 t ha^{-1}in the LDS (Table 6.1), whereas that reported by Russell-Smith *et al.* (2009) for the higher rainfall methodology assessment was estimated at 6.8 t ha^{-1} in the EDS and 8.95 t ha^{-1} in the LDS. These observations are consistent with the general lower productivity of savanna vegetation under lower, more seasonal, and annually variable rainfall conditions (Whitehead *et al.*, Chapter 2).

As noted by Williams *et al.* (2003) and Cook (2003) from observations made at Kapalga in present-day Kakadu National Park in the higher rainfall region, there is a strong seasonality element in accession of litter, such that fine fuels in the EDS were observed to be dominated by grasses (72%) whereas litter averaged more than half (and up to 70% in some years) in their LDS treatments. Given that most sampling for the higher rainfall abatement methodology was conducted in the early- to mid-dry season, in the absence of other observations this seasonality differential was addressed simply by adding 1.8 t ha^{-1} to mean fine fuels based directly on observations from Kapalga (Russell-Smith *et al.* 2009). In the present study, we account for the seasonal differential in litter fuels applying observations reported by Cuff and Brocklehurst (Chapter 7) who found that litter accession in the LDS was 73% of the annual total. For the purposes of parameterising the lower rainfall abatement methodology, we have applied this adjustment to derive separate fine fuel parameter values for EDS and LDS periods (Table 6.14).

As with the higher rainfall methodology, few statistically robust trends were observed between FLA and TSF for coarse, heavy and shrub fuel components. In these instances we have simply used mean values to characterise respective FLA parameters irrespective of TSF (Table 6.14). The paucity of observed statistical relationships between coarse and heavy woody fuel components and TSF suggest that accumulation of both of these components is likely to be disturbed by pulsed inputs following severe fires or fire regimes (Williams *et al.* 1998; Cook *et al.* 2005), as well as localised and subregional wind disturbance events such as pre-monsoon storms, tornadoes and cyclones (Hutley *et al.* 2013). The lack of trend in shrub fuel accumulation with time is also likely to reflect very substantial inter-site variability in historical processes (Russell-Smith *et al.* 1998).

Olson curve relationships yielded lower R^2 values than corresponding linear models across all fuel types, but especially fine fuels. While Olson curves are useful for depicting biological processes describing accumulation modulated by decay, we note that our field data, based on time-for-space sampling and incorporating inherent variability (e.g. different domestic stock and feral herbivore grazing and environmental histories), do not strictly meet the necessary assumptions of the formal Olson curve and need to be interpreted accordingly. Conversely, linear regressions presented here make no such assumptions. Nevertheless, fuel accumulation relationships based on the Olson curve are applied in both Australia's national greenhouse accounts (DoE 2014) and the Carbon Farming Initiative (Commonwealth of Australia 2013).

For the purposes of implementing the methodology, we note that FLA relationships for the grass component of fine fuels were derived specifically under observed low-to-moderate grazing conditions. Given that emissions estimates should not be distorted by inclusion of heavily grazed paddocks, such situations need to be excluded from defined savanna burning project areas, as specified in the higher rainfall methodology (Commonwealth of Australia 2013).

Further, we note however that an extensive northern pastoral industry literature addresses sustainable grazing management issues under a range of soil type and rainfall conditions, with general agreement that annual 'safe' forage utilisation rates (AURs) provide a more realistic and robust measure of sustainability than stocking rates (e.g. Hunt 2008; Walsh and Cowley 2011). Given that the great majority of recurrent fire in savanna landscapes occurs on relatively infertile soil types, and hence savanna burning projects will focus predominantly on such situations but also include some areas of relatively fertile land types, based on available pastoral industry standards (Novelly and Baird 2001; Chilcott *et al.* 2005; Walsh and Cowley 2011) it is plausible to suggest that savanna burning projects should exclude sites/paddocks with AURs >20% (Dr Dionne Walsh, pers. comm.). However, from a practical assessment perspective, the requirements of project proponents to demonstrate AURs at paddock scale would be both very onerous and presumably reliant on sophisticated modelling (e.g. Walsh and Cowley 2011).

Fire patchiness

Based on a very large sample size, the fine-scale fire patchiness observed here (79% burnt in the EDS, 97% burnt in the LDS: Table 6.14) is more than, but otherwise consistent with, fire patchiness values (71% EDS, 89% LDS) reported under the higher rainfall conditions (Russell-Smith *et al.* 2009). Different assessment methods produced comparable results and there was limited variation among groundcover types. As noted in earlier discussion, we attribute the higher proportion of area burnt under lower rainfall conditions to reduced canopy, and associated greater continuity in, fuel cover.

Burning efficiency

Fuel combustion in the 90 lower rainfall fire treatments was observed to vary little between fires of low, moderate, and high severity for fine fuels and shrubs, and by factors of three or more for coarse and heavy woody fuels. These observations differ from those reported for the higher rainfall regions in that there was significant differential combustion in fires of different severity of all fuel components in the higher rainfall zone; and the proportion of fuel components consumed was substantially greater in fires of high severity in the higher rainfall area. For example, heavy fuel consumption in higher rainfall treatments was 15.5% and 58% under fires of low/moderate, and high severity (Russell-Smith *et al.* 2009), whereas comparable figures under lower rainfall treatment conditions reported here are 6.51%, and 12.42%, respectively. This implies that fire intensities experienced in lower rainfall methodology treatments were generally significantly lower than those under high rainfall savanna conditions, presumably reflecting the influence of generally lower available fuel loads, and, given the definition used here of high severity fires (those which scorch the canopy), the potential for exaggerating fire severity effects in vegetation types of relatively low stature (Russell-Smith and Edwards 2006).

In assessment of residual combusted materials under higher rainfall conditions (Russell-Smith *et al.* 2009), it was found that a mean 11.6% of pre-fire fuels was converted to 'ash'. In the assessment presented here for fire treatments conducted under lower rainfall conditions, we calculated that a mean of 3.79% of pre-fire fuels remained as residual combusted material following fire treatments. A first issue to address concerns terminology, noting that 'ash' in the previous study equates to our broader definition of 'residual material' (charred remains plus white ash) applied here. Our field observations, as well as those reported in Russell-Smith *et al.* (2014) for an equivalent assessment of residual materials following burning under tussock grassland conditions, indicate that the residual white ash component can be very small in at least some situations. Hence, here we have preferred to use the broader descriptor 'residual material'.

Although the proportion of residual materials observed here is less than that observed in the previous higher rainfall assessment, present values are generally consistent with studies from Kakadu National Park (7%) (Cook 1994) and other tropical savanna and forest ecosystems, which typically range from 3% (e.g. Kauffman *et al.* 2003; Barbosa and Fearnside 2005) to 11% (e.g. Kauffman *et al.* 1994). It is feasible that the smaller mean proportion of residual material observed in the current assessment likewise reflects lower fire intensities experienced in our fire treatments.

Taking into account the relatively large contribution of fine fuels to the total fuel which is combusted, and minor differentiation in fine fuel combustion under different fire severity conditions, calculated BEFs were found consequently to exhibit little seasonal differentiation for the respective fuel components (Table 6.14). This is in contrast to BEFs reported for the higher rainfall methodology assessment, where seasonal differences ranged between ~10–20% for consumption in respective fuel components (Russell-Smith *et al.* 2009). As noted above, we attribute the lack of seasonal differentiation in BEFs to generally lower fire intensities, and reduced seasonal effects, under lower rainfall conditions.

In sum, this assessment, in combination with other work (Chapter 4) indicates that, for the purposes of implementing a savanna burning abatement project under lower rainfall conditions, two parameters are of particular importance for differentiating seasonal effects on fuels consumed and hence emissions—higher litter fuel inputs in the LDS, and more patchy fires in the EDS.

Acknowledgements

Thanks are extended to the numerous people who helped with the extensive field data collection, in particular the Indigenous ranger groups Waanyi Garawa Rangers, Mimal Rangers and Yugul Mangi Rangers (in the Northern Territory), Carpentaria Rangers (in Queensland), and the Uunguu Rangers and the Wunambal Gaambera Aboriginal Corporation, the Wunggurr Rangers and the Wilinggin Aboriginal Corporation, the Balanggarra Rangers and the Balanggarra Aboriginal Corporation, the Dambimangari Rangers and the Dambimangari Aboriginal Corporation, and the Kimberley Land Council (in Western Australia). Thanks are also extended to Jon Schatz (CSIRO) and Ian Dudgeon (Australian Wildlife Conservancy) for help collecting fuel accumulation data included in this chapter.

References

Adobe (2002) Adobe Photoshop Elements version 2.0.

Barbosa RI, Fearnside PM (2005) Above-ground biomass and the fate of carbon after burning in the savannas of Roraima, Brazilian Amazonia. *Forest Ecology and Management* **216**, 295–316. doi:10.1016/j.foreco.2005.05.042

Chilcott C, Oxley T, Dyer R, MacDonald R (2005) 'Grazing Land Management – Katherine version workshop notes.' Meat & Livestock Australia: Sydney.

Commonwealth of Australia (2013) Carbon Credits (Carbon Farming Initiative) Reduction of Greenhouse Gas Emissions through Early Dry Season Savanna Burning—1.1) Methodology Determination 2013. www.comlaw.gov.au/Series/F2013L01165

Cook GD (1994) The fate of nutrients during fires in a tropical savanna. *Australian Journal of Ecology* **19**, 359–365. doi:10.1111/j.1442-9993.1994.tb00501.x

Cook GD (2003) Fuel dynamics, nutrients and atmospheric chemistry. In *Fire in Tropical Savannas: The Kapalga Experiment.* (Eds AN Andersen, GD Cook, RJ Williams) pp. 47–58. Springer: New York.

Cook GD, Liedloff AC, Eager RW, Chen X, Williams RJ, O'Grady AP, Hutley LB (2005) The estimation of carbon budgets of frequently burnt tree stands in savannas of northern Australia, using allometric and isotopic discrimination. *Australian Journal of Botany* **53**, 621–630. doi:10.1071/BT04150

DoE (Department of Environment) (2014) National Inventory Report 2012, vol. 2. www.environment.gov.au/climate-change/greenhouse-gas-measurement/publications/national-inventory-report-2012

Eamus D, McGuiness K, Burrows W (2000) 'Review of allometric relationships for estimating woody biomass for Queensland, the Northern Territory and Western Australia.' Charles Darwin University. Report to the Australian Greenhouse Office: Canberra.

Hunt LP (2008) Safe pasture utilisation rates as a grazing management tool in extensively grazed tropical savannas of northern Australia. *The Rangeland Journal* **30**, 305–315. doi:10.1071/RJ07058

Hutley LB, Evans BJ, Beringer J, Cook GD, Maier SW, Razon E (2013) Impacts of an extreme cyclone event on landscape-scale savanna fire, productivity and greenhouse gas emissions. *Environmental Research Letters* **8**. doi:10.1088/1748-9326/8/4/045023

Kauffman JB, Cummings DL, Ward DE (1994) Relationships of fire, biomass and nutrient dynamics along a vegetation gradient in the Brazilian cerrado. *Journal of Ecology* **82**, 519–531. doi:10.2307/2261261

Kauffman JB, Steele MD, Cummings DL, Jaramillo VJ (2003) Biomass dynamics associated with deforestation, fire, and conversion to cattle pasture in a Mexican tropical dry forest. *Forest Ecology and Management* **176**, 1–12. doi:10.1016/S0378-1127(02)00227-X

Novelly P, Baird P (2001) 'A process for re-assessing the potential carrying capacity of Kimberley pastoral leases.' In *Proceedings of the Northern Australian Beef Industry Conference.* Kununurra, WA. 8–9 November 2001. pp. 41–47. Department of Agriculture: Perth.

Olson JS (1963) Energy storage and the balance of producers and decomposers in ecological systems. *Ecology* **44**, 322–331. doi:10.2307/1932179

Price O, Russell-Smith J, Edwards A (2003) Fine-scale patchiness of different fire intensities in sandstone heath vegetation in northern Australia. *International Journal of Wildland Fire* **12**, 227–236. doi:10.1071/WF03040

Russell-Smith J, Edwards A (2006) Seasonality and fire severity in savanna landscapes of monsoonal northern Australia. *International Journal of Wildland Fire* **15**, 541–550. doi:10.1071/WF05111

Russell-Smith J, Murphy BP, Meyer CP, Cook GD, Maier S, Edwards AC, Schatz J, Brocklehurst P (2009) Improving estimates of savanna burning emissions for greenhouse accounting in northern Australia: limitations, challenges, applications. *International Journal of Wildland Fire* **18**, 1–18. doi:10.1071/WF08009

Russell-Smith J, Ryan PG, Klessa D, Waight G, Harwood R (1998) Fire regimes, fire-sensitive vegetation, and fire management of the sandstone Arnhem Plateau, monsoonal northern Australia. *Journal of Applied Ecology* **35**, 829–846. doi:10.1111/j.1365-2664.1998.tb00002.x

Russell-Smith J, Yates CP, Brock C, Westcott V (2010) Fire regimes and interval-sensitive vegetation in semi-arid Gregory National Park, Northern Territory. *Australian Journal of Botany* **58**, 300–317.

Russell-Smith J, Yates CP, Evans J, Desailly M (2014) Developing a savanna burning emissions abatement methodology for tussock grasslands in high rainfall regions of northern Australia. *Tropical Grasslands – Forrajes Tropicales* **2**, 154–166.

Walsh D, Cowley RA (2011) Looking back in time: can safe pasture utilisation rates be determined using commercial paddock data in the Northern Territory? *The Rangeland Journal* **33**(2), 131–142. doi:10.1071/RJ11003

Williams RJ, Gill AM, Moore PHR (1998) Seasonal changes in fire behaviour in a tropical savanna in northern Australia. *International Journal of Wildland Fire* **8**, 227–239. doi:10.1071/WF9980227

Williams RJ, Gill AM, Moore PHR (2003) Fire behavior. In *Fire in Tropical Savannas: The Kapalga Experiment.* (Eds AN Andersen, GD Cook, RJ Williams) pp. 33–46. Springer: New York.

7

Leaf and coarse fuel accumulation and relationships with vegetation attributes in 'evergreen' tropical eucalypt savannas

Nicholas Cuff and Peter Brocklehurst

Summary

The distribution and proportional composition of leaf litter (<0.6 mm diameter) and coarse litter (woody debris, 0.6–5 cm diameter) play an important role in fire fuel dynamics in tropical savanna ecosystems. This chapter details geographic and seasonal variation in the rate and volume of leaf litter and coarse litter fuel accumulation in evergreen *Eucalyptus miniata* and *E. tetrodonta*-dominated open forests and woodlands along a rainfall gradient (750–1600 mm) in the Northern Territory. The aim is to provide a basis for correlating remotely sensed measures of vegetation attributes (height, basal area, cover) with fine fuel accumulation, a key variable for emissions calculation.

Leaf litter accumulation rates were found to vary significantly with rainfall and tree basal area. The seasonality of leaf litter accumulation was found to vary most significantly with rainfall and latitude. Under higher rainfall (>1000 mm p.a.), mean leaf litter input was 5.01 ± 0.48 Mg ha^{-1} yr^{-1}, with 54% occurring in the late dry season (LDS) period after July. Under lower rainfall (<1000 mm p.a.), mean leaf litter input was 2.87 ± 0.41 Mg ha^{-1} yr^{-1}, with 74% occurring in the LDS period. Mean coarse litter inputs were 0.55 Mg ha^{-1} $yr^{-1}$and 0.22 Mg ha^{-1} yr^{-1}, at higher and lower rainfall sites, respectively. Although seasonal differences were observed in coarse litter inputs, these were not associated with vegetation structural characteristics and it is suggested that, at plot scales, such inputs rather reflect disturbance events (e.g. high winds, severe fires).

Seasonal differences in leaf litter fuel loads are likely to contribute to significant differences in the total fine fuel loads between the early and late dry season, especially in the <1000 mm rainfall zone. Generally, these results reflect plant community compositional and hypothesised plant ecophysiological differences along the rainfall gradient. Rates of leaf litter accumulation in Top End savannas appear to be consistently greater than in other evergreen eucalypt formations outside the wet–dry tropics. Difficulties associated with the accurate quantification of coarse-woody debris accumulation at the plot scale are also highlighted, and recommendations for further studies to better quantify the range of variation in fine fuel accumulation rates within tropical savanna ecosystems are discussed.

Introduction

Savanna ecosystems cover some 1.9 million km^2, ~25% of the continental landmass of Australia between ~11°S and 20°S but extending in parts as far south as 26°S (Eamus and Prior 2001). These ecosystems are typified by a strongly seasonal wet–dry climatic cycle in which >90% of total annual rainfall is generally received during the summer months October to April (the 'wet' season). This is followed by an extended period of seasonal moisture deficit (the 'dry' season) where little, if any, rainfall is received in most months and large-scale wildfires are a prominent feature of the landscape.

Although the general nature of climatic conditions and seasonal rainfall patterns are consistent across the Australian savannas, the geographic range of these systems across the continent imparts strong longitudinal differences in the drivers and timing of phenological patterns (e.g. ENSO fluctuations between Indian and Pacific Ocean coastlines), and similarly strong gradients in rainfall with latitude. This pronounced seasonality and wide climatic range has significant implications for plant production, plant physiological strategies and, in turn, the quantity and cycling of fine and coarse fuels and savanna fire regimes. Detailed descriptions of the biophysical conditions driving savanna fire regimes are discussed in Chapters 1 and 2.

It has long been recognised that accurate quantification of emissions from savanna burning is dependent upon various spatial and temporal aspects of climate (e.g. time of year, local meteorology), fire behaviour (e.g. intensity and patchiness), vegetation characteristics (e.g. fuel 'types' and quantity), and the interactions between these factors.

Previous studies have identified fine fuels (particles of <6 mm diameter in size including grass, tree/shrub leaf litter and woody debris) as a major contributor to fuel loads in Australian tropical savannas (Russell-Smith *et al.* 2009). Extensive work in the 'Top End' savannas of the Northern Territory has aimed at better defining the gross quantities of fine fuels produced in these structural formations, although the proportions of the different fine fuel components in these systems has not been extensively investigated. It has previously been reported that in higher rainfall (>1000 mm) mesic Australian savannas, leaf litter (including twigs) may comprise up to 50% of the total fine fuel load (Cook 2003). However, general quantification of leaf litter as a proportion of the total fine fuel load and variation in leaf litter production across rainfall zones in the tropical savannas has not been widely investigated. It could be reasoned that given the wide range in rainfall and potentially plant primary production, there may be equally pronounced variation in the quantities of leaf litter produced by different vegetation types.

Similarly, these distinct seasonal rhythms have a profound influence on the physiological strategies of plants (particularly phenophases) within tropical savannas (e.g. Egan and Williams 1996; Duff *et al.* 1997), which may potentially influence seasonal proportions of fine fuel components produced over the course of the wet and dry seasons. Patterns of woody species leaf phenology, including initiation, expansion, senescence and abscission (i.e. leaf fall), have been well documented in many Australian ecosystems (e.g. Pook 1984; Lowman 1992) and in savannas from around the world, including tropical Australia (e.g. Bowman *et al.* 1991; Medina and Silva 1990; Williams *et al.* 1997). Leaves have been shown to exhibit patterns of deciduousness and/or abscission linked to various factors related to seasonal water stress, including soil and atmospheric moisture declines as well as internal-water deficits, characteristic of different times during the dry season (Duff *et al.* 1997; Myers *et al.* 1997; Williams *et al.* 1997). However, most of these studies have dealt primarily with savanna ecosystems or several vegetation associations primarily occurring within one rainfall zone. Few tropical Australian studies (e.g. Ma *et al.* 2013) appear to have attempted to quantify variation in field phenological patterns within a floristic formation or association across rainfall zones through the extent of their geographic range. Leaf phenological patterns of woody species determine the timing, magnitude and rate of accumulation as well as the proportional composition of fuel loads at different times of the year. Thus fires at different times of the year may be expected to have available different proportions of grassy or litter fine fuels dependent on the phenological stages of the constituent vegetative material contributing to this total fuel load. This in turn will affect the types and proportions of constituent gases emitted from a fire during combustion.

The more complete identification and quantification of these proportions of vegetation fuel production and dynamics across the range of savanna climate zones and vegetation types has previously been identified as a significant uncertainty in the estimation of fuel loads and subsequent calculation of emissions parameters from savanna burning (Cook and Meyer 2009). Understanding the variation in the nature and magnitude of leaf phenological patterns with climatic (e.g. rainfall), geographic (e.g. latitude), vegetation structural (e.g. foliage cover, basal area etc.) or plant physiological traits (e.g. leaf sclerophylly) will significantly contribute to the improved parameterisation of savanna burning emissions calculations used for emissions accounting. Furthermore, accurate parameterisation of models used to calculate emissions from savanna burning are essential if a sustainable and equitable economic benefit is to be gained from reducing greenhouse gas emissions through alteration of seasonal burning patterns in line with existing and future carbon farming initiative savanna burning abatement methodologies (DCCII 2013).

This study aimed to:

1. quantify variation in leaf litter accumulation rates for extensive *Eucalyptus miniata* and *E. tetrodonta*-dominated formations in the tropical Northern Territory
2. provide improved estimates of leaf litter accumulation rates across the latitudinal range of these dominant structural and floristic formations in the Top End of the Northern Territory
3. document the vegetation structural characteristics of this dominant formation across its latitudinal range in the context of both 'on-ground' and 'top-down' measures useful for remote sensing applications, see Chapters 11 and 15
4. investigate relationships between leaf litter accumulation rates in these associations with vegetation structural characteristics and with latitudinal and rainfall variation
5. provide a preliminary investigation of parameters useful for on-ground estimation of leaf litter accumulation to be used in potential explanatory models of fine fuel accumulation across the broader landscape, see Chapter 6
6. develop models of seasonal leaf litter and gross canopy phenological cycling within these dominant structural and broad floristic formations across the latitudinal range
7. provide improved estimates of proportional leaf litter inputs to total fine fuel loads across the wet–dry seasonal cycle over the latitudinal range of these systems in the tropical Northern Territory.

Methods

Study area and site selection

Eighteen monitoring sites were established in the monsoonal savanna region of the Top End of the Northern Territory, Australia (Fig. 7.1a). The main objective of site placement was to capture the structural and environmental range of the regionally dominant fire-prone vegetation association, namely *Eucalyptus miniata–E. tetrodonta* forests and open

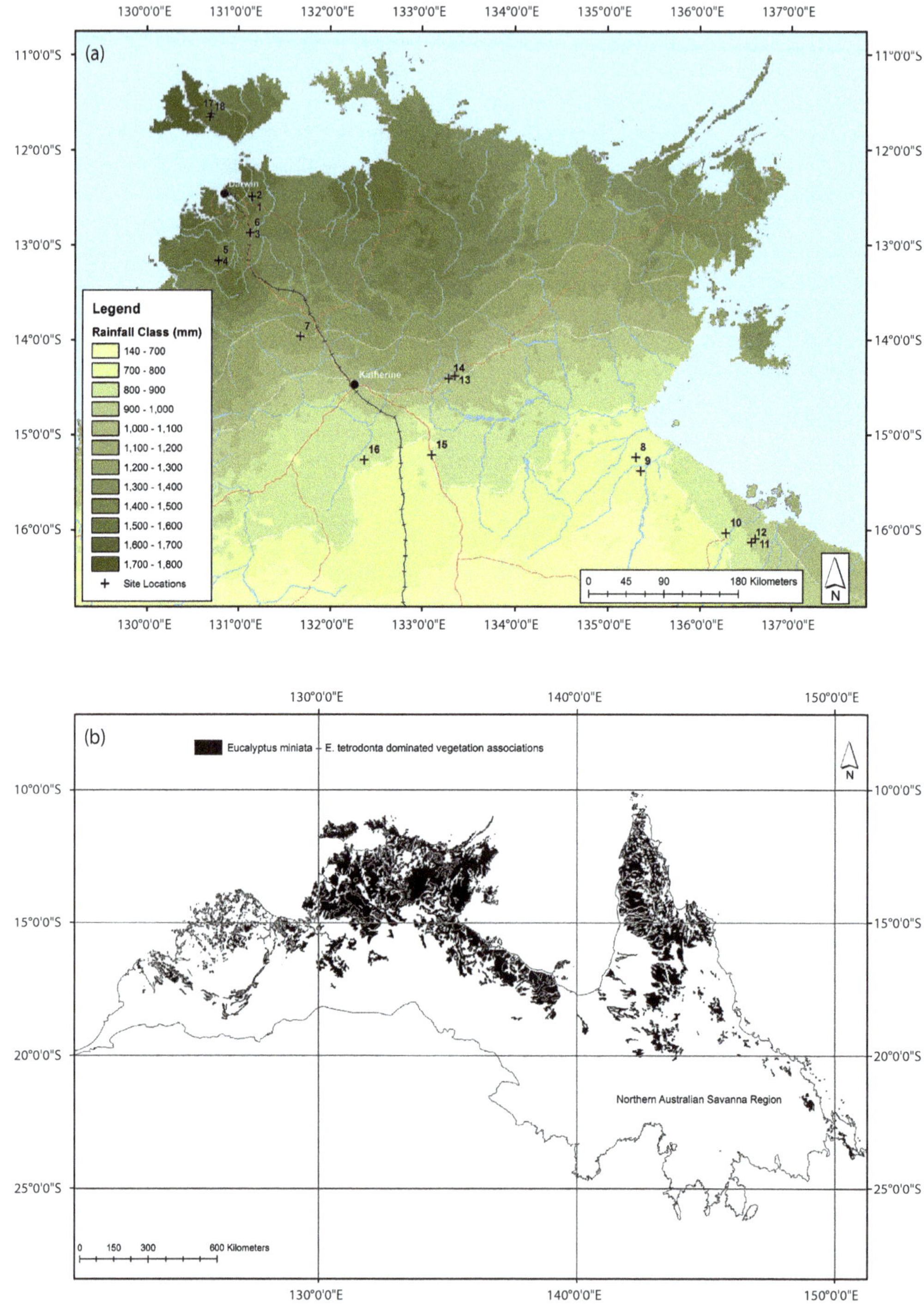

Figure 7.1. (a) Site locations and median annual rainfall for the Northern Territory north of ~15°S (Bureau of Meteorology 2003). (b) Spatial extent of *Eucalyptus miniata–E. tetrodonta*-dominated vegetation associations across northern Australia (Fox *et al.* 2001).

woodlands (Wilson *et al.* 1990), along a rainfall gradient from the highest rainfall regions to regions with less than 1000 mm p.a.

Floristic variation in *E. miniata–E. tetrodonta* associations could generally be considered to reflect rainfall, substrate type and parent material. However, previous studies have suggested that floristic variation is relatively limited despite the wide geographic range the association occupies within the Northern Territory (Williams *et al.* 1996). This vegetation association is common across the whole northern Australian region (Fig. 7.1b) and has been extensively studied in terms of fire regimes, fuel loads and greenhouse gas emissions, and is considered representative of the functionally evergreen eucalypt associations that dominate the tropical savannas of northern Australia (~350 000 km^2) (e.g. Edwards and Russell-Smith 2009; Russell-Smith *et al.* 2009). Potential site localities were initially appraised via GIS following analysis of Moderate Resolution Imaging Spectroradiometer (MODIS) (250 m resolution) and Landsat (30 m resolution) imagery to produce a homogeneity/uniformity layer (mean, minimum, maximum and standard deviation of focal range), which was cross-referenced against existing vegetation mapping and site information to provide the range of this particular vegetation association. The purpose of this was to highlight regions which would be suitably uniform (spatially and floristically) and large enough to be of use for Landsat resolution imagery analysis.

Data on geological parent material, landform and soil type (where available at an appropriate spatial resolution) were assessed against potential site locations in order to control as much as possible the edaphic growing conditions available to plants at each site location. However, it is worth noting that while no detailed nutrient analyses were undertaken as part of this study, there was subtle variation in general substrate types from lateritic kandosols in the northern half of the study area to sandier kandosols and tenosols on sandstone plateaux and plains in the south of the study area. This may have an influence on soil conditions at individual sites, although it is worth considering that, despite these mapped differences, the parent materials are generally of a similar deeply weathered quartzose sandstone origin. Furthermore, the resolution of the available data to assess these differences is too coarse for application at the site-scale where micro-topographic variability is likely to obscure much of the difference indicated by the available spatial information. Sites were established in the field as far as was practicable in areas of similar landform and soil texture properties upon assessment during the initial site evaluation.

Also considered in site selection was availability (tenure considerations), ease of vehicle access, and minimal disturbance regimes from grazing herbivores. Additionally, where possible, areas subject to a less frequent fire regime were selected in an attempt to ensure the continuity of data collection over the study period and to better represent intact vegetation systems. Further consideration was the ability to put two sites in relatively close proximity and/or to collaborate on existing site networks (e.g. Monash University/Charles Darwin University flux towers, Hutley *et al.* 2011). Sites were field-verified for suitable access, landscape position and vegetation characteristics before establishment in an endeavour to select locations with the most similar floristic composition and which best met all the selection criteria. Sites were established on Melville Island (~1800 mm p.a.), in the vicinity of Darwin (~1700 mm p.a.), the Sturt Plateau region (~850 mm p.a.), and in the Gulf region (~750 mm p.a.).

Sampling design

Sampling design and site layout were based on a modified point-intercept approach (Muir *et al.* 2011) with the addition of fine fuel (leaf and twig) accumulation quadrats and

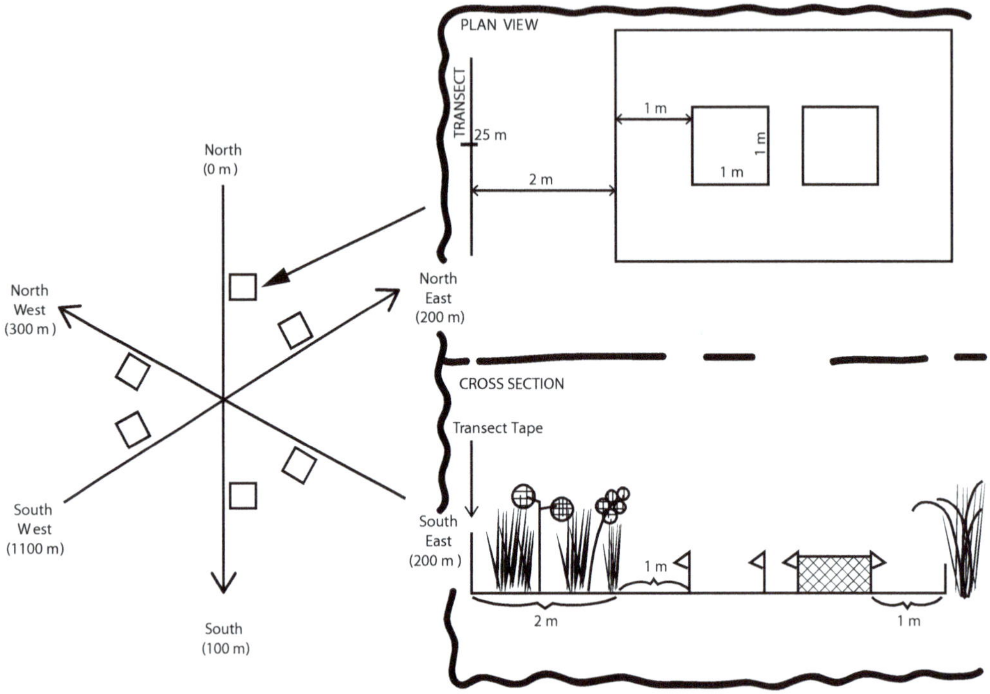

Figure 7.2. Example transects and plot arrangement.

hemispherical camera measurements of foliage cover. A sampling site consisted of 3 × 100 m transects and 12 × 1 m^2 leaf accumulation quadrats. The central transect was located in a north–south direction with the two additional transects intercepting the 50 m point at ±60° from north (60° and 300° respectively, Fig. 7.2). On establishment, each transect endpoint and centre point was permanently marked and a GPS location taken at the centre and end points of the central transect.

Leaf accumulation quadrats were established, one fenced and one unfenced, located at the 25 and 75 m mark along each transect. Plots were fenced in order to determine if there was any significant difference in leaf accumulation as a result of wind removal. An area of 15 m^2 was cleared for the accumulation quadrats, 1 m offset from transect line. The two quadrats were established with marker flags, 1 m apart and surrounded by 1 m of cleared ground. On the first site visit, woody vegetation was retained, and leaf litter and grass were removed. One set of quadrat marker flags were surrounded by a 25 cm high sediment fence or shade-cloth. Fenced and unfenced quadrats were established.

Field sampling techniques

Vegetation structure

Dominant species and structure for each stratum was recorded as per Brocklehurst *et al.* (2007) on establishment in order to describe the vegetation community.

Transect measurements

Vegetation and land-cover observations following Muir *et al.* (2011) were made at 1 m intervals along each transect tape with the following modifications:

1. Vegetation observations were collected within strata defined following standard methods in Australian vegetation communities (NCST 2009) to allow a more detailed analysis of variation in crown attributes within layers and functional species groups.
2. Ground observations were further divided into shrub and graminoid classes for both live and dead leaf material to gain a more detailed understanding of total woody plant leaf productivity within traditionally defined vegetation strata at the sites.
3. Ground observations were further divided into substrate size classes following NCST (2009).

Basal area

Basal area was measured using Bitterlich gauge sweeps (Loetsch *et al.* 1973) at seven locations along each transect (at 25 m, 75 m and centre). Species and species counts were recorded.

Halfway through the measurement campaign, the basal gauge was reused and trees counted 'in' by the basal gauge were tagged, measured for over-bark diameter at breast height and top height, with species identifications recorded.

Hemispherical camera (each visit for first 1.5 years)

Seven digital hemispherical photos (DHPs) were taken at the same seven locations as the basal sweeps. Photos were taken vertically upward, after levelling the lens, with the camera ~1.2 m from ground level. Photos assessed the mid- and upper-stratum. Downward (i.e. ground) photos were not taken. Photos were taken with a Nikon D40 digital camera equipped with a sigma hemispherical (fisheye) lens.

Site photos were processed using Hemispherical Photo Analyser (Opie and Newnham 2011) with the seven measures then averaged for the site for each visit (Fig. 7.3). This data was used to compare with the vegetation values derived from the 1 m transect data (e.g. foliage projective cover) as well as to determine correlations with satellite indices such as MODIS leaf area index (LAI). Given the logistic limitations of this study, time of day at which the photographs were captured was unable to be controlled.

Leaf accumulation quadrats

For each site visit, all leaf fuel (leaf, twig and woody debris <0.6 cm diameter) and coarse-woody derived fuel (0.6–5 cm diameter branches) was collected from both fenced and unfenced quadrats. This was weighed to ±1 g in the field. A sub-sample of each fuel type

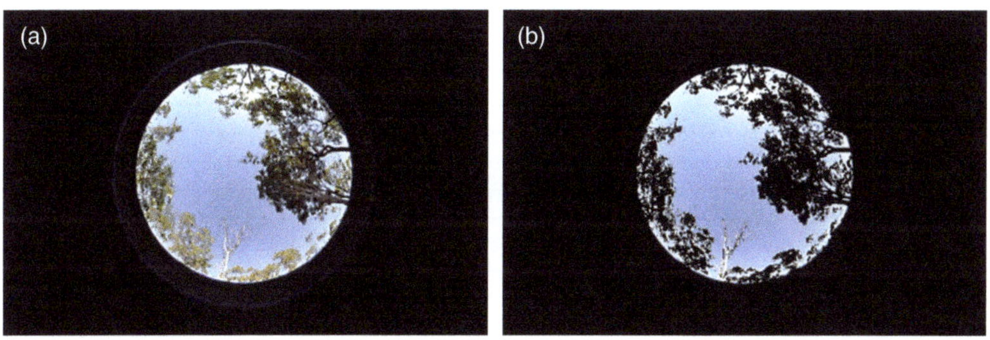

Figure 7.3. Raw (a) and processed (b) hemispherical camera images used to derive measures of canopy attributes.

was also collected, weighed and subsequently dried in an oven to determine total dry weight of fuels. Grass litter was removed from quadrats and discarded (not weighed) on each visit.

Coarse fuels

Coarse fuels were originally collected from six 1 × 1 m quadrats located in conjunction with the basal area and hemispherical camera photo points.

Site observations

Sites were established on the mainland from Darwin to the Gulf in the early dry season 2010. Sites were established on Melville Island in the 2011 early dry season. Most sites were measured at least six times over the course of the study, with the exception of site 16 (two measurements) and the Melville Island sites (four measurements) (Fig. 7.4). Site 16, at Nenen Station on the Sturt Plateau, was established next to an adjacent CSIRO flux tower. However, cattle damaged most of the leaf fall quadrats and the site was abandoned after the first two measures. Melville Island sites were established later in the study (high rainfall zone), although the relatively high financial costs involved in visiting the sites limited the number of measurements.

Rainfall 2010–2012

Yearly rainfall varied across the sites during the course of the survey when compared to mean annual rainfall (international standard 30 year period, 1961–1990). Rainfall was higher than average for both the 2009–10 and 2010–11 wet seasons, particularly in the Gulf region. The 2010–11 wet season was also the wettest recorded in much of the Top End

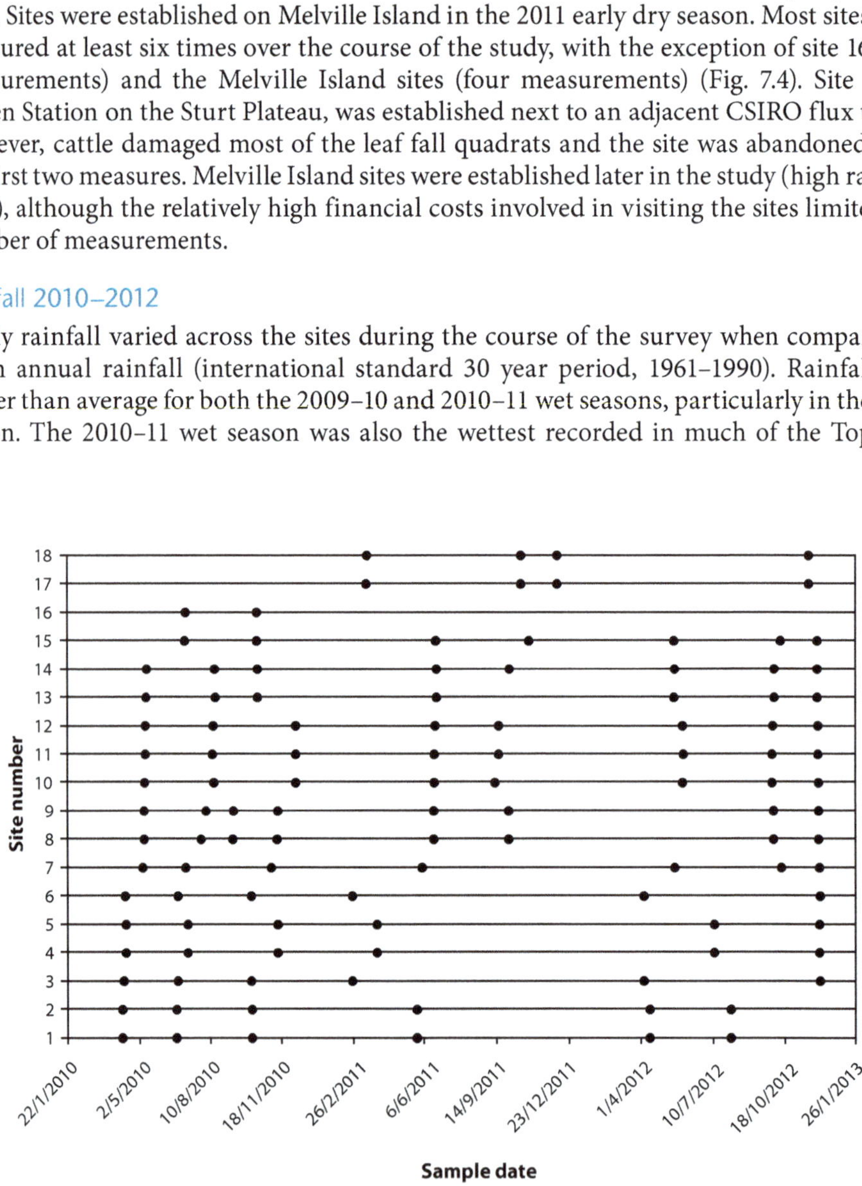

Figure 7.4. Timing of site measurements across 3 years of study. Site numbers as given in Fig. 7.1a.

(Bureau of Meteorology 2013). Conversely, rainfall for the 2009 wet season was below average in the Gulf, and the 2011–12 wet season was below average for all study sites although the study was completed by December, early into the 2011–12 wet season.

Normalised difference vegetation index standardised anomaly mapping 2010–12

The normalised difference vegetation index (NDVI) standardised anomaly is the departure of NDVI from the long-period average, normalised by the long-period variability. It indicates whether the vegetation 'greenness' (active photosynthetic material in all strata) at a particular location is typical for a particular period of the year, or whether the vegetation is more or less green than average. Although the NDVI also includes the ground layer greenness, it is a direct measure of the growing state of the vegetation and provides some insight into the conditions experienced over the course of this study in terms of longer-term conditions (international standard 30 year period 1961–1990).

Reference to the NDVI anomaly mapping indicates that, for most of the study period, vegetation greenness was near average or slightly above average for most regions, across all sites. The mapping also indicates an increased greenness in the Gulf region for almost the whole study period, particularly from 1 January 2011 to 30 June 2012. Average or slightly lower greenness occurred in northern regions in the first 6 months of 2011 and last 6 months of 2012.

Data preparation

Transect data

Data for each measured parameter were tallied for each transect. Individual transect measures were summed and averaged for an individual observation date at a site, with site averages from all re-visits calculated over the study period to produce mean measures of vegetation parameters for the site in the following classes (Table 7.1).

Foliage projective cover (FPC), foliage cover (FC) and crown cover (CC) were calculated for both the canopy and combined woody strata on each site visit (Hnatiuk *et al.* 2009; Muir *et al.* 2011). Additionally, first intercept measures of 'satellite green', 'satellite brown' and 'satellite bare' were derived for the purposes of comparison with satellite imagery. These measures represent the 'top down' view of the land surface as seen by a satellite sensor and the fractions of total land cover associated with the site from this perspective. These cover fractions are consistent with the approach of Scarth *et al.* (2010) for calibrating and validating remotely sensed landcover products.

Basal area data were averaged per site into live and dead species basal area and one total live and dead stand basal area figure. This was used to investigate relationships with FPC, FC, fine fuel accumulation etc., as well as to estimate above-ground biomass. Basal area was assumed to be static throughout the project period. For the purposes of subsequent statistical analyses, Site 16 (Nenen Station) was excluded given large-scale and repeated disturbance of leaf litter accumulation plots by cattle.

Leaf litter accumulation: fine and coarse fuels

Dry weight of fine fuels was averaged per visit per site for fenced and unfenced quadrats. Raw data weights were initially used to determine relationships between fenced and unfenced quadrats, with no significant difference detected between leaf litter accumulation volumes in fenced and unfenced plots within sites.

Table 7.1. Vegetation structural measures collected from point-intercept transects and their use in this study

Class	How	Used for analysis of relationships with					
		Basal area	AGB	Fuels	Remote sensed data	Rain	Camera
Foliage projective cover (FPC) canopy	Per observation			Yes	Yes		Yes
	Average all visits	Yes	Yes	Yes			Yes
Foliage cover (FC) canopy (leaf and branch)	Per observation		Yes	Yes		Yes	
	Average all visits	Yes	Yes	Yes			Yes
Woody FPC	Per observation	Yes		Yes	Yes		Yes
	Average all visits			Yes			Yes
Woody FC	Per observation	Yes		Yes	Yes		Yes
	Average all visits			Yes		Yes	
Satellite green	Per observation				Yes		
Satellite brown	Per observation				Yes		
Satellite bare	Per observation				Yes		

Average figures per site per visit were subsequently used to determine accumulation rates (g day^{-1}) between visits (seasonality). Sites were assessed using NAFI hotspot and fire-scar data (refer to Chapter 5) to determine if plots were burnt between visits, allowing the accumulation period start date to be reset to the post-fire period (rather than last visit). A final average accumulation rate in Mg ha^{-1} yr^{-1} per site was derived from all site visit data for comparison with other final averaged data (e.g. basal area, rainfall, average FPC). It should be noted that several transects were affected by fire over the duration of the study, although only a limited number of fuel accumulation quadrats were actually burnt. Coarse fuel accumulation from transect data was treated in a similar manner to fine fuels.

Results and discussion

Summary transect vegetation attributes

Figure 7.5 and Appendix 1 present a summary of average vegetation parameter measures for the sites over the course of the study. All sites fall within the range of open forest to open woodland structural formations previously documented for *E. miniata–E. tetrodonta* vegetation types in the Northern Territory (Wilson *et al.* 1990). The majority (10 out of 18 sites) were classified as woodland on both foliage and crown cover criteria in the dominant upper stratum, with a further three sites classified as borderline woodland (Howard Springs-2, Manton-3 and Limmen-9) based on one of these criteria (ESCAVI 2003; Hnatiuk *et al.* 2009). A further four sites (Howard Springs-1, Manton-6, Litchfield-5 and Nenen-16) were classified as open forest or borderline open forest. One site (Litchfield-4) was classified as open woodland (ESCAVI 2003; Hnatiuk *et al.* 2009).

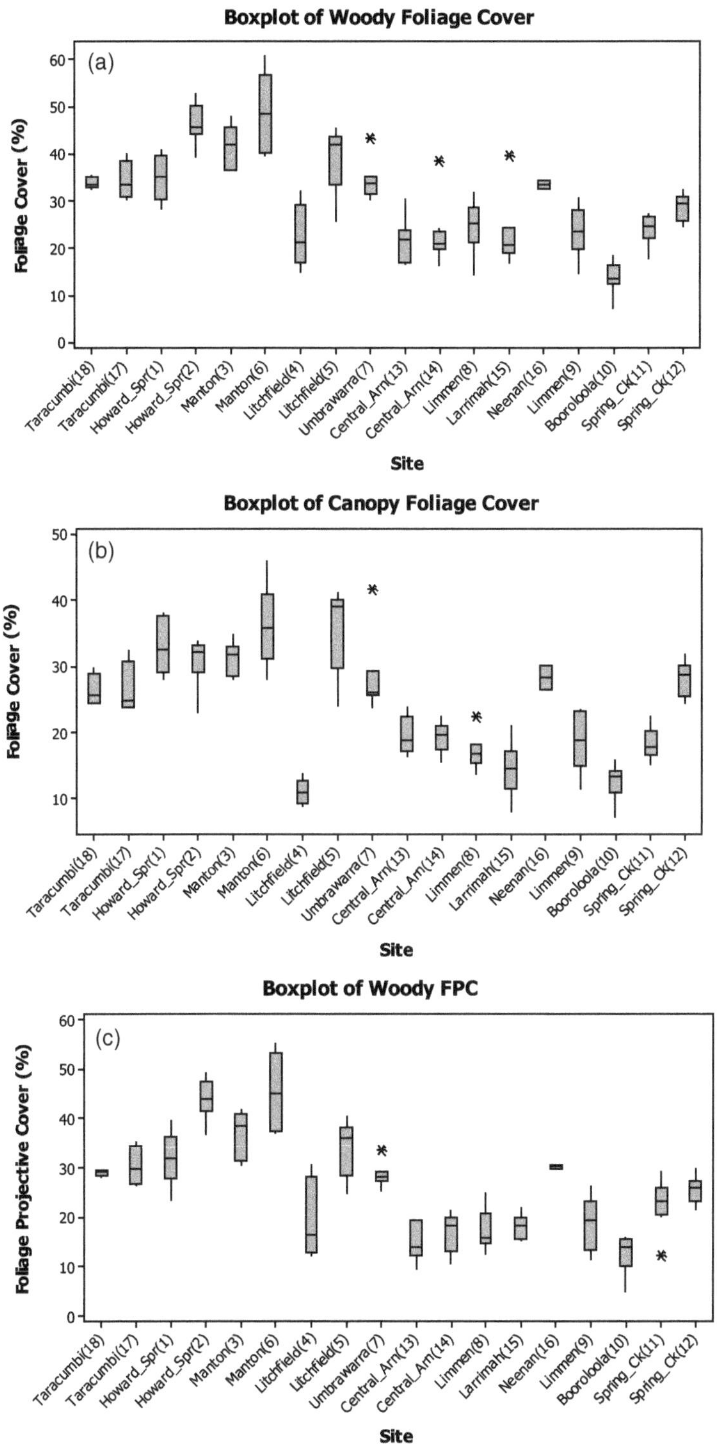

Figure 7.5. (a–f): Bar-charts and summary boxplots of vegetation cover and basal area characteristics from 18 sites in *Eucalyptus miniata–E. tetrodonta* vegetation communities in the Northern Territory. Sites are ordered north to south on the *x*-axis.

(Continued)

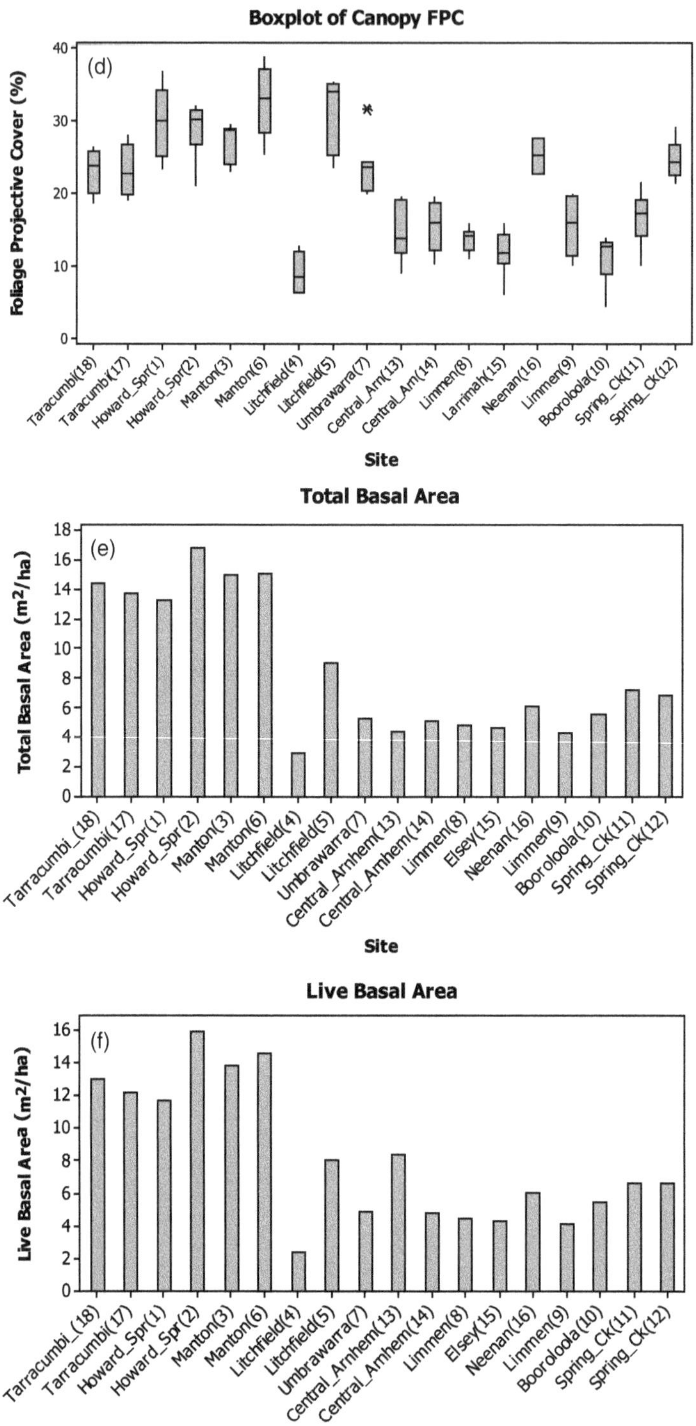

Figure 7.5. (Continued).

Figure 7.5 displays box plots of selected derived vegetation measures over the course of the study, illustrating the variability (95% confidence interval (CI) of mean) in measures across the study area between revisits. Some of this variability may be attributed to observer variability between operators and site visits, particularly for crown cover estimates where the delineation of canopy perimeters is particularly susceptible to operator subjectivity. Broadly, structural attributes ('cover' and basal area) appear to be positively correlated with rainfall and more northerly latitude. However, markedly increased variability in combined (woody) FPC and foliage cover at sites with lower latitude (i.e. northern sites) can be observed in the data. These variations are discussed further in relation to the latitudinal/rainfall gradient later in this chapter.

Similar patterns in variability with latitude and rainfall were evident in the landcover attributes derived from the on-ground measures, primarily reflecting the known seasonal patterns in vegetation flushing and senescence over the seasonal wet–dry cycle.

Basal area varied across the study area with rainfall and latitude, and generally reflected site structural formation classification characteristics. Average total basal area ranged from 10.85 m^2 ha^{-1} in open forests to <3 m^2 ha^{-1} at the single open woodland monitoring site. It should be noted that the range of variability in both total and live basal area values recorded for stands in both open forest and woodland vegetation types overlap by a large margin.

Vegetation parameters and the rainfall gradient

As observed in Appendix 1 and Fig. 7.5, derived vegetation parameters varied substantially across the study area. Furthermore, trends depicted in Fig. 7.6 suggest that variability *within* sites appears to increase with decreasing latitude. It is suggested that this variability possibly reflects climatic productivity differences across the study area, particularly annual rainfall along the latitudinal gradient, and this in turn influences community structure and composition.

Table 7.2 gives Pearson correlation coefficients between derived vegetation parameters and median grid rainfall across the study area. Median rainfall is strongly positively correlated with most vegetation parameters measured in this study, and also appears to play an important role in the production of leaf litter. Total basal area, woody FPC and FC were the vegetation attributes most strongly correlated with rainfall across the study area, suggesting an important role of annual rainfall on woody vegetation cover in *E. miniata–E. tetrodonta* associations. Analysis of variance for repeated measures of total foliage characteristics at our sample sites exhibits a highly significant relationship with rainfall over the course of this study (FPC: DF = 116; S = 20.42; $r^2 = 0.69$; $P = 0.000$; FC: DF = 116; $F = 16.51$; $r^2 = 0.64$, $P = 0.000$).

Table 7.2. Pearson correlation coefficients for mean annual rainfall (mm) vs. a range of vegetation parameters

TBA[1] (m^2 ha^{-1})	LBA[2] (m^2 ha^{-1})	Canopy-FPC[3]	Canopy-FC[4]	Canopy-CC[5]	Woody-FPC	Woody-FC	Leaf litter accumulation (Mg ha^{-1} yr^{-1})
0.730 ($P = 0.001$)	0.692 ($P = 0.001$)	0.512 ($P = 0.03$)	0.502 ($P = 0.03$)	0.335 ($P = 0.17$)	0.624 ($P = 0.006$)	0.624 ($P = 0.006$)	0.769 ($P = 0.000$)

1 TBA denotes total stand basal area (m^2 ha^{-1} of all live + dead standing stems).
2 LBA denotes live stand basal area (m^2 ha^{-1}) only.
3 FPC denotes percentage foliage projective cover (Hnatiuk *et al.* 2009).
4 FC denotes percentage foliage cover (Hnatiuk *et al.* 2009).
5 CC denotes percentage crown cover (Hnatiuk *et al.* 2009).

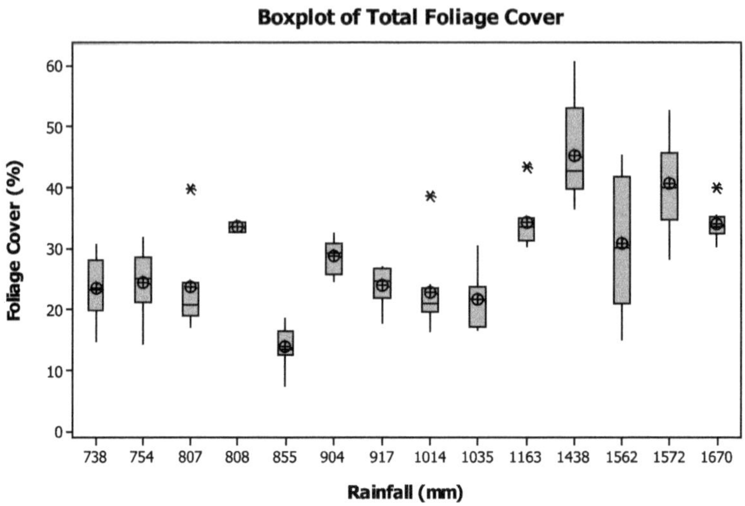

Figure 7.6. Boxplot of total woody foliage cover in all strata versus rainfall at 18 sites in Northern Territory *Eucalyptus miniata–E. tetrodonta* vegetation associations.

Further, regression analyses between rainfall and important vegetation structural parameters identified almost identical relationships for mean woody FPC and mean woody FC (DF = 17; $F = 10.2$; $r^2 = 0.39$; $P = 0.006$). Mean basal area was also strongly positively correlated with increasing rainfall (DF = 17; $F = 18.2$; S = 3.30; $r^2 = 0.53$, $P = 0.001$). When restricted to woodland and open forest structural formations (i.e. omitting the one open woodland site), predictive relationships between rainfall for mean woody FPC (DF = 16; $F = 16.52$; S = 6.68; $r^2 = 0.53$; $P = 0.001$), mean woody FC (DF = 16; $F = 17.71$; S = 6.39; $r^2 = 0.54$, $P = 0.001$), and total basal area (TBA) (DF = 16; $F = 49.14$; S = 2.30; $r^2 = 0.77$; $P = 0.000$), improved significantly.

These results are consistent with previous studies (Williams *et al.* 1996) which have found strong relationships between rainfall, cover and basal area in the savannas of the Northern Territory. This suggests that annual rainfall plays a key role in the determination of eucalypt savanna structure and productivity, and that median rainfall could be used as a parameter in models to reliably estimate total basal area within mature eucalypt-dominated woodland structural formations in some tropical savanna ecosystems. Given the strong relationship between stand basal area and leaf litter accumulation, it could be reasoned that regionally refined models for leaf fuel accumulation could similarly be derived with median rainfall as a key explanatory variable contributing to the model (see *Vegetation parameters and leaf litter accumulation*).

Temporal variability of foliage projective cover

As discussed previously, total FPC increased with increasing rainfall. However, the magnitude and timing of woody plant leaf phenology appears relatively uniform regardless of rainfall quantum. Median rainfall was seen to be the most important determinant of FPC at the sites across the geographic range of the study (Eqn 1 and Fig. 7.7).

$$\text{logit}(\text{FPC}) = 0.1008\sin(2\pi(\text{pDOY})) - 0.0865\cos(2\pi(\text{pDOY})) + 0.964\ln(\text{RF}) - 7.6573 \quad (1)$$

*pDOY = Day of Year as a proportion (i.e. Julian Day/365); RF = median annual rainfall.

Peak FPC levels were attained in the late wet season (March–April) with minima generally occurring in the late dry season (September–October). There was considerable variation in woody foliage measures within sites. However, the average percentage change in woody FPC minima and maxima was generally consistent across the latitudinal and rainfall gradient, with values ranging from 38–41% both north and south of the 1000 mm isohyet/15°S disjunction. This superficially seems at odds with the results for leaf litter accumulation (Fig. 7.6). It would seem logical that the timing and magnitude of variations in woody plant phenology should correlate with the observed seasonal variations in leaf litter accumulation. However, this is consistent with the observations of Williams *et al.* (1997) and O'Grady *et al.* (2000) for LAI in *E. miniata–E. tetrodonta* associations in high rainfall areas of the Northern Territory. These studies indicated that despite the prominence of deciduousness as a strategy for the management of seasonal water stress in the subordinate tree layers and mid-storey, the actual contribution of these strata to total LAI is small (~10%) regardless of season, and the large variations in canopy fullness in these vegetation strata are likely to be obscured by the smaller variations in LAI of the dominant evergreen taxa in the upper tree layer when averaged across the community. Although not directly comparable to FPC, the relationship between LAI and FPC is such that temporal trends in the two parameters may be expected to be similar (Scarth *et al.* 2008). This would appear to be confirmed by the timing of seasonal minima and maxima for canopy fullness, which coincides with the cycles identified for the dominant evergreen overstorey taxa in *E. miniata–E. tetrodonta* woodland and open forest associations (Williams *et al.* 1997; O'Grady *et al.* 2000; Ma *et al.* 2013).

Basal area as a predictor of vegetation foliage measures

All measures of foliage characteristics (FPC, foliage cover and crown cover) were positively correlated with basal area. Total basal area was seen to be most strongly related to total FPC in all strata (woody FPC; S = 5.094; $r^2 = 0.70$; $P = 0.000$) and total FC in all strata (woody foliage cover; S = 5.283; $r^2 = 0.67$; $P = 0.000$). Figure 7.8 gives the linear regression relationship between woody FPC and total basal area. These results are consistent with other studies suggesting a strong relationship between foliage characteristics and total basal area measures (Danaher *et al.* 1992; Kuhnell *et al.* 1998; Danaher *et al.* 2004; Armston *et al.* 2009).

However, it should be noted that a linear model explained the greatest proportion of variation in the data from this study, as opposed to the higher order relationships determined in previous studies. This may be a result of this study focusing on one particular vegetation community with a restricted range of structural characteristics, as opposed to the previous studies which aimed to develop a model for a much broader range of floristic and structural vegetation types. Figure 7.9 also shows the woody FPC/total basal area relationship plotted along with the generalised model of Armston *et al.* (2009). Monitoring sites from the Northern Territory sit within the expected range of the generalised model at the higher range of basal area values and slightly above the expected range at lower values (<10 m^2 ha^{-1}). This may be a result of the general model incorporating a wider variety of eucalypt and non-eucalypt associations (and thus canopy foliage densities) across a wider geographic range at the lower end of the basal area range than the single association considered in this study.

Vegetation foliage measures have long been operationally derived from remote sensing platforms (e.g. leaf area index (LAI) from MODIS; foliage projective cover (FPC) from

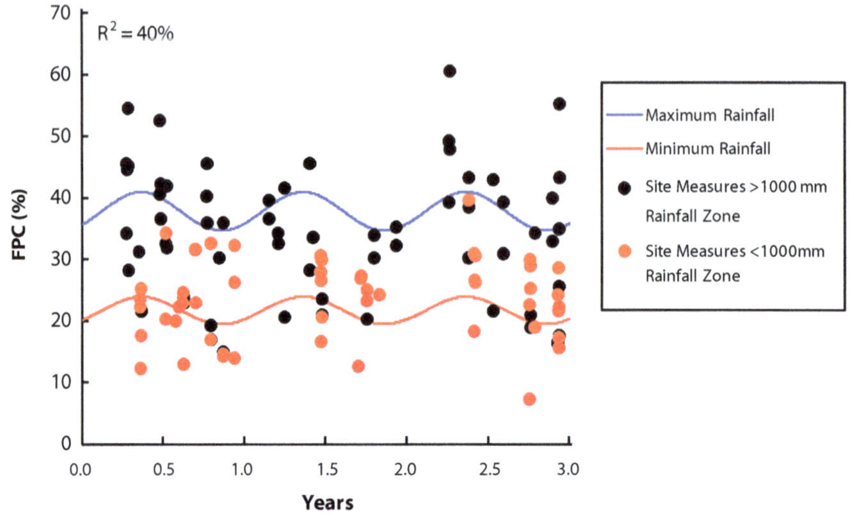

Figure 7.7. Foliage projective cover observations and models from repeated measures at 18 sites along a rainfall gradient in *E. miniata* and *E. tetrodonta* vegetation associations from the tropical Australian monsoonal savannas between March 2010 and December 2012.

Landsat 5/7). The allometric relationship between stand basal area and FPC is utilised to predict on-ground FPC which can then be used to model Landsat-FPC and associated fractional cover products. The refinement of this relationship between TBA and site FPC for Northern Territory conditions is important for the up-scaling and ultimately the mapping of vegetation cover classes and fuel accumulation relationships for savanna systems in the Northern Territory.

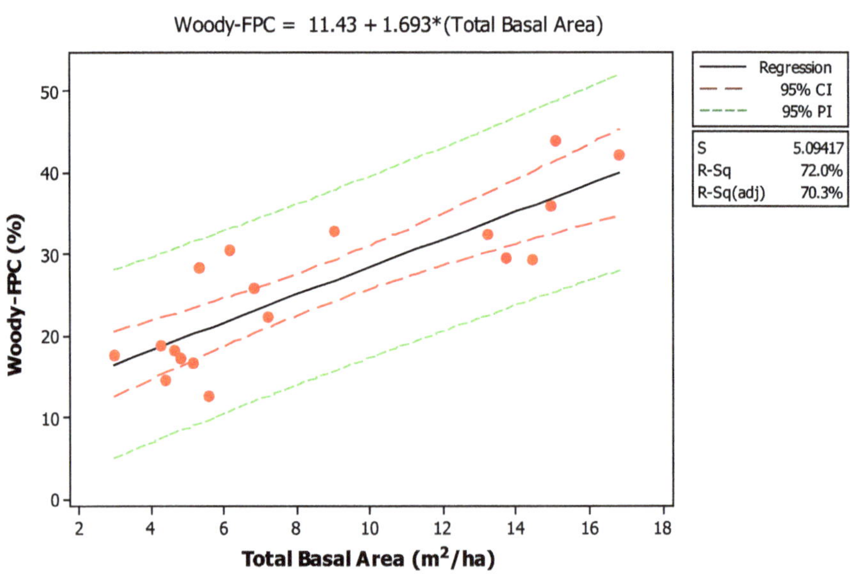

Figure 7.8. Linear regression of woody foliage projective cover (FPC) and total stand basal area (TBA) at 17 sites in Northern Territory *Eucalyptus miniata–E. tetrodonta* vegetation associations.

Vegetation parameters and leaf litter accumulation

Figures 7.10a,b present mean boxplot and interval plots of leaf litter accumulation rates and Appendix 2 provides descriptive statistics for the 18 monitoring sites over the course of the study. Mean leaf litter accumulation rates varied substantially across the study area from a maximum average of 7.11 Mg ha^{-1} yr^{-1} on Melville Island (Tarracumbi-18) to 1.64 Mg ha^{-1} yr^{-1} at Limmen National Park (LIM-9), although marked variation was observed within sites between revisits. Raw descriptive statistics suggested average leaf litter accumulation at sites 13 and 14 were significantly affected by a small number of unusually high values ($n = 3$) in single sample quadrats on these sites. These plots were located beneath large trees at the north-western and southern ends of the sites respectively, and revisits subsequent to fire events at these sites exhibited excessively large (>3*inter-quartile range) variance from the mean rates of litter accumulation. For the purposes of statistical analyses these measures were considered outliers and removed from the dataset.

Table 7.3 summarises Pearson correlation coefficients between leaf litter accumulation rate and derived vegetation and rainfall attributes for the 18 sites represented in this study. Leaf litter accumulation was found to be strongly positively correlated with especially stand basal area measure, total woody foliage measures and rainfall. The general declining trend in mean litter accumulation rates from north to south (Figs 7.10a,b) further suggests median annual rainfall is playing an important role in determining total litter loads.

Simple linear regression of these parameters with leaf litter accumulation indicated that total basal area (s = 0.887193; r^2 = 0.74; P = 0.000) and live stand basal area (s = 0.924691; r^2 = 72.2%; P = 0.000) described the strongest relationships, although strong relationships were also evident with woody foliage cover (s = 0.99514; r^2 = 67.8%; P = 0.000) and woody foliage projective cover (s = 0.995524; r^2 = 67.8%; P = 0.000).

TBA and rate of leaf litter accumulation were plotted against comparable measures of these parameters from other studies (Fig. 7.11). Our observations were generally consistent with data from other studies in New South Wales, Queensland, Western Australia and the Northern Territory as summarised in Cook (2003), However, the predictive power of this

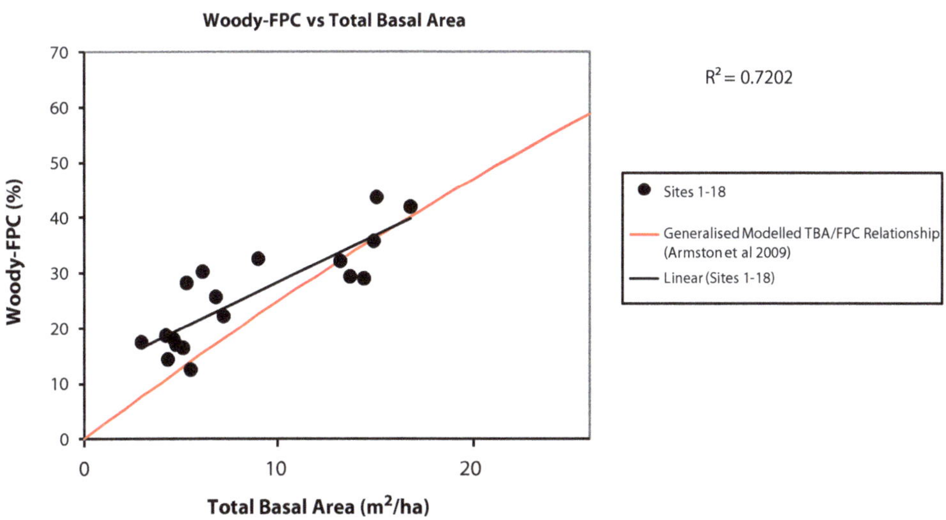

Figure 7.9. Linear regression of woody foliage projective cover and total stand basal area with generalised model of Armston et al. 2009 (red line).

Table 7.3. Pearson correlation coefficients of leaf litter accumulation with vegetation parameters

	TBA[1] (m^2 ha^{-1})	LBA[2] (m^2 ha^{-1})	Canopy-FPC[3]	Canopy-FC[4]	CC[5]	Woody-FPC	Woody-FC	Rainfall (mm)
Leaf litter accumulation (Mg ha^{-1} yr^{-1})	0.871 (*P* = 0.000)	0.859 (*P* = 0.000)	0.742 (*P* = 0.000)	0.712 (*P* = 0.001)	0.602 (*P* = 0.008)	0.835 (*P* = 0.000)	0.835 (*P* = 0.000)	0.769 (*P* = 0.000)

1 TBA denotes total stand basal area (m^2 ha^{-1} of all live + dead standing stems).
2 LBA denotes live stand basal area (m^2 ha^{-1}) only.
3 FPC denotes percentage foliage projective cover (Hnatiuk *et al.* 2009).
4 FC denotes percentage foliage cover (Hnatiuk *et al.* 2009).
5 CC denotes percentage crown cover (Hnatiuk *et al.* 2009).

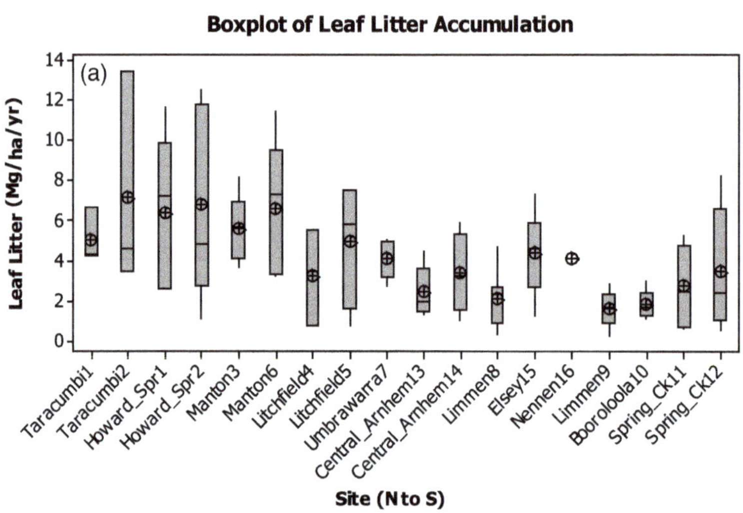

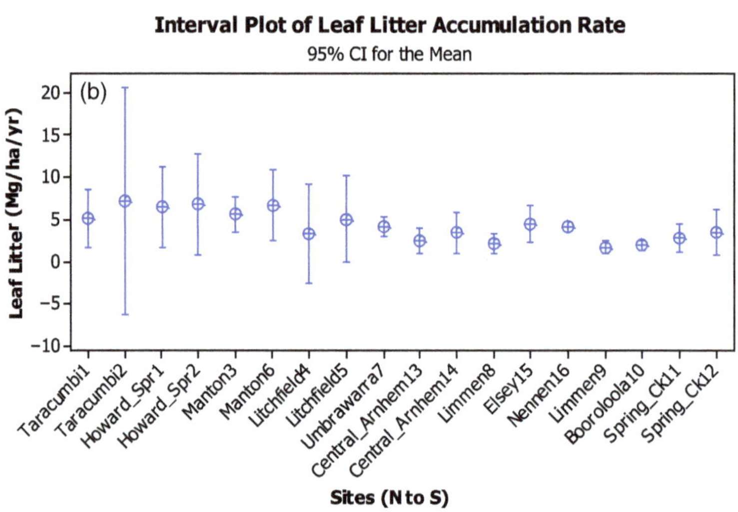

Figure 7.10. (a) Boxplot of leaf litter accumulation rates at 18 sites (ordered north to south) in Northern Territory *Eucalyptus miniata–E. tetrodonta* vegetation associations. (b) Interval plot of leaf litter accumulation rates at 18 sites (ordered north to south) in Northern Territory *Eucalyptus miniata–E. tetrodonta* vegetation associations.

relationship was substantially lower than the original model (DF = 44; F = 54.65; r^2 = 0.56, P < 0.000 *v.* r^2 = 0.76, P < 0.000).

It should be noted that, as for the Kapalga (NT) sites detailed in Cook (2003), the sites from this study sit above the expected levels of leaf litter accumulation for a given total basal area as represented by the general model. Leaf litter accumulation rates in both these studies was substantially higher than that reported by Chen *et al.* (2003) for similar savanna sites at 2 Mg ha^{-1} yr^{-1}. Figure 7.11 also shows the data with fitted relationships representing the total set of data as well as the NT sites from both this and the Kapalga studies. The gradient of this relationship diverges markedly from the general relationship for all vegetation types.

Fitted regressions of TBA and leaf litter accumulation rate for this subset of Northern Territory savanna structural formations are slightly more predictive than the overall relationship (DF = 25; F = 35.9; S = 0.985; r^2 = 0.60; P <0.000). As with the sites from the current study, the sites from the Kapalga study are *E. miniata-* and *E. tetrodonta*-dominated associations ranging from woodlands to open forests. This suggests that there may be underlying characteristics of tropical Australian savanna ecosystems that result in higher rates of leaf litter accumulation for a given stand basal area than for other eucalypt-dominated associations in Australia. As discussed below, the physiological mechanisms (e.g. deciduousness) employed by tropical Australian savanna trees and shrubs to cope with seasonal water stress may be playing a significant role in the way in which leaf litter accumulates within these systems and could explain some of the variation observed between Northern Territory sites and those from southern and eastern Australia.

Deciduousness and leaf litter accumulation

Table 7.3 demonstrated significant variability in mean rates of leaf litter accumulation across the geographic and precipitation range of the study area. Previous studies have

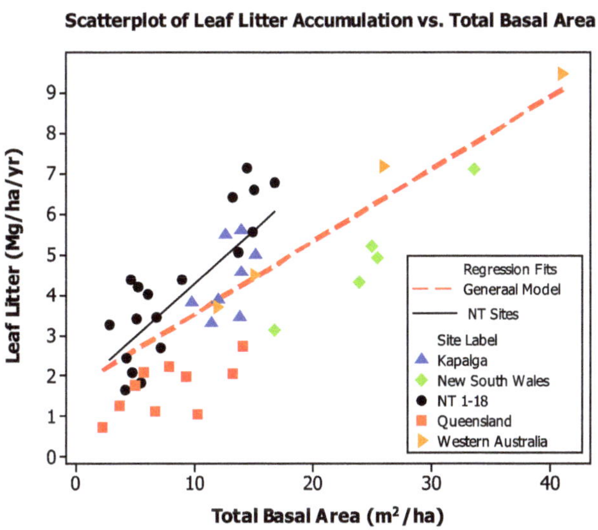

Figure 7.11. Overall (after Cook 2003 – red dashed line) and Northern Territory (black line) relationships between leaf litter accumulation and total basal area at 45 sites in eucalypt-dominated vegetation associations fitted to intercept. Sites indicated by: Queensland (red square), New South Wales (green diamond), Western Australia (orange triangle) and the Northern Territory (blue triangle and black circle).

suggested that annual rainfall plays an important role in the determination of savanna structure, floristics and function (e.g. Williams *et al.* 1996). The results detailed in the preceding section are consistent with these summations. Similarly, total annual leaf fall has previously been linked to tree species physiology and this in turn to annual rainfall in tropical Australian savanna ecosystems (Williams *et al.* 1996, 1997).

Similarly, other studies (e.g. Williams *et al.* 1997; Prior *et al.* 2004) have discussed the annual cycling of evergreen tree species canopies in northern Australian savannas, suggesting that species such as *E. tetrodonta* and *E. miniata* undergo almost total canopy leaf replacement over the course of a wet–dry–wet season cycle in the higher rainfall areas. This differs from eucalypt species in temperate and arid southern and eastern Australia, which have been shown to turn canopy leaves over on a bi-annual or longer time scale (e.g. Pook 1984). This could be interpreted as a response to the relatively more uniform patterns of resource availability within these environments when compared with the 'boom–bust' seasonal cycles of the tropical savannas.

Increasing intensity and magnitude of seasonal water deficit has long been implicated in the development of both deciduousness and the sclerophyllous leaf traits dominant in savanna taxa, including increased leaf mass per area (LMA – the inverse of specific leaf area (SLA); e.g. Prior *et al.* 2004). Development of such traits, particularly increased LMA, has been demonstrated to correlate positively with leaf longevity (LL) (Casper *et al.* 2001; Wright *et al.* 2001, 2004a,b; Wright and Westoby 2002) and is thought to reflect adaptation to essentially 'droughty' climatic conditions, affording a degree of thermal protection to leaves in such environments (Westoby 1998; Wright *et al.* 2002; Curtis *et al.* 2012). Among species, this is thought to reflect the trade-off between lower LMA allowing for lower investment in leaf construction and therefore higher potential growth and turnover rates, versus increased LL allowing for a longer unit return on the investment in leaf production but a reduced potential for growth (Westoby 1998; Wright *et al.* 2002). Australian temperate and arid zone woody species have been generally shown to conform to this model of resource economics.

The implications of this for leaf litter accumulation rates are likely to be significant given the strong correlations with both vegetation structural attributes (total basal area, woody foliage cover) and rainfall as a determinant of floristic composition, tree performance, leaf retention strategy and community deciduousness. These points warrant future investigation with reference to fine fuel accumulation in the Australian savannas.

Leaf litter accumulation rate (i.e. leaf fall) was generally observed to have a negative relationship with latitude and consequently a positive relationship with increasing rainfall. Again, raw leaf litter accumulation data were observed to vary considerably within sites, but a generally positive trend was evident when plotted against annual median rainfall. Figure 7.12 shows the mean and range of leaf litter accumulation ordered by annual rainfall at each of the sites monitored in this study compiled from individual measures over an ~3 year period.

Yearly averages for leaf fall in the high rainfall regions (>1000 mm) were 5.01 Mg ha^{-1} yr^{-1} as compared with lower rainfall region (<1000 mm), where the mean accumulation was 2.87 Mg ha^{-1} yr^{-1}. Combined average leaf fall for all regions was 4.17 Mg ha^{-1} yr^{-1}.

Mean leaf litter accumulation rates varied significantly with rainfall in regression analysis, indicating a positive relationship between leaf litter accumulation rates and the rainfall gradient (DF = 17; F = 14.44; P = 0.002 S = 1.327; r^2 = 44.2). This relationship improved considerably when considering woodland structural formations in isolation (DF = 16; F = 25.30; P = 0.000; S = 1.11414; r^2 = 60.3%). These results suggest that there are

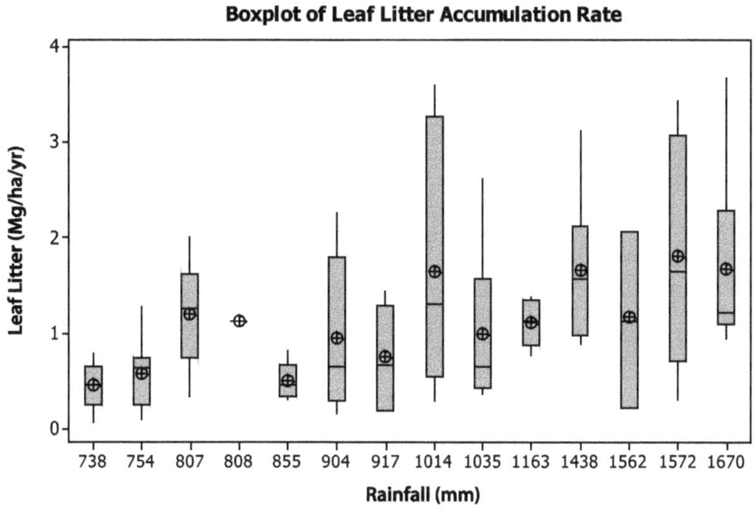

Figure 7.12. Boxplot of leaf litter accumulation versus rainfall at 18 sites in Northern Territory *Eucalyptus miniata–E. tetrodonta* vegetation associations.

fundamental differences between *E. miniata–E. tetrodonta* associations in different rainfall zones that are influencing the rate of leaf litter production in these locations.

It has previously been reported that, within the Northern Territory, deciduousness has a generally negative relationship with latitude and thus a positive relationship with rainfall (Egan and Williams 1996; Williams *et al.* 1996). This trend towards deciduousness is a feature of savanna ecosystems of the wet–dry tropics globally, although this is somewhat at odds with many other savanna systems where deciduousness tends to increase with decreasing rainfall within the biome. Although considered dominantly evergreen, *E.miniata–E. tetrodonta*-dominated associations may exhibit a high proportion of deciduous tree and shrub species in the secondary strata (e.g. shrub or mid-strata) or as sub-dominant components of the upper tree layers, and these species may contribute significantly to the volume and rates of leaf litter accumulation.

Consequently, the species records from the structural assessments at each site were partitioned according to the four leaf phenological groups described by Williams *et al.* (1997) and the proportions of deciduousness analysed in both floristic and stand structural contexts. The criteria employed by Williams *et al.* (1997) were used to assess deciduousness. We observed a generally consistent, albeit relatively weak, pattern of deciduousness similar to those reported by previous authors (Egan and Williams 1996; Williams *et al.* 1996) (Fig. 7.13). The proportion of deciduous species increased with decreasing latitude, except in the very highest rainfall zones (>1600 mm), where the proportion of deciduous species in the regional flora declined slightly. The occurrence of deciduousness as a proportion of each site's woody floristic complement ranges from ~22% of species at the lowest mainland latitudes (~12°S), to an average of ~9% at the most southerly, lower rainfall locations (~16°S).

Similarly, trends in deciduousness as a proportion of total basal area and thus stand dominance were also observed to reflect these latitudinal and rainfall gradients in the floristic composition of the association. Highest levels of deciduousness as a proportion of stand basal area were recorded on mainland sites at the lowest latitude (17% of TBA) declining to lowest levels south of 16°S (5%).

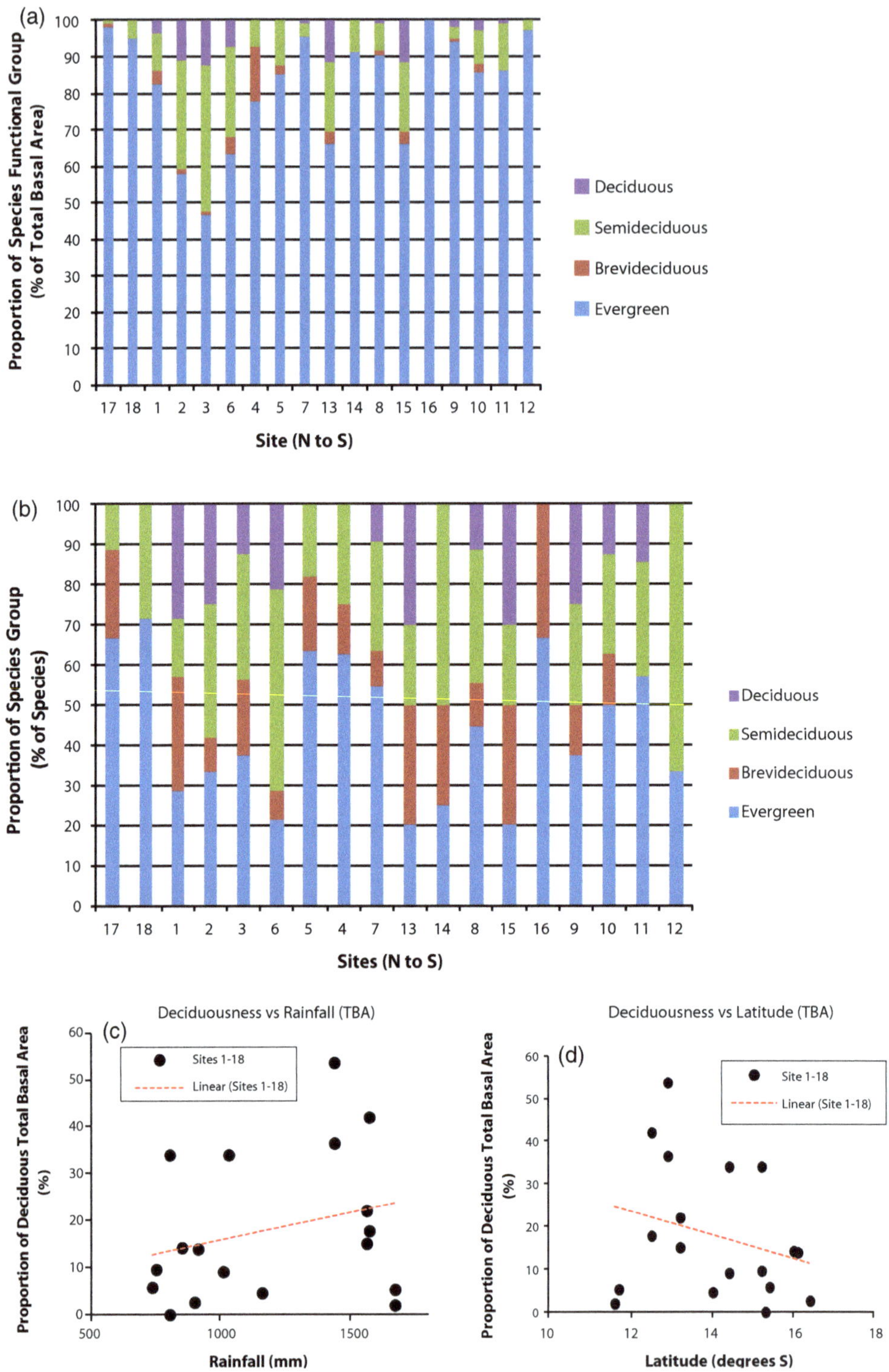

Figure 7.13. Proportions of deciduousness classes as a function of both total basal area (a, c, d) and number of species (b, e, f) of woody trees and shrubs versus latitude and rainfall at 18 sites in *Eucalyptus miniata–E. tetrodonta* vegetation associations.

(Continued)

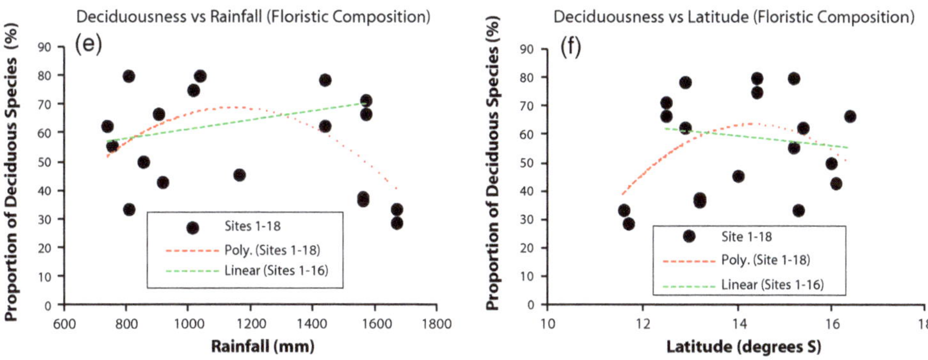

Figure 7.13. (Continued).

Egan and Williams (1996) reported marked differences in the incidence of deciduousness in the NT regional flora with increasing latitude and decreasing rainfall. Variation between sites in adjacent 100 mm rainfall increments in this study was on average less than 10% and in the order of 8% between adjacent 1° increments of latitude, comparable to the more northerly, less arid results reported by Egan and Williams (1996). However, a marked increase in the dominance of evergreen species as a proportion of TBA was observed south of ~15°S, coinciding with the approximate location of the 1000 mm median rainfall isohyet. Evergreen species comprise 92.3% of the TBA in sites south of 15° while contributing on average only 77% to the TBA of mainland sites north of this latitude. This is likely to be of particular importance to the expression of seasonal patterns in community leaf litter accumulation, given the distinct differences previously reported for the timing of seasonal foliage minima in species with differing leaf retention strategies within this vegetation type.

The relationships of leaf traits such as LMA (SLA) and in turn LL to rainfall and soil moisture gradients are also of particular interest. These data suggest that underlying leaf trait and physiological strategies in essentially evergreen species such as the dominant eucalypt species may be influencing the rates at which leaf litter is accumulating within these systems. The investment in producing leaves at more southerly latitudes, subject to seasonally longer periods of moisture deficit and possibly less reliable inter-annual rainfall, may be influencing the rates of annual canopy leaf turnover in these systems, producing longer LL and thus reducing rates of observed litter accumulation for essentially similar canopy foliage characteristics. This would be in keeping with the resource investment/rate of return over time trade-off models proposed for the explanation of leaf-trait variability between species (including phylogenetically independent contrasts), and habitats along resource gradients (e.g. Westoby 1998; Wright *et al.* 2001, 2002).

It could therefore be expected that leaves within a species in lower rainfall habitats may have higher LMA that yields a longer LL for given canopy densities as a result of seasonal water deficit. This could in turn influence the rate of canopy leaf turnover in functionally evergreen species and in turn the leaf litter accumulation rates and total fine fuel composition along the rainfall gradient.

Final leaf litter accumulation model

Simple univariate relationships between leaf litter accumulation and vegetation parameters explain a significant proportion of the observed variation within the data.

However, inclusion of additional variables to the model, particularly rainfall, results in improvements in the overall variation explained by the model (S = 0.841266; $P = 0.000$; $r^2 = 0.77$) as would be expected given the significant role the rainfall gradient appears to play in the determination of foliage cover, plant deciduousness and thus leaf fall (Fig. 7.14). Eqn 2 details the modelled relationship between leaf litter accumulation, total basal area and rainfall. Interestingly, models incorporating the proportion of deciduousness at each site as an explanatory variable did not significantly improve the results.

$$\textit{Leaf Litter} = 0.376 + 0.248^{*}(\textit{Total Basal Area}) + 0.00140^{*}(\textit{Rainfall}) \qquad (2)$$

The development of this model potentially allows the on-ground estimation of leaf litter at a location from several simply derived measures.

Coarse litter accumulation

Appendix 3 and Fig. 7.15 present coarse litter accumulation data from the 18 monitoring sites across the study area. Coarse litter accumulation rates varied from a mean maximum of greater than 1 Mg ha^{-1} yr^{-1} on Melville Island to levels consistently less than 0.35 Mg ha^{-1} yr^{-1} in all but the most northerly mainland locations. As these data indicate, the variability in amount and rate of coarse litter production across the study area is high, with some sites not producing measurable quantities of coarse litter during the period of the study (e.g. Nenen Station, Site 16).

Table 7.4 illustrates that coarse litter accumulation is not correlated strongly with structural vegetation parameters. Of the parameters derived from this study, basal area is most strongly correlated (but non-significantly) with coarse litter accumulation. Rainfall was significantly correlated with coarse litter accumulation rate ($P = 0.014$).

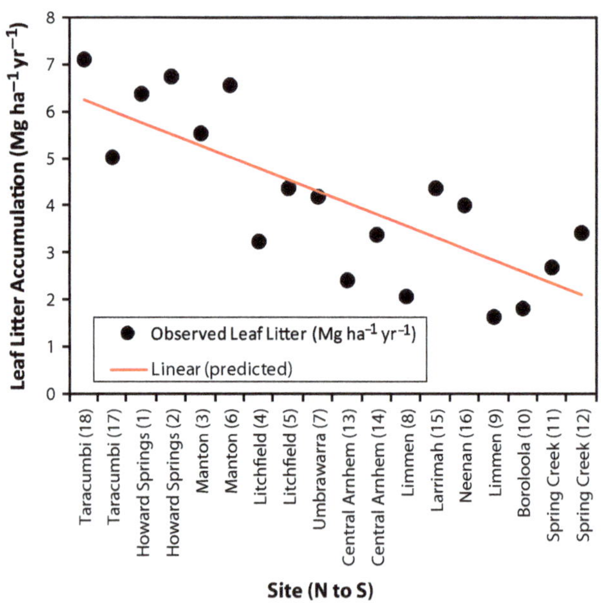

Figure 7.14. Observed leaf litter accumulation (points) plotted with the modelled relationship (line from Eqn 2) for Total Basal Area and Rainfall (mm) 6 at 18 sites in *Eucalyptus miniata–E. tetrodonta* vegetation associations.

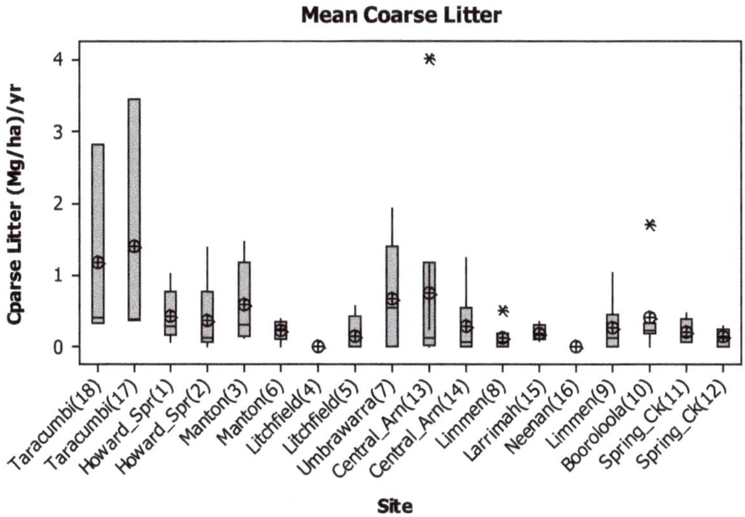

Figure 7.15. Boxplot of coarse litter accumulation (Mg ha^{-1} yr^{-1}) at 18 sites in *Eucalyptus miniata* and *E. tetrodonta* vegetation associations in the Northern Territory.

Regression analysis of coarse litter accumulation rates and rainfall suggests that rainfall and thus plant productivity is playing an important role in the production of coarse fuel, but this is not a strong predictive relationship ($r^2 = 34.2$; $P = 0.014$).

These results would appear to suggest that some parameter(s) other than those measured in this study are regulating the production of coarse litter at the landscape scale. Observations from this study suggest that localised and/or stochastic weather events (e.g. strong winds, storms and cyclones) and stand senescence may play a more important role. Several of the field locations monitored during this study were affected by a combination of strong winds and intense rainfall events over the 2010–11 wet season. Subsequent observations from these sites and surrounding areas suggested a non-uniform patterning of coarse litter production in the landscape operating at a scale beyond the observations collected in this study.

Preliminary investigation of alternative methods for collecting coarse litter accumulation data were commenced during the course of this study with a larger sample area to adequately account for the variability and larger spatial scale of coarse woody fuel production. Proceeding from the 0 m end of each transect, a 1 m-wide strip on the right hand side

Table 7.4. Pearson correlation coefficients of coarse litter accumulation rates (Mg ha^{-1} yr^{-1}) with derived vegetation measures and rainfall

	TBA[1] (m^2 ha^{-1})	LBA[2] (m^2 ha^{-1})	Canopy-FPC[3]	Canopy-FC[4]	CC[5]	Woody-FPC	Woody-FC	Rainfall (mm)
Coarse litter accumulation (Mg ha^{-1} yr^{-1})	0.408 (*P* = 0.104)	0.369 (*P* = 0.145)	0.015 (*P* = 0.953)	0.050 (*P* = 0.849)	–0.098 (*P* = 0.708)	0.070 (*P* = 0.790)	0.109 (*P* = 0.678)	0.585 (*P* = 0.014)

1 TBA denotes total stand basal area (m^2 ha^{-1} of all live + dead standing stems).
2 LBA denotes live stand basal area (m^2 ha^{-1}) only.
3 FPC denotes percentage foliage projective cover (Hnatiuk *et al.* 2009).
4 FC denotes percentage foliage cover (Hnatiuk *et al.* 2009).
5 CC denotes percentage crown cover (Hnatiuk *et al.* 2009).

of each tape was cleared of all coarse (1–5 cm diameter) woody material. This created a sampling area of 300 m^2 rather than the original 12 m^2. On subsequent re-visits the coarse fuel was collected, weighed and sub-samples taken for drying to determine moisture content and dry weight. However, preliminary results suggest that even this sampling area does not adequately scale to the spatial patterning of coarse fuel production in the landscape.

Temporal variability and seasonality of leaf fall and coarse fuel accumulation

Temporal variability in fuel production is likely to have a profound influence on the behaviour and severity of a fire at a given time of the year. Similarly, the relative proportions of different fuel classes present at the time a fire occurs will influence the characteristics of both the fire itself and the products of combustion (e.g. the types and amounts of gases emitted). Thus the characterisation of potential temporal variation in the accumulation of fuel components could prove important in the formulation and parameterisation of fire models at the landscape scale.

Previous studies have identified distinct cycles in the production, retention and shedding of tree and shrub leaves in tropical savanna ecosystems (e.g. Williams *et al.* 1997; Myers *et al.* 1998; O'Grady *et al.* 2000). As discussed previously, tree/shrub deciduousness is a prominent feature of tropical Australian savanna ecosystems and a common physiological adaptation to seasonal water stress in savannas worldwide. Previous studies have highlighted significant differences in the onset and attainment of maximum canopy reduction (i.e. leaflessness) in some woody tropical savanna species based on canopy deciduousness class (Williams *et al.* 1997). This study demonstrated a significant offset of these canopy parameters earlier in the year (May–October) for deciduous, as opposed to semi-deciduous and evergreen species, which began canopy reduction and attained canopy minima later in the year (September–November). Similarly, the trends in overall deciduousness with latitude and rainfall gradients have been discussed. Consequently, when considered at the association level and in a regional context, it could be reasoned that differences should be evident in the patterns of seasonal litter accumulation and foliage cover along the rainfall gradient, possibly reflecting the changes in floristic composition associated with the changes in proportion of community deciduousness.

Measures from the 18 sites monitored in this study highlighted several broad and repeatable patterns of seasonal leaf litter accumulation (i.e. leaf fall) in the tropical savannas of the Northern Territory. Median rainfall was determined to be the most important determinant of variability in temporal leaf litter accumulation rates across the study area (Eqn 3). Generally, the amount of leaf litter accumulation peaks during the dry season months of May to October when seasonal water stress is most pronounced. However, there was notable variation in the magnitude of leaf litter accumulation peaks between years (Fig. 7.16), with generally lower rates of leaf litter accumulation observed in the second year (2011) of this study, particularly in areas with a median annual rainfall exceeding 1000 mm. This could be attributable to the record rainfall received in the 2010–11 wet season across much of northern Australia and the subsequent extended period of above-average water availability in the landscape after this event. This is evidenced by the generally 'positively green' nature of NDVI anomalies recorded for the period 1 July 2010 through 31 December 2011.

It is worth noting the distinct differences observable between leaf litter accumulation peaks for high (>1000 mm) vs. low (<1000 mm) rainfall locations. Peak leaf

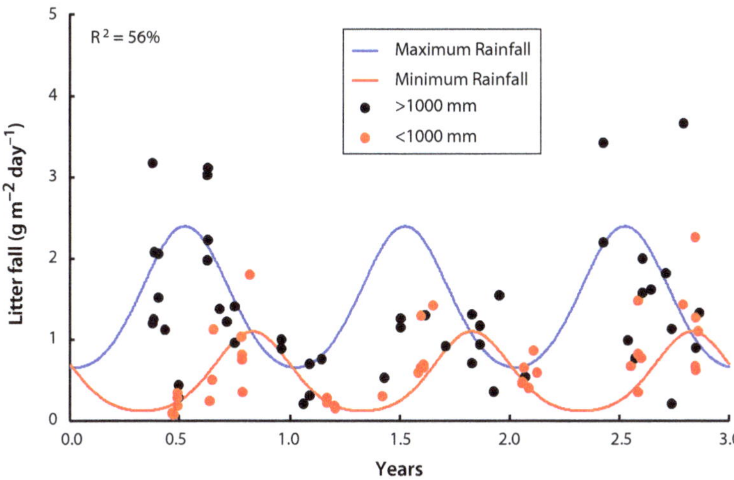

Figure 7.16. Leaf litter accumulation models (Eqn 3) from repeated-measures at 18 sites along a rainfall gradient in *E. miniata–E. tetrodonta* associations from the tropical Australian monsoonal savannas between March 2010 and December 2012.

litter accumulation occurs in June–July for locations receiving >1000 mm, as opposed to September–October for locations receiving <1000 mm median annual rainfall. This pattern of community 'leaf fall' strongly correlates with the observations of Williams *et al.* (1997) for woody taxa leaf phenology in tropical savanna ecosystems. Given the pattern of increasing deciduousness with increasing rainfall, it is reasonable to suggest that much of the difference in timing of peak leaf fall could be attributed to the higher proportions of deciduous species found in sites north of the 1000 mm rainfall isohyet. Conversely, patterns of leaf fall in sites south of the 1000 mm isohyet reflect the increased dominance of evergreen and lower proportion of deciduous taxa in the composition of the *E. miniata–E. tetrodonta* community, as has been reported previously (Williams *et al.* 1996; Egan and Williams 1996).

$$LF^{0.3} = -1.7588sin(2\pi(pDOY)) + 2.8008cos(2\pi(pDOY)) + 0.3888\ln(RF) + 0.2325\ln(RF)[sin(2\pi(pDOY)] - 0.4054\ln(RF)[cos(2\pi(pDOY)] - 1.7974 \qquad (3)$$

*pDOY = Day of Year as a proportion (i.e. Julian Day/365); RF = Median annual rainfall (mm).

In order to aid the estimation of the different emissions resulting from early or late dry season fires, leaf and coarse fuel accumulation data were averaged for the early dry season (January–July) and late dry season (August–December) for both the high rainfall (>1000 mm p.a.) and lower rainfall regions (<1000 mm p.a.) (Table 7.5).

Yearly averages for leaf fall was 5.01 Mg ha^{-1} yr^{-1} in the high rainfall regions and 2.87 Mg ha^{-1} yr^{-1} in the lower rainfall region. These differences and the variation in proportion of total leaf litter accumulation between season and with rainfall are proposed to relate primarily to the documented differences in dominant leaf phenophases and deciduousness along the north–south latitudinal gradient. A further logical extension of this study would be to sort accumulated litter on the basis of deciduousness groups in order to confirm the relative contributions of these groups to total leaf fall.

Table 7.5. Leaf litter and coarse litter accumulation by proportion in early and late dry season

Rainfall zone	Leaf litter accumulation		
Leaf litter	**Mean accumulation Mg ha^{-1} yr^{-1}**	**Proportion early dry season (%)**	**Proportion late dry season (%)**
>1000 mm p.a.	5.01 (±0.48)	46	54
<1000 mm p.a.	2.87 (±0.41)	26	74
Coarse litter			
>1000 mm p.a.	0.551 (±0.13)	29	71
<1000 mm p.a.	0.221 (±0.04)	60	40

Differences in the accumulation of coarse fuel accumulation with rainfall and season are less easily interpretable. As discussed previously, the production and accumulation of coarse fuel within the landscape does not appear to correlate strongly with vegetation parameters at the scales measured in this study. However, rainfall variability does appear to have a significant influence on the amount of coarse fuel produced in *E. miniata–E. tetrodonta* associations in the Top End. Intuitively, this would appear to make ecological sense given the positive relationships previously described between plant productivity and moisture availability along the north–south rainfall gradient. However, reliable correlations between increased production and coarse fuel loads remain unresolved at this time.

Seasonality of coarse fuel production also appears to vary with rainfall regime, although this may in actuality reflect latitudinal variability in the frequency and timing of intense weather events (e.g. thunderstorms) likely to produce fuel at the landscape scale. The categorisation of early and late dry season from January–July and August–December respectively is likely to correlate with the periods of most intense storm activity in the <1000 mm and >1000 mm rainfall zones respectively. This may account for the observed shift in seasonal distribution of coarse litter production between low and high rainfall zones.

Conclusion

This study has provided significant additional information on the volume and seasonality of leaf litter contributions to total fine fuel load along the rainfall gradient in the 'evergreen' eucalypt savannas of the Northern Territory. Leaf litter accumulation has been shown to vary significantly with rainfall and vegetation structural attributes, particularly stand basal area, in *E. miniata–E. tetrodonta* woodlands and open forests. Similarly, the seasonality of leaf fall in these systems varies with latitude and rainfall, with a distinct leaf fall maximum in June–July for savannas receiving >1000 mm median annual rainfall as opposed to the September–October peak observed in savannas receiving <1000 mm rainfall. When considered in terms of community deciduousness, this results in very different rates and patterns of leaf litter production across the season between higher and lower rainfall locations. These results allow for the refinement of models used to account for the emissions resulting from wildfire at different times of the year and, potentially, more flexible application of abatement models to determine the results of fire management actions by land managers on total wildfire emissions.

Several significant components of leaf litter contribution to fine fuel accumulation remain to be addressed in these vegetation types. These include: how decomposition rates may vary in conjunction with the rainfall gradient and how this affects total fine fuel loads; the relative contributions of deciduous versus evergreen species to leaf litter loads in relation to the rainfall gradient; and the influence of leaf attributes such as longevity and leaf mass per area on the rates of canopy turnover.

More generally, relationships between leaf litter accumulation, vegetation structure and rainfall remain to be addressed in the broader range of vegetation types encountered in the tropical savannas. Of particular interest is quantification of these relationships in deciduous eucalypt communities found in lower rainfall areas of the eastern Northern Territory and throughout large areas of northern Queensland. This could potentially allow further regional refinement of emissions accounting models for the purposes of fire abatement projects across the northern Australian savannas.

Acknowledgements

Access to sites was provided or facilitated by Alan Withers, John McCartney, Kristen Appel (Parks and Wildlife Commission NT), Max Gorringe (Elsey Station – NLC), Cyril Kalippa, Willie Rioli, Kate Hadden, Gibsen Farmer, Roger Smith (Tiwi Land Council), Jimmy Morrison, Jack Green, Steven Anderson (NLC – Booroloola and Station), Ray Whear (Jawoyn Association) and Nicholas Kostowski (Nenen Station).

Assistance with field data collection was provided by Mark Desally, Dominique Lynch, Jay Evans, Andrew Edwards (Darwin Centre for Bushfire Research, CDU (formerly Bushfires NT)), Chris Mangion, Grant Staben, Bart Edmeades, Jason Barnetson, Agnes Lahille (DLRM NT), Tahnee Thompson, Stuart Woerle, Shelley Inglis (Parks and Wildlife Commission NT), Ben Sparrow (University of Adelaide, SA), Tony Rosling (Department of Environment, Australian Government), Willie Rioli, Colin Kerinauia, Vivian Kerinauia, Brian Tipungwuti, Desmond Bruppacher, José Puruntatameri, Kim Brooks, Leon Puruntatameri (Tiwi Land Council – Tiwi Islands/Tiwi Rangers), Jimmy Morrison (Waanyi-Gaarawa Ranger Group – NLC), Jon Shatz, Maeli Cooper and Celine Christophe (CSIRO, Darwin).

Technical advice, project management and funding were provided by J Russell-Smith, Cameron Yates, Andrew Edwards (Darwin Centre for Bushfire Research, CDU (formerly Bushfires NT)), Grant Staben, Jason Hill, Mark Tarrant, Agnes Lahille (DLRM NT), Brett Murphy (University of Melbourne), Jon Schatz, Gary Cook, Dick Williams, Anna Richards (CSIRO, Darwin), Lindsay Hutley, Stefan Maier and Mila Bristow (Charles Darwin University, Darwin).

References

Armston JD, Denham RJ, Danaher TJ, Scarth PF, Moffie TN (2009) Prediction and validation of foliage projective cover from Landat-5 TM and Landsat-7 ETM+ imagery. *Journal of Applied Remote Sensing* **3**, 033540. doi:10.1117/1.3216031

Bowman DMJS, Wilson BA, Woinarski JCZ (1991) Floristic and phenological variation in a northern Australian rocky *Eucalyptus* savanna. *Proceedings of the Royal Society of Queensland* **101**, 79–90.

Brocklehurst P, Lewis D, Napier D, Lynch D (2007) 'Northern Territory Guidelines and Field Methodology for Vegetation Survey and Mapping.' Technical Report No. 02/2007D. Department of Natural Resources, Environment and the Arts: Palmerston, NT.

Bureau of Meteorology (2003) 'Median Monthly and Annual Rainfall Percentile Data.' Commonwealth of Australia: Canberra.

Bureau of Meteorology (2013) Climate Data Online. Commonwealth of Australia: Canberra. www.bom.gov.au/climate/data

Casper BB, Forseth IN, Kempenich H, Seltzer S, Xavier K (2001) Drought prolongs leaf life span in the herbaceous desert perennial *Cryptantha flava*. *Functional Ecology* **15**, 740–747.

Chen X, Hutley L, Eamus D (2003) Carbon balance of a tropical savanna of northern Australia. *Oecologia* **137**, 405–416.

Cook GD (2003) Fuel dynamics, nutrients and atmospheric chemistry. In *Fire in Tropical Savannas: The Kapalga Experiment.* (Eds AN Andersen, GD Cook, RJ Williams) pp. 47–58. Springer: New York.

Cook GD, Meyer CP (2009) Fire, fuels and greenhouse gases. In *Culture, Ecology and Economy of Fire Management in North Australian Savannas: Rekindling the* Wurrk *Tradition.* (Eds J Russell-Smith, P Whitehead, P Cooke) pp. 313–328. CSIRO Publishing: Melbourne.

Curtis EM, Leigh A, Rayburg S (2012) Relationships among leaf traits of Australian arid zone plants: alternative models of thermal protection. *Australian Journal of Botany* **60**, 471–483.

Danaher T, Carter J, Brook K, Peacock A, Dudgeon G (1992) Broadscale vegetation mapping using NOAA-AVHRR imagery. *6th Australasian Remote Sensing and Photogrammetry Conference (ARSPC).* Auckland, NZ. pp. 128–137.

Danaher T, Armston J, Collett L (2004) A regression approach for mapping woody foliage projective cover using Landsat imagery in Queensland, Australia. *Proceedings of the Geoscience and Remote Sensing Symposium, 2004. IGARSS '04.* 2004 IEEE International, October 2004.

DCCII (Australian Government Department of Climate Change, Industry and Innovation) (2013) 'Carbon Credits (Carbon Farming Initiative) (Reduction of Greenhouse Gas Emissions Through Early Dry Season Savanna Burning–1.1) Methodology Determination 2013.' Commonwealth of Australia: Canberra.

Duff GA, Myers BA, Williams RJ, Eamus D, O'Grady AO, Fordyce IR (1997) Seasonal patterns in soil moisture, vapour pressure deficit, tree canopy cover and pre-dawn water potential in a northern Australian savanna. *Australian Journal of Botany* **45**, 211–224. doi:10.1071/BT96018

Eamus D, Prior LD (2001) Ecophysiology of trees of seasonally dry tropics: comparisons among phenologies. *Advances in Ecological Research* **32**, 113–197. doi:10.1016/S0065-2504(01)32012-3

Edwards AC, Russell-Smith J (2009) Ecological thresholds and the status of fire-sensitive vegetation in western Arnhem Land, northern Australia: implications for management. *International Journal of Wildland Fire* **18**, 127–146. doi:10.1071/WF08008

Egan JL, Williams RJ (1996) Life form distribution of woodland plant species along a moisture availability gradient in Australia's monsoonal tropics. *Australian Systematic Botany* **9**, 205–217. doi:10.1071/SB9960205

ESCAVI (Executive Steering Committee for Australian Vegetation Information) (2003) 'Australian Vegetation Attribute Manual.' National Vegetation Information System, Version 6.0. Department of the Environment and Heritage: Canberra.

Fox ID, Neldner VJ, Wilson GW, Bannink PJ, Wilson BA, Brocklehurst PS, Clark MJ, Dickinson KJM, Beard JS, Hopkins AJM, Beeston GR, Harvey JM, Thompson EJ, Ryan TS, Thompson SL, Butler DW, Cartan H, Addicott EP, Bailey LP, Cumming RJ, Johnson DC, Schmeider M, Stephens KM, Bean AR (2001) 'The Vegetation of the Australian Tropical Savannas, (1:2 000 000 scale map in 3 sheets).' Queensland Herbarium, Environmental Protection Agency, Brisbane and the Cooperative Research Centre for the Sustainable Development of Tropical Savannas, Darwin.

Hnatiuk RJ, Thackway R, Walker J (2009) Vegetation. In *Australian Soil and Land Survey Field Handbook*. 3rd edn. National Committee on Soil and Terrain. CSIRO Publishing: Melbourne.

Hutley LB, BeringerJ, Isaac PR, Hacker JM, Cernusak LA (2011) A sub-continental scale living laboratory: Spatial patterns of savanna vegetation over a rainfall gradient in northern Australia. *Agricultural and Forest Meteorology* **151**, 1417–1428.

Kuhnell CA, Goulevitch BM, Danaher TJ, Harris DP (1998) Mapping woody vegetation cover over the state of Queensland using Landsat TM imagery. *9th Australasian Remote Sensing and Photogrammetry Conference (ARSPC)*. Sydney, NSW.

Loetsch F, Zohrer F, Haller KE (1973) *Forest Inventory* Vol. 2, pp. 348–368. BLV Verlagsgesells-chaft, mbH: München.

Lowman MD (1992) Leaf growth dynamics and herbivory in five species of Australian rain-forest canopy trees. *Journal of Ecology* **80**, 433–447. doi:10.2307/2260689

Ma X, Huete A, Yu Q, Coupe NR, Davies K, Broich M, Ratana P, Beringer J, Hutley LB, Cleverly J, Boulain N, Eamus D (2013) Spatial patterns and temporal dynamics in savanna vegetation phenology across the North Australian Tropical Transect. *Remote Sensing of Environment* **139**, 97–115. doi:10.1016/j.rse.2013.07.030

Medina E, Silva JF (1990) Savannas of northern South America: a steady state regulated by fire-water interactions on a background of low nutrient availability. *Journal of Biogeography* **17**, 403–413.

Muir J, Schmidt M, Tindall D, Trevithick R, Scarth P, Stewart JB (2011) 'Field measurement of fractional ground cover: a technical handbook supporting ground cover monitoring in Australia.' Prepared by the Queensland Department of Science, Information Technology, Innovation and the Arts for the Australian Bureau of Agricultural and Resource Economics and Sciences: Canberra.

Myers BA, Duff GA, Eamus D, Fordyce IR, O'Grady A, Williams RJ (1997) Seasonal variation in water relations of trees of differing leaf phenology in a wet-dry tropical savanna near Darwin, northern Australia. *Australian Journal of Botany* **45**, 225–240. doi:10.1071/BT96015

Myers BA, Williams RJ, Fordyce I, Duff GA, Eamus D (1998) Does irrigation affect leaf phenology in deciduous trees of the savanna of northern Australia? *Australian Journal of Ecology* **23**, 329–339. doi:10.1111/j.1442-9993.1998.tb00738.x

NCST (National Committee on Soil and Terrain) (2009) *Australian Soil and Land Survey Field Handbook*. 3rd edn. CSIRO Publishing: Melbourne.

O'Grady APO, Chen X, Eamus D, Hutley LB (2000) Composition, leaf area index and standing biomass of eucalypt open forests near Darwin in the Northern Territory, Australia. *Australian Journal of Botany* **48**, 629–638. doi:10.1071/BT99022

Opie K, Newnham G (2011) Hemispherical Photo Analyser Ver. 1. CRC Forestry and CSIRO.

Pook EW (1984) Canopy dynamics of *Eucalyptus maculata* Hook. I. Distribution and dynamics of leaf populations. *Australian Journal of Botany* **32**, 387–403. doi:10.1071/BT9840387

Prior LD, Eamus D, Bowman DMJS (2004) Tree growth rates in north Australian savanna habitats: seasonal patterns and correlations with leaf attributes. *Australian Journal of Botany* **52**, 303–314. doi:10.1071/BT03119

Russell-Smith J, Murphy BP, Meyer CP, Cook GD, Maier S, Edwards AC, Schatz J, Brocklehurst P (2009) Improving estimates of savanna burning emissions for greenhouse accounting in northern Australia: limitations, challenges, applications. *International Journal of Wildland Fire* **18**, 1–18. doi:10.1071/WF08009

Scarth P, Armston J, Danaher T (2008) On the relationship between crown cover, foliage projective cover and leaf area index. *15th Australasian Remote Sensing and Photogrammetry Conference*. Darwin NT. SSSI.

Scarth P, Roder A, Schmidt M (2010) Tracking grazing pressure and climate interaction – the role of Landsat fractional cover in time series analysis. *15th Australasian Remote Sensing and Photogrammetry Conference (ARSPC)*. Alice Springs, NT. SSSI.

Westoby M (1998) A leaf-height-seed (lhs) plant ecology strategy scheme. *Plant and Soil* **199**, 213–227. doi:10.1023/A:1004327224729

Williams RJ, Duff GA, Bowman DMJS, Cook GD (1996) Variation in the composition and structure of tropical savannas as a function of rainfall and soil texture along a large-scale climatic gradient in the Northern Territory, Australia. *Journal of Biogeography* **23**, 747–756. doi:10.1111/j.1365-2699.1996.tb00036.x

Williams RJ, Myers BA, Muller WJ, Duff GA, Eamus D (1997) Leaf phenology of woody species in a north Australian tropical savanna. *Ecology* **78**(8), 2542–2558. doi:10.1890/0012-9658(1997)078[2542:LPOWSI]2.0.CO;2

Wilson BA, Brocklehurst PS, Clark MJ, Dickinson KJM (1990) 'Vegetation of the Northern Territory, Australia.' Technical Report No. 49. Conservation Commission of the Northern Territory: Darwin.

Wright IJ, Westoby M (2002) Leaves at low versus high rainfall: coordination of structure, lifespan and physiology. *New Phytologist* **155**, 403–416. doi:10.1046/j.1469-8137.2002.00479.x

Wright IJ, Reich PB, Westoby M (2001) Strategy shifts in leaf physiology, structure and nutrient content between species of high- and low-rainfall and high- and low-nutrient habitats. *Functional Ecology* **15**, 423–434. doi:10.1046/j.0269-8463.2001.00542.x

Wright IJ, Westoby M, Reich PB (2002) Convergence towards higher leaf mass per area in dry and nutrient-poor habitats has different consequences for leaf life span. *Journal of Ecology* **90**, 534–543. doi:10.1046/j.1365-2745.2002.00689.x

Wright IJ, Groom PK, Lamont BB, Poot P, Prior LD, Reich PB, Schulze E-D, Veneklaas EJ, Westoby M (2004a) Leaf-trait relationships in Australian plant species. *Functional Plant Biology* **31**, 551–558. doi:10.1071/FP03212

Wright IJ, Reich PB, Westoby M, Ackerly DD, Baruch Z, Bongers F, Cavender-Bares J, Chapin T, Cornelissen JHC, Diemer M, Flexas J, Garnier E, Groom PK, Gulias J, Hikosaka K, Lamont BB, Lee T, Lee W, Lusk C, Midgley JJ, Navas M-L, Niinemets U, Oleksyn J, Osada N, Poorter H, Poot P, Prior L, Pyankov VI, Roumet C, Thomas SC, Tjoelker MG, Veneklaas EJ, Villar R (2004b) The worldwide leaf economics spectrum. *Nature* **428**, 821–827. doi:10.1038/nature02403

Appendix 1. Averaged vegetation measured vegetation parameters, latitude and median rainfall figures for 18 monitoring sites in *Eucalyptus miniata* and *E. tetrodonta* dominated vegetation associations in the Northern Territory

* denotes that Nenen Station was discontinued as a monitoring site after a single revisit.

Site	Location	Latitude	Rainfall (mm)	TBA (L+D)[1]	LBA (L)[2]	Mean canopy-FPC[3]	Mean canopy-FC[4]	Mean CC[5]	Mean woody-FPC	Mean woody-FC	Mean satellite-green[6]	Mean satellite-brown[7]	Mean satellite-bare[8]
1	**Howard Springs**	–12.5	1572	13.21	11.64	30.67	33.33	49.00	32.27	35.07	42.60	52.40	4.87
2	**Howard Springs**	–12.5	1572	16.79	15.93	29.73	31.87	44.33	42.07	46.13	51.53	38.73	10.00
3	**Manton Dam Recreation Area**	–12.9	1438	14.93	13.86	26.80	31.13	44.00	35.87	41.00	61.87	33.20	5.53
4	**Litchfield National Park**	–13.2	1562	2.96	2.41	8.47	10.67	14.47	17.53	21.40	31.40	40.13	28.33
5	**Litchfield National Park**	–13.2	1562	9.00	8.00	30.53	34.93	51.47	32.67	37.67	57.20	33.80	8.80
6	**Manton Dam Recreation Area**	–12.9	1438	15.07	14.57	31.87	35.47	50.07	43.80	47.67	60.60	34.53	6.13
7	**Umbrawarra Gorge Nature Park**	–14.0	1163	5.32	4.86	23.39	28.78	43.50	28.28	34.89	36.61	40.83	22.56
8	**Limmen National Park**	–15.2	754	4.79	4.50	13.42	17.04	28.54	17.10	24.60	21.13	50.38	28.75
9	**Limmen National Park**	–15.4	738	4.25	4.14	15.54	19.21	29.92	18.75	23.42	24.46	47.83	27.50
10	**Booroloola**	–16.0	855	5.54	5.46	11.14	12.52	17.52	12.57	13.95	23.10	31.05	41.10
11	**Spring Creek Station**	–16.1	917	7.21	6.64	16.38	18.29	28.05	22.29	24.86	30.38	45.57	24.24
12	**Spring Creek Station**	–16.1	904	6.82	6.61	25.10	29.10	43.71	25.76	29.86	37.38	41.10	20.76
13	**Central Arnhem Rd. 1**	–14.4	1035	4.36	8.36	14.33	19.17	30.67	14.44	21.61	21.00	38.89	40.22
14	**Central Arnhem Rd. 2**	–14.4	1014	5.14	4.79	15.05	19.00	30.24	16.52	22.94	33.24	46.90	20.14
15	**Elsey Station**	–15.2	807	4.64	4.29	12.78	15.22	24.06	18.17	24.17	23.89	62.22	13.78
16	**Nenen Station ***	–15.3	808	6.14	6.07	27.67	30.00	52.33	30.33	32.67	32.67	57.33	10.00
17	**Melville Island – Tarracumbi 1**	–11.6	1670	13.71	12.14	23.11	26.78	36.00	29.44	34.22	48.67	25.67	26.44
18	**Melville Island – Tarracumbi 2**	–11.7	1670	14.43	13.00	23.00	26.11	36.22	29.11	34.11	40.44	47.00	12.56

1 Total stand basal area (m^2 ha^{-1} of all live + dead standing stems).
2 Live stand basal area (m^2 ha^{-1}).
3 Foliage projective cover (%) (Hnatiuk *et al.* 2009).
4 Foliage cover (%) (Hnatiuk *et al.* 2009).
5 Crown cover (%) (Hnatiuk *et al.* 2009).
6 Total live photosynthetic plant material, as percentage of plot area, and as seen from above (satellite).
7 Total non-photosynthetic/senescent plant material (litter, branch, dry grass etc.), as percentage of plot area, and as seen from above (satellite).
8 Total inorganic material (bare soil, rock etc.), measured as percentage of plot area, and as seen from above (satellite).

Appendix 2. Descriptive statistics for leaf litter accumulation (Mg ha^{-1} $year^{-1}$) at 18 monitoring sites across the Northern Territory, Australia

Site	Location	Number of revisits	Mean	Median	Minimum	Maximum	SE mean	Standard deviation	Variance	Sum of squares
1	Howard Springs	5	6.3875	7.22	2.57	11.62	1.73	3.87	14.94	264.96
2	Howard Springs	5	6.7525	4.79	1.13	12.51	2.15	4.82	23.20	321.98
3	Manton Dam Recreation Area	5	5.548	5.668	3.650	8.13	0.751	1.679	2.818	165.561
4	Litchfield National Park	3	3.2485	3.49	0.78	5.54	1.38	2.39	5.71	43.42
5	Litchfield National Park	4	4.38	5.82	0.77	7.51	1.61	3.23	10.43	130.43
6	Manton Dam Recreation Area	5	6.57	7.29	3.24	11.39	1.52	3.40	11.58	262.13
7	Umbrawarra Gorge Nature Park	5	4.1975	4.102	2.780	5.019	0.412	0.922	0.850	86.247
8	Limmen National Park	8	2.0805	2.301	0.342	4.639	0.480	1.357	1.842	47.941
9	Limmen National Park	8	1.6425	1.647	0.265	2.823	0.303	0.857	0.734	26.251
10	Booroloola	7	1.825	1.679	1.090	2.947	0.265	0.701	0.491	26.262
11	Spring Creek Station	7	2.701	2.437	0.649	5.213	0.705	1.865	3.477	73.296
12	Spring Creek Station	7	3.431	2.38	0.55	8.25	1.09	2.87	8.25	132.94
13	Central Arnhem Rd. 1	5	2.427	1.932	1.331	4.436	0.561	1.254	1.572	35.763
14	Central Arnhem Rd. 2	5	3.392	3.274	1.052	5.902	0.884	1.978	3.911	73.184
15	Elsey Station	6	4.38	4.590	1.234	7.331	0.855	2.095	4.389	137.290
16	Nenen Station	1	4.015	4.015	4.015	4.015	NA	NA	NA	16.7117
17	Melville Island – Tarracumbi 1	3	5.037	4.278	4.2.5	6.631	0.797	1.380	1.904	79.951
18	Melville Island – Tarracumbi 2	3	7.1175	4.60	3.45	13.38	3.14	5.44	29.55	212.12

Appendix 3. Descriptive statistics for coarse litter accumulation (Mg ha^{-1} $year^{-1}$) at 18 sites in *Eucalyptus miniata* and *E. tetrodonta* dominated vegetation associations in the Northern Territory

Site	Location	Number of revisits	Mean	Median	Minimum	Maximum	SE mean	Standard deviation	Variance	Sum of squares
1	Howard Springs	5	0.436	0.294	0.057	1.020	0.164	0.367	0.135	1.488
2	Howard Springs	5	0.358	0.124	0.000	1.376	0.256	0.573	0.328	1.956
3	Manton Dam Recreation Area	5	0.598	0.316	0.129	1.454	0.257	0.575	0.331	3.110
4	Litchfield National Park	3	0.000	0.000	0.000	0.000	0.000	0.000	0.000	0.000
5	Litchfield National Park	4	0.141	0.000	0.000	0.566	0.141	0.283	0.080	0.320
6	Manton Dam Recreation Area	5	0.2267	0.253	0.000	0.383	0.065	0.145	0.021	0.341
7	Umbrawarra Gorge Nature Park	5	0.676	0.547	0.000	1.933	0.355	0.794	0.630	4.805
8	Limmen National Park	8	0.1228	0.068	0.000	0.501	0.060	0.169	0.029	0.321
9	Limmen National Park	8	0.257	0.118	0.000	1.039	0.127	0.360	0.130	1.438
10	Booroloola	7	0.413	0.216	0.000	1.716	0.220	0.583	0.340	3.236
11	Spring Creek Station	7	0.2025	0.152	0.054	0.459	0.061	0.161	0.026	0.442
12	Spring Creek Station	7	0.141	0.131	0.000	0.291	0.046	0.120	0.015	0.226
13	Central Arnhem Rd. 1	6	0.759	0.128	0.000	4.021	0.654	1.601	2.563	16.28
14	Central Arnhem Rd. 2	6	0.279	0.056	0.000	1.246	0.199	0.488	0.238	1.656
15	Elsey Station	4	0.1860	0.155	0.086	0.348	0.057	0.113	0.013	0.177
16	Nenen Station	1	0.000	0.000	0.000	0.000	NA	NA	NA	0.000
17	Melville Island – Tarracumbi 1	3	1.40	0.38	0.36	3.46	1.03	1.78	3.18	12.25
18	Melville Island – Tarracumbi 2	3	1.189	0.409	0.326	2.831	0.822	1.423	2.025	8.288

8

Measuring and mapping fire severity in the tropical savannas

Andrew C Edwards, Jeremy Russell-Smith and Stefan W Maier

Summary

This chapter addresses the application of, and challenges associated with, satellite-based fire severity mapping in north Australian savanna landscapes. Fire severity describes the effects of fire on vegetation and can be classified by simple semiquantitative schemata. In savannas, fire severity typically is classified as low (fires are patchy, consuming only the grass and litter layer), moderate (fires affect the woody mid-storey foliage), or high (fires affect the upper canopy). We present important applications of fire severity mapping in fire management operations, ecological assessments, and greenhouse gas emissions accounting. We first describe recent research undertaken to characterise fire severity mapping using satellite imagery, informed by substantial ground- and aerial-based calibration and validation assessments.

While current fire severity models show considerable promise, a particular challenge remains to discriminate high severity fires reliably. We then consider how available fire severity data can help inform the development of a savanna burning abatement methodology in the lower rainfall (600–1000 mm mean annual rainfall) region. We propose that fire severity in lower rainfall savanna situations is usefully described by a two-class model (not-severe, severe), and that, given current challenges with fire severity mapping, early (EDS) and late (LDS) dry season fire severity characteristics are best described using a site-based approach, in line with the methods applied in the higher rainfall methodology. In time, reliable automated satellite-based fire severity mapping processes will allow for application of spatially explicit, dynamic, and hence more flexible savanna burning emissions accounting frameworks.

Introduction

Fire severity describes the direct impact of fire on combustible material (Keeley 2009) as exhibited in immediate post-fire effects (Fig. 8.1). In north Australian savanna management and research contexts, fire severity has been applied in ecological studies (Russell-Smith and Edwards 2006; Edwards and Russell-Smith 2009; Russell-Smith *et al.* 2010),

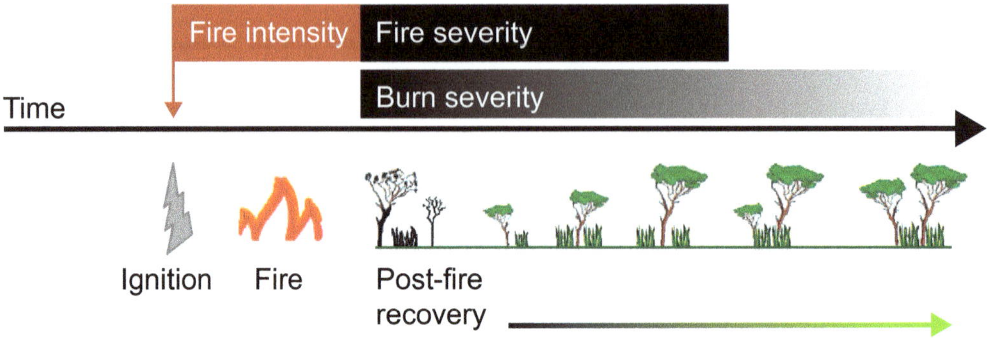

Figure 8.1. The fire continuum. The illustrations below the time line represent the fire: the ignition; the fire event; and the post-fire recovery. The boxes above the line represent the fire terminology with respect to time, where fire intensity is the measureable thermal energy emitted from the active fire; fire severity is the measureable post-fire effect on the combustible material; and burn severity is the degree of environmental change caused by fire.

models for calculating greenhouse gas emissions (Russell-Smith *et al.* 2006; Meyer *et al.* 2008), and bio-sequestration assessments (Murphy *et al.* 2009, 2010, 2013). It also has practical applications for aiding biodiversity management, for landscape fire managers to describe characteristics of individual fires and fire regimes (Edwards 2009), and as a key component of fire planning and implementation operations.

In this chapter we first describe studies that have been directed at developing automated fire severity mapping products derived from MODIS satellite imagery for public dissemination, such as those proposed for inclusion on the North Australia Fire Information (NAFI) website (http://www.firenorth.org.au). On the basis of the state of development of this fire severity mapping tool, and comprehensive field validation, we then address its potential for describing fire severity as a parameter for inclusion in savanna burning greenhouse gas (GHG) emissions abatement and bio-sequestration methodology applications.

Assessing fire severity in savanna environments

Fire severity can be quantified both horizontally and vertically, both through and within the canopy (Russell-Smith and Edwards 2006). The areal component reflects fire patchiness, while the vertical component reflects the degree of charring and scorching of photosynthetic (PV) and non-photosynthetic (NPV) plant material in different canopy strata. Patchiness also exists in the vertical dimension, but is not as amenable to optical remote sensing. A practical field guide illustrating fire severity categories for characteristic savanna woodland has been developed to provide simple metrics for characterising fire severity (Edwards 2009). It uses the scorch height (i.e. the height to non-pyrolised fire-affected foliage (Williams *et al.* 1998) and the proportion of ground patchiness. Although fire severity is a continuous variable, five severity categories are readily defined in practice (Table 8.1). A slightly enhanced system (Fig. 8.2) can be readily used by skilled operators to assess fire severity using two, three, or seven classes. This system is used extensively in north Australian aerial landscape-scale assessments, and provides the basis for calibration and validation data reported later in this chapter.

In general, studies attempting to characterise fire severity are habitat specific and/or involve a single measure of impact on vegetation. Several variables have been used to

Table 8.1. Metrics and descriptions of fire severity classes, from consultation with fire managers and researchers

Fire severity	Leaf scorch[1]	Patchiness	Fire intensity[2] (kW m^{-1})	Description
Patchy	≤2 m	Yes	<<100	A patchy fire of very low intensity; only the ground layer is affected
Low	≤2 m	No	<1000	A low intensity fire; only the ground layer is fire-affected
Moderate	>2 m	No	1000–5000	A moderate intensity fire; the mid-storey is fire-affected
High	Complete	No	5000–10 000	A high intensity fire; the upper storey is fire-affected
Extreme	All charred	No	>10 000	A high intensity fire; all the foliage is charred

[1] (Williams *et al.* 2003).
[2] (Russell-Smith and Edwards 2006).

describe severity, including minimum branch diameter (Pérez and Moreno 1998; Whight and Bradstock 1999; Keeley 2006; Williams *et al.* 2006), tree mortality (Chappell and Agee 1996), crown damage (Russell-Smith and Edwards 2006), scorch and char height (Williams *et al.* 1998), soil burn depth (Schimmel and Granstrom 1996; Chafer 2008), fuel moisture

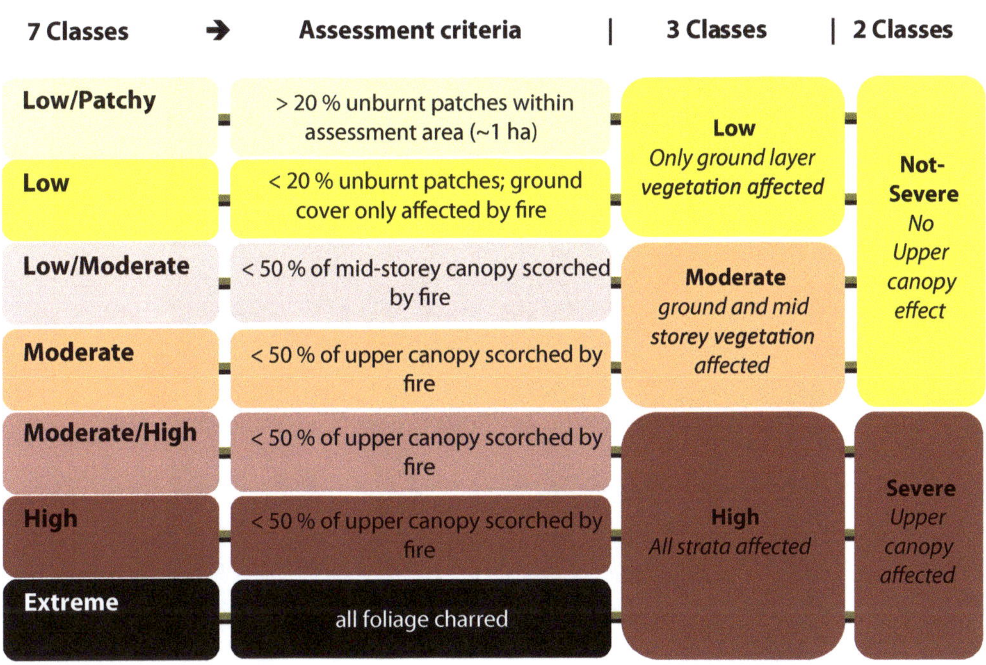

Figure 8.2. Fire severity classes (seven classes) and a description of the assessment criteria used to attribute fire effects during an aerial survey, which may be further grouped into classification schemes for ecological assessment (three classes) and a simple classification to determine general fire effects (two classes); this also has the greatest classification accuracy.

conditions (Ferguson *et al.* 2002), and white ash production (Smith and Hudak 2005; Smith *et al.* 2005). These observations can also serve the purposes of image calibration, where ground samples are used as training sites for classifying fire severity and also for validation.

Short-term persistence of fire scars under tropical savanna conditions (Edwards *et al.* 2001; Bowman *et al.* 2003; Pereira 2003) requires frequent image capture for burnt area mapping (BAM). Rapid change in albedo (Beringer *et al.* 2003) and other observations from ground sites indicate that the window for post-fire image sampling for measuring fire effects on vegetation is brief (5 to 6 days), as opposed to months or years for other biomes and sampling techniques (Hammill and Bradstock 2006; Allen and Sorbel 2008; De Santis and Chuvieco 2009; Holden *et al.* 2009). The main contributors to this are rapid vegetative re-flushing (Fig. 8.3) and leaf fall from fires affecting mid and upper storeys

Figure 8.3. Photos illustrating the change in albedo caused by leaf fall after a high severity fire. (a) 3 days and (b) 11 days after fire at the same site.

Figure 8.4. The recovery of *Cycas armstrongii* after a high severity fire: (a) after 1 day, with complete foliage removal, and (b) after 7 days, completely re-foliated.

(Fig. 8.4). Fire severity may be measured after this period but spectral responses will be affected; for image calibration purposes, temporal sampling consistency is required.

MODIS satellite sensors

Available satellite sensor data meeting the requirement of 5–6-day post-fire sampling are currently of relatively coarse scale (pixels ≥250 m). Data at this resolution from the MODIS sensors are the most cost-effective available for the whole of northern Australia (Maier and Russell-Smith 2012). The MODIS geo-location approach has reduced large on-orbit geo-location errors to better than 45 m (Qu 2006), a crucial factor for multitemporal analyses. The MODIS sensors had an expected 5-year life from 1999 (Terra) and 2002 (Aqua) (http://modis.gsfc.nasa.gov: accessed 14 February 2014). The Suomi National Polar-orbiting Partnership or Suomi NPP satellite with its visible infrared imaging radiometer suite (VIIRS) has provided an alternative since 2012. To date MODIS imagery is still being used to undertake fire severity mapping, although algorithms will need to be re-configured for VIIRS on Suomi NPP and the future satellites of the Joint Polar Satellite System (JPSS).

Methods and results

Methodological approach for MODIS-derived fire severity mapping

In preliminary analyses, significant correlations were obtained between field-based measurements and reflectance spectra averaged for the optical MODIS channels 5 (R_{1240}), 6 (R_{1640}) and 7 (R_{2130}), the normalised difference vegetation index (NDVI) and the normalised burn ratio (NBR) (Edwards 2011). The indices NDVI and NBR were able to separate severe from not-severe fire effects. R^2 for NBR was greater than R^2 for NDVI; therefore NBR was selected for assessment. A candidate set of MODIS channels, based on the significance of correlation and inference from the literature, was analysed with Akaike's information criteria (Burnham and Anderson 2002). Separation within the not-severe (low and moderate) class was inconclusive. The most parsimonious model, simply using MODIS channel 6, was selected.

A result of 94% overall accuracy was obtained in the binary classification (not-severe vs. severe) using ΔNBR. However, technical difficulties with MODIS channel 6 yielded unsatisfactory (48%) overall accuracy, with considerable overlap in discriminating low from moderate severity fire effects. A revised set of candidate models was derived incorporating MODIS channels 2 (R_{860}) and 5 (R_{1240}) and 7 (R_{2130}). The overall accuracy, 57% (2+5) and 60% (2*5*7), and class overlap improved; however, kappa $\hat{K}$ statistics (Congalton 1991) were still poor (+0.22 and +0.19).

The overlap in classification suggested that detecting fire effects in lower severity categories was challenging, unlike in severe fires which predominantly involve effects on the upper canopy that contain most of the PV. The smaller proportion of plant material in lower and mid canopies implies that the major influence on reflectance change in lower fire severity categories derives from the ground storey, with a greater proportion of NPV and soil effects. As noted previously, field categorisation of fire severity in savannas is influenced by assessment of scorch height and only a coarse representation of ground patchiness (Table 8.1). To produce more accurate calibration/validation data we determined that, for characterisation of the low and moderate fire severity categories, aerial assessment data needed to include estimates of patchiness. The aerial assessment data also required calibration from a ground-based survey having greater sampling intensity and accuracy at fine scales.

Initial calibration

An extensive calibration dataset increases the accuracy of the model and its ability to make better predictions of fire severity. Aerial transects were pre-determined to maximise intersection with mapping of fire-affected areas provided on the NAFI website. Transects were provided to a helicopter pilot via a series of waypoints. Following the published methodology, the helicopter flew at ~500 feet (150 m) above ground level at an approximate velocity of 60 kn (108 km h^{-1}) to provide maximum visibility and data point collection frequency (Edwards *et al.* 2001). Waypoints were collected using a GPS-enabled PDA Trimble Juno. The PDA was pre-loaded with a database sequence (Steventon and Liebenberg 2009) that collected a track file and waypoints. Waypoints were attributed with seven categories describing fire severity (Fig. 8.2).

In 2011, transects were sampled on foot (33 km), by ground vehicle (320 km) and by helicopter (4900 km) throughout the fire season, and across the extent of the tropical savannas (Fig. 8.5). Fire severity, including unburnt areas, was assessed at each waypoint along aerial and vehicular-based transects. Scorch height and basal area were also recorded along transects assessed on foot. The unburnt category was collected to provide an accuracy assessment of the BAM. The fire severity algorithm is applied only within fire-affected areas determined from the BAM.

A total of 8097 waypoints were collected in 2011 (Fig. 8.5). Of these, 3272 waypoints were used for fire severity calibration, while the complete dataset was used for BAM validation. Of the 3272 points, those assessed from the air were further filtered to remove points for which confidence in their accuracy was lower than 100%. This produced a high quality dataset of 1048 points incorporating substantial spatial and temporal variation and the entire range of severity classes.

Initial validation

In 2012, aerial transects (total length 2400 km) were undertaken for validation and assessment purposes in four different regions (Fig. 8.5). Waypoints (n = 3148) were intersected with the fire severity mapping. Following Congalton (1991), an error matrix was derived to determine mapping accuracy in terms of omission error (the amount missing from the mapping of a category); commission error (the amount wrong added to the mapping of a category); and overall accuracy.

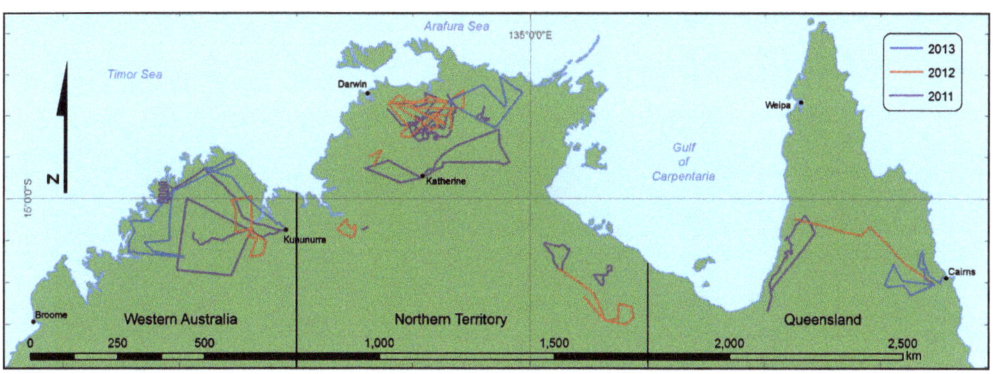

Figure 8.5. Fire severity waypoints collected along aerial transects, 2011–2013.

Table 8.2. Error matrix of the binary classification of fire severity, 2012

	Field data			
Classified image	**Not-severe**	**Severe**	**Total**	**Commission accuracy**
Not-severe	1455	361	1816	0.80
Severe	863	469	1332	0.35
Total	2318	830	3148	
Omission accuracy	0.63	0.57		**Overall accuracy 0.61**

The results of the binary classification (not-severe vs. severe) are presented in Table 8.2 to illustrate the performance of the algorithm. Overall accuracy for the 2012 fire season was found to be 61% with the major error associated with misclassification of severe fires, whereas more reliable mapping was achieved for not-severe fires. The resulting fire severity map for the whole region for 2011 is given in Fig. 8.6.

Our assessments also showed that fire severity mapping accuracy in the EDS was far higher than in the LDS. This is counter to accuracy assessments undertaken of fire extent mapping, where EDS fires, being smaller in extent and generally patchier, consume much less of the biomass, have less impact on the reflectance signal, and therefore are more difficult to detect. In contrast, while LDS fire-affected areas tend to be more extensive, there is typically much greater variability in combustion completeness and thereby fire severity (e.g. Russell-Smith and Edwards 2006). Increasing deciduousness and curing of ground-storey vegetation, especially in the latter part of the dry season, can add to this complexity by reducing the reflectance contrast between lower and more severely burnt areas. Given that fire severity mapping takes the fire extent mapping product as its spatial basis, it follows that not-severe fires are more reliably mapped in the EDS, whereas there is greater fire severity complexity later in the dry season.

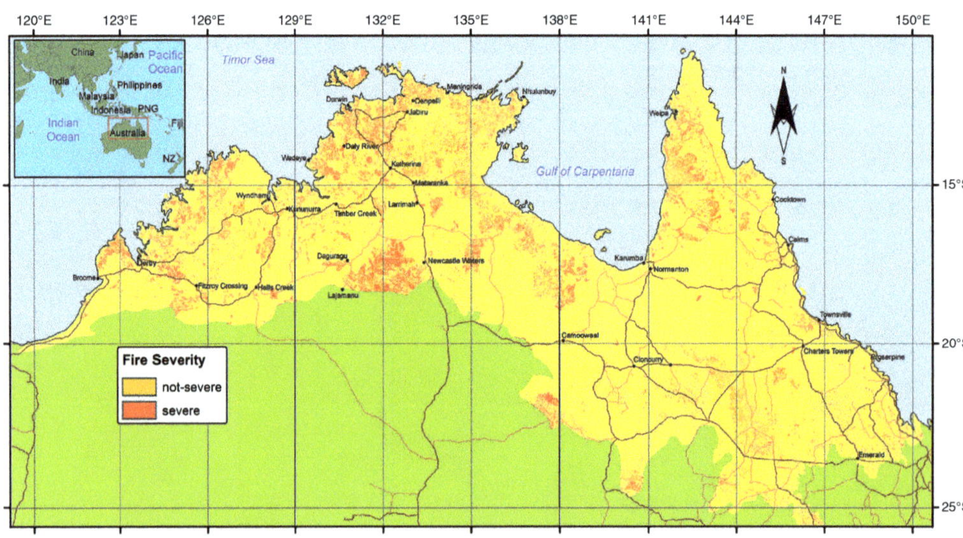

Figure 8.6. MODIS-derived fire severity mapping for northern Australian savannas, for 2012. (Fire severity: not-severe = upper canopy not fire-affected; severe = upper canopy fire-affected).

Current algorithm

On the basis of the preceding work, key current methodological improvements have included:

- replacing the automated burnt area mapping with the NAFI fire extent mapping monthly product. This improves the visual quality of the map and provides the flexibility to vary the post-fire period of image acquisition.
- calculating reflectance change from images 4 to 7 days apart where possible
- calculating reflectance changes as a proportional change, rather than absolute, to reduce the effects of topography, substrate and pre-fire PV and NPV vegetative cover.

New calibration

All spatial analyses were undertaken as raster files with 250 m pixels in ArcInfo Spatial Analyst (ESRI 2011). A calibration analysis of the combined 2011–12 field data was used to produce a new algorithm. The calibration data points (CDPs) were converted to raster (cell size = 250 m), and attributed uniquely with the date of acquisition. MODIS raster images ($n = 792$) of channels 1, 2, 5 and 7 for 2012 were clipped using the NAFI BAM. The proportional reflectance change (%RC) was calculated for each CDP:

$$\%RC = (\text{reflectance\_}t - \text{reflectance\_}t{+}1) / \text{reflectance\_}t) * 100 \quad (1)$$

where t and t+1 equal the earlier and later image dates, respectively.

Threshold values of MODIS channel 2 for a 3 class severity classification were used. The proportional change values for each class are: Low = 1 to 18%; Moderate = 19 to 30%; High = 31–51%. Negative values and values greater than 51% usually represented cloud cover or cloud shadow in either of the images and were classified as zero in the reclassification. Each image was masked for a given month/year by the NAFI BAM. The maximum proportional reflectance change value in the series of results for a given month was attributed for each pixel. The months were then combined to create the fire severity map for each year.

Validation

The 2013 fire severity mapping was derived from an algorithm using the calibration from the 2011–12 data. An error matrix was derived from the 2013 aerial validation data (Table 8.3 and Fig. 8.5). Overall classification accuracy improved slightly from 61% in 2012 (Table 8.2), to 65% in the lower rainfall and 76% in the higher rainfall regions in 2013 (Table 8.3). The majority of fires assessed in 2013 were LDS fires. This assessment demonstrates that high severity fires are more difficult to classify in the LDS. Nevertheless, the accuracy of fire severity mapping, although not yet in the realms of simple burnt area mapping (see Chapter 3), shows considerable promise and already approaches adequate levels of reliability, especially under higher rainfall conditions.

Calibration and validation are ongoing processes, as demonstrated, with incremental improvements in the 2013 severity mapping results. The 2013 field data will be collated with the 2011–12 data to improve the fire severity algorithm further. Calibration to date has relied upon threshold reflectance change values in the near-infrared. Previous research has demonstrated that the near-infrared does not capture all of the variability required to classify severity (Pereira 2003; De Santis *et al.* 2010). However, as the improvement in

Table 8.3. Error matrix for (a) the lower and (b) the higher rainfall regions for fire severity mapping in northern Australia, for 2013

	Field data			
Mapping data	**Not-severe**	**Severe**	**Total**	**Commission accuracy**
(a) Lower rainfall regions				
Not-severe	84	39	123	0.68
Severe	75	129	204	0.63
Total	159	168	327	
Omission accuracy	0.53	0.77		Overall accuracy **0.65**
(b) Higher rainfall regions				
Not-severe	60	13	73	0.82
Severe	47	130	177	0.74
Total	107	143	250	
Omission accuracy	0.56	0.91		Overall accuracy **0.76**
Total accuracy of (a) and (b) = 0.72				

mapping accuracy from 2012 to 2013 has shown, continued expansion of the empirical calibration and validation datasets will potentially make enhanced detection more feasible. This applies particularly to the 750 000 km^2 lower rainfall region where currently the available calibration dataset stands at a modest 2644 points.

Seasonality of fire severity

As described previously, work is ongoing to develop a remotely sensed fire severity mapping product that discriminates severe fires reliably. However, for the purposes of characterising the seasonality of fire severity for application in the lower rainfall (600–1000 mm) savanna burning abatement methodology, potentially we could use the substantial aerial survey validation dataset ($n = 2644$) assembled over the period 2011–2013 as an alternative. These data describe the proportion of low, moderate and high severity fires sampled in both the higher (>1000 mm) and lower (600–1000 mm) mean annual rainfall regions in EDS and LDS periods (Table 8.4). The data indicate that in the lower rainfall region there is little observed difference in the proportion of not-severe fires between seasons (64% early, 60% late), whereas the proportion of low severity fires in the LDS is about double that in the EDS (Table 8.4). The appropriateness of applying these aerial validation data for the purposes of describing seasonal fire severity distributions is considered in the Discussion.

Discussion

Fire severity mapping

Fire severity mapping products have a variety of practical management and applied research applications. Such products currently require the development of a reliable remotely sensed model incorporating simple thresholds for describing ranges of the proportional change in reflectance of fire-affected areas in the near-infrared. Given rapid initial changes in fire effects, consistent temporal sampling of images is also required. Such models can perform well in comparison with other fire severity modelling approaches (Chafer 2008; French *et al.* 2008; De Santis *et al.* 2010; Holden *et al.* 2010; Veraverbeke and

Table 8.4. Seasonal proportions (%) of fires occurring in fire severity classes (low, moderate or high) for (a) higher (>1000 mm) and (b) lower (600–1000 mm) rainfall savanna regions, derived from different assessment sources as indicated

	Fire season					
	Early dry season (EDS)			Late dry season (LDS)		
	Low	Moderate	High	Low	Moderate	High
(a) Higher rainfall region						
Fire severity mapping (472 450 km^2, 2009–2013)[a]	40	29	31	31	26	43
Aerial validation assessments (17 487 km transects flown collecting 2925 applicable waypoints)[a]	36	40	24	22	37	41
(b) Lower rainfall region						
Fire severity mapping (744 689 km^2, 2009–2013)[a]	36	30	34	36	29	35
Aerial validation assessments (3243 km transects flown collecting 2644 applicable waypoints)[a]	28	37	35	14	46	40

[a] this study.

Hook 2013), even compared with mapping derived from higher pixel resolution imagery (e.g. Landsat). In savanna management contexts, fire severity mapping is particularly useful for determining the efficacy of EDS prescribed burning activities, especially since these are of typically lesser severity. As described here, a particular challenge under seasonally changing savanna vegetation structural conditions concerns reliably differentiating relatively severe fires.

Fire severity mapping in north Australian savannas has thus far utilised two processes. As described here, a first method has involved manual selection of pre- and post-fire images 4–7 days apart from an archive, followed by automated procedures for masking, calculating proportional change, and then merging mapping products for desired mapping intervals (e.g. by months, years). A second procedure fully automates the image acquisition and difference imaging process using the first post-fire image, determined from an automated process (Maier and Russell-Smith 2012), and then calculates the difference image solely for burnt pixels. This method has demonstrated similar reliability with far less manual effort; however, it requires a substantial amount of highly valuable high-end processing.

Fire severity mapping studies conducted in other landscape conditions and vegetation types have similarly reported difficulties in defining thresholds between classes, and overlap in mapping of moderate and high severity classes (Key and Benson 2006; Smith *et al.* 2007; Boer *et al.* 2008; French *et al.* 2008; Kasischke *et al.* 2008; De Santis *et al.* 2010). In the longer term, the reliability of regional fire severity mapping products is achievable through ongoing iterative refinements to regional models informed by independent validation and calibration data.

Two or three severity classes?

Sparse foliage (i.e. <10% canopy cover) provides minimal pre-fire detectable PV material. Therefore its post-fire removal is often not detectable, thus reducing mapping accuracy as demonstrated. Vegetation structure in the lower rainfall region falls predominantly into this category, with distinct ground and low open tree/shrub layers. Therefore, we suggest it is preferable to apply a classification that more appropriately describes fire effects in this manner with a binary, severe versus not-severe, classification. The binary classification also has the advantage of being more accurate; Fig. 8.2.

Application to savanna burning methodologies

Outputs of the fire severity mapping model (2009–13) reported here for both higher rainfall and lower rainfall regions imply that seasonal distributions of fire severity are broadly similar for both (Table 8.4). Assuming a two-class fire severity model for the lower rainfall region (see above), 65% of mapped fires in both EDS and LDS were of lesser (low + moderate) severity. These proportions are similar also to corresponding mean values (64% and 60% respectively) derived from independent aerial validation data (2011–13) for the lower rainfall region (Table 8.4). While this level of correspondence between model and validation fire severity class proportions is reassuring, we note from previous discussion that reliable mapping of high severity fires remains an ongoing technical challenge. Therefore, for present purposes we consider that the seasonal proportions of fire severity classes derived from available aerial validation data should not be applied to the proposed lower rainfall savanna burning emissions abatement methodology until these errors have been further reduced. Instead, consistent with the process of characterising the proportions of each fire severity class seasonally for the higher rainfall savanna burning emissions abatement methodology, we suggest a site-based assessment.

Utility of fire severity data for the greenhouse gas emissions abatement methodology

The seasonal distribution of fire severity classes derived from decadal observations (1995–2004; Table 8.5a) of fire effects at permanent sandstone monitoring plots ($n = 79$) in Kakadu and Nitmiluk national parks (KNNP; Russell-Smith and Edwards 2006) is manifestly different from that observed in both fire severity mapping and associated validation data for the higher rainfall savanna region (compare with data in Table 8.4). In particular, EDS fires in the KNNP dataset are markedly less severe. Significantly, the KNNP sandstone plot data inform the parameterisation of fire severity seasonality as applied in the higher rainfall savanna burning abatement methodology (DCCEE 2013; Russell-Smith *et al.* 2009). Broadly equivalent seasonal fire severity class distributions derived from plot-based observations are observed from analysis of the same set of KNNP sandstone plots for the subsequent nine-year period, 2005–13 (Table 8.5b), and lower rainfall zone plots ($n = 267$, refer Chapter 6 for details) over the 5-year period 2009–13 (Table 8.5c).

Vertical and horizontal fire effects on vegetation are readily assessed in plot-based assessments (Russell-Smith and Edwards 2006). By contrast, the aerial validation assessments reported in this study are relatively spatially heterogeneous (both horizontally and vertically), where individual assessment waypoints integrate fire effects over an area approximating one MODIS pixel (~6 ha). Further, whereas plot observations effectively provide a random sample, for sampling efficiency and cost-effectiveness the aerial assessment relies on making multiple observations of fire severity within large (hence likely relatively severe) fire-affected areas identified in advance from the NAFI website. As such,

Table 8.5. Seasonal proportions (%) of fires in respective severity classes observed in (a) KNNP sandstone and lowland plots, 1995–2004, (b) KNNP sandstone and lowland plots, 2005–2013; (c) lower rainfall plots, 2009–2013 (refer Chapter 6)

Season severity	Low	Moderate	High
(a) Higher rainfall (KNNP sandstone) 1995–2004			
EDS	72	20	8
LDS	19	47	34
(b) Higher rainfall (KNNP sandstone) 2005–2013			
EDS	77	22	1
LDS	15	66	19
(c) Lower rainfall 2009–2013			
EDS	73	25	2
LDS	13	47	40

while the two assessment methods provide qualitatively different observations, both are valid when applied to the specific purposes they are designed for.

Conclusion

This chapter demonstrates the need and potential for incorporating remotely sensed, validated, savanna-wide assessments of fire severity in future iterations of savanna burning methodologies at such time as models are deemed sufficiently robust to discriminate severe fire effects effectively. Whereas currently fire severity is inferred from seasonality, its replacement with reliable fire severity mapping will simplify the accounting process.

References

Allen JL, Sorbel B (2008) Assessing the differenced normalized burn ratio's ability to map burn severity in the boreal forest and tundra ecosystems of Alaska's national parks. *International Journal of Wildland Fire* **17**, 463–475. doi:10.1071/WF08034

Beringer J, Hutley LB, Tapper NJ, Coutts A, Kerley A, O'Grady AP (2003) Fire impacts on surface heat, moisture and carbon fluxes from a tropical savanna in northern Australia. *International Journal of Wildland Fire* **12**, 333–340. doi:10.1071/WF03023

Boer MM, Macfarlane C, Norris J, Sadler RJ, Wallace J, Grierson PF (2008) Mapping burned areas and burn severity patterns in SW Australian eucalypt forest using remotely-sensed changes in leaf area index. *Remote Sensing of Environment* **112**, 4358–4369. doi:10.1016/j.rse.2008.08.005

Bowman DMJS, Zhang Y, Walsh A, Williams RJ (2003) Experimental comparison of four remote sensing techniques to map tropical savanna fire-scars using Landsat-Tm imagery. *International Journal of Wildland Fire* **12**, 341–348. doi:10.1071/WF03030

Burnham KP, Anderson DR (2002) *Model Selection and Multi-Model Inference: A Practical Information Theoretic Approach*. Springer: New York.

Chafer CJ (2008) A comparison of fire severity measures: an Australian example and implications for predicting major areas of soil erosion. *Catena* **74**, 235–245. doi:10.1016/j.catena.2007.12.005

Chappell CB, Agee JK (1996) Fire severity and tree seedling establishment in *Abies magnifica* forests, Southern Cascades, Oregon. *Ecological Applications* **6**, 628–640. doi:10.2307/2269397

Congalton RG (1991) A review of assessing the accuracy of classifications of remotely sensed data. *Remote Sensing of Environment* **37**, 35–46. doi:10.1016/0034-4257(91)90048-B

DCCEE (Australian Government Department of Climate Change and Energy Efficiency) (2013) Carbon Credits (Carbon Farming Initiative) (Reduction of Greenhouse Gas Emissions through Early Dry Season Savanna Burning—1.1) Methodology Determination 2013. Carbon Credits (Carbon Farming Initiative) Act 2011. Federal Register of Legislative Instruments F2013L01165. Parliamentary Secretary for Climate Change Industry and Innovation. Australian Government: Canberra.

De Santis A, Chuvieco E (2009) Geocbi: a modified version of the composite burn index for the initial assessment of the short-term burn severity from remotely sensed data. *Remote Sensing of Environment* **113**, 554–562. doi:10.1016/j.rse.2008.10.011

De Santis A, Asner GP, Vaughan PJ, Knapp DE (2010) Mapping burn severity and burning efficiency in California using simulation models and Landsat imagery. *Remote Sensing of Environment* **114**, 1535–1545. doi:10.1016/j.rse.2010.02.008

Edwards AC (2009) Fire Severity Categories for the Tropical Savanna Woodlands of Northern Australia. Bushfire Cooperative Research Centre: Melbourne.

Edwards AC (2011) I Spy with My Little Eye Something Beginning With … Fire. PhD Thesis. Faculty of Engineering, Health, Science and the Environment. Charles Darwin University: Darwin, Northern Territory.

Edwards A, Russell-Smith J (2009) Ecological thresholds and the status of fire-sensitive vegetation in western Arnhem Land, northern Australia: implications for management. *International Journal of Wildland Fire* **18**, 127–146. doi:10.1071/WF08008

Edwards A, Hauser P, Anderson M, McCartney J, Armstrong M, Thackway R, Allan G, Hempel C, Russell-Smith J (2001) A tale of two parks: contemporary fire regimes of Litchfield and Nitmiluk National Parks, monsoonal northern Australia. *International Journal of Wildland Fire* **10**, 79–89. doi:10.1071/WF01002

ESRI (2011) Arcgis Desktop: Release 10. Environmental Systems Research Institute: Redlands, CA.

Ferguson SA, Ruthford JE, McKay SJ, Wright D, Wright C, Ottmar R (2002) Measuring moisture dynamics to predict fire severity in longleaf pine forests. *International Journal of Wildland Fire* **11**, 267–279. doi:10.1071/WF02010

French NHF, Kasischke ES, Hall RJ, Murphy KA, Verbyla DL, Hoy EE, Allen JL (2008) Using Landsat data to assess fire and burn severity in the North American boreal forest region: an overview and summary of results. *International Journal of Wildland Fire* **17**, 443–462. doi:10.1071/WF08007

Hammill KA, Bradstock RA (2006) Remote sensing of fire severity in the Blue Mountains: influence of vegetation type and inferring fire intensity. *International Journal of Wildland Fire* **15**, 213–226. doi:10.1071/WF05051

Holden ZA, Morgan P, Evans JS (2009) A predictive model of burn severity based on 20-year satellite-inferred burn severity data in a large southwestern US wilderness area. *Forest Ecology and Management* **258**, 2399–2406. doi:10.1016/j.foreco.2009.08.017

Holden ZA, Morgan P, Smith AMS, Vierling L (2010) Beyond Landsat: a comparison of four satellite sensors for detecting burn severity in ponderosa pine forests of the Gila Wilderness, NM, USA. *International Journal of Wildland Fire* **19**, 449–458. doi:10.1071/WF07106

Kasischke ES, Turetsky MR, Ottmar RD, French NHF, Hoy EE, Kane ES (2008) Evaluation of the composite burn index for assessing fire severity in Alaskan black spruce forests. *International Journal of Wildland Fire* **17**, 515–526. doi:10.1071/WF08002

Keeley JE (2006) Fire severity and plant age in postfire resprouting of woody plants in sage scrub and chaparral. *Madroño* **53**, 373–379.

Keeley JE (2009) Fire intensity, fire severity and burn severity: a brief review and suggested usage. *International Journal of Wildland Fire* **18**, 116–126. doi:10.1071/WF07049

Key CH, Benson NC (2006) Landscape Assessment (LA) sampling and analysis methods; Composite Burn Index field methods. FIREMON—Fire Effects Monitoring and Inventory Protocol. System for Environmental Management: Montana, USA. www.fire.org/firemon

Maier SW, Russell-Smith J (2012) Measuring and monitoring of contemporary fire regimes in Australia using satellite remote sensing. In *Flammable Australia: Fire Regimes, Biodiversity and Ecosystems in a Changing World.* (Eds RA Bradstock, AM Gill, RJ Williams) pp. 79–95. CSIRO Publishing: Melbourne.

Meyer MC, Russell-Smith J, Murphy B, Cook GD, Edwards A, Yates C, Schatz J, Brocklehurst P, Luhar A, Mitchell R, Parry D (2008) Accounting and Verification of Greenhouse Gas Emissions from Fire Management Programs in Northern Australia. International Bushfire Research Conference 2008 – incorporating The 15th annual AFAC Conference, Adelaide, South Australia, Australia.

Murphy BP, Russell-Smith J, Watt FA, Cook GD (2009) Fire management and woody biomass carbon stocks in mesic savannas. In *Culture, Ecology and Economy of Fire Management in North Australian Savannas: Rekindling the* Wurrk *Tradition.* (Eds J Russell-Smith, P Whitehead, P Cooke), pp. 361–378. CSIRO Publishing: Melbourne.

Murphy BP, Russell-Smith J, Prior LD (2010) Frequent fires reduce tree growth in northern Australian savannas: implications for tree demography and carbon sequestration. *Global Change Biology* **16**, 331–343. doi:10.1111/j.1365-2486.2009.01933.x

Murphy BP, Lehmann CER, Russell-Smith J, Lawes MJ (2013) Fire regimes and woody biomass dynamics in northern Australian savannas. *Journal of Biogeography* **41**, 133–144.

Pereira JMC (2003) Remote sensing of burned areas in tropical savannas. *International Journal of Wildland Fire* **12**, 259–270. doi:10.1071/WF03028

Pérez B, Moreno JM (1998) Methods for quantifying fire severity in shrubland-fires. *Plant Ecology* **139**, 91–101. doi:10.1023/A:1009702520958

Qu JJ (2006) Modis geolocation. In *Earth Science Satellite Remote Sensing: Science and Instruments.* (Eds JJ Qu, W Gao, M Kafatos, RE Murphy, VV Salomonson) Chapter 4. Tsinghua University Press and Springer: Beijing and Berlin.

Russell-Smith J, Edwards AC (2006) Seasonality and fire severity in savanna landscapes of monsoonal northern Australia. *International Journal of Wildland Fire* **15**, 541–550. doi:10.1071/WF05111

Russell-Smith J, Edwards AC, Cook GD, Schatz J, Brocklehurst P (2006) Estimating greenhouse gas emissions from savanna burning in northern Australia. Fifth International Conference on Forest Fire Research, Figuiera da Foz, Portugal *Forest Ecology and Management* **234** (Supplement 1), S152. doi:10.1016/j.foreco.2006.08.203

Russell-Smith J, Murphy BP, Meyer CP, Cook GD, Maier S, Edwards AC, Schatz J, Brocklehurst P (2009) Improving estimates of savanna burning emissions for greenhouse accounting in northern Australia: limitations, challenges, applications. *International Journal of Wildland Fire* **18**, 1–18. doi:10.1071/WF08009

Russell-Smith J, Price O, Murphy BP (2010) Managing the matrix: decadal responses of eucalypt-dominated mesic savanna to ambient fire regimes *Ecological Applications* **20**, 1615–1632. doi:10.1890/09-1553.1

Schimmel J, Granstrom A (1996) Fire severity and vegetation response in the boreal Swedish forest. *Ecology* **77**, 1436–1450. doi:10.2307/2265541

Smith AMS, Hudak AT (2005) Estimating combustion of large downed woody debris from residual white ash. *International Journal of Wildland Fire* **14**, 245–248. doi:10.1071/WF05011

Smith AMS, Wooster MJ, Drake NA, Dipotso FM, Falkowski MJ, Hudak AT (2005) Testing the potential of multi-spectral remote sensing for retrospectively estimating fire severity in African savannahs. *Remote Sensing of Environment* **97**, 92–115. doi:10.1016/j.rse.2005.04.014

Smith AMS, Lentile LB, Hudak AT, Morgan P (2007) Evaluation of linear spectral unmixing and ΔNBR for predicting post-fire recovery in a North American ponderosa pine forest. *International Journal of Remote Sensing* **28**, 5159–5166. doi:10.1080/01431160701395161

Steventon J, Liebenberg L (2009) Cybertracker V 3.181.

Veraverbeke S, Hook SJ (2013) Evaluating spectral indices and spectral mixture analysis for assessing fire severity, combustion completeness and carbon emissions. *International Journal of Wildland Fire* **22**, 707–720. doi:10.1071/WF12168

Whight S, Bradstock R (1999) Indices of fire characteristics in sandstone heath near Sydney, Australia. *International Journal of Wildland Fire* **9**, 145–153. doi:10.1071/WF00012

Williams RJ, Gill AM, Moore PHR (1998) Seasonal changes in fire behaviour in a tropical savanna in northern Australia. *International Journal of Wildland Fire* **8**, 227–239. doi:10.1071/WF9980227

Williams RJ, Gill AM, Moore PHR (2003) Fire behavior. In *Fire in Tropical Savannas: The Kapalga Experiment.* (Eds AN Andersen, GD Cook, RJ Williams) pp. 33–46. Springer: New York.

Williams RJ, Wahren C-H, Bradstock RA, Müller WJ (2006) Does alpine grazing reduce blazing? A landscape test of a widely-held hypothesis. *Austral Ecology* **31**, 925–936. doi:10.1111/j.1442-9993.2006.01655.x

9

Biomass combustion and emission processes in the northern Australian savannas

CP (Mick) Meyer and Garry D Cook

Summary

The savanna woodlands of northern Australia contribute annually ~3% of Australia's net accountable greenhouse gas emissions. In recent years the development of fire management techniques based on the traditions of the Indigenous land owners has demonstrated that these emissions can be reduced by carefully planned and implemented prescribed burning. This led to the development of a certified methodology for claiming tradable carbon credits under Australia's Carbon Farming Initiative (CFI). The methodology was applicable only to savannas with a mean annual rainfall greater than 1000 mm. Greenhouse gas emission rates are determined by fuel composition; the fine fuels in this region are predominantly tree litter, with grasses dominating in only a few vegetation communities, and therefore the accounting methodology is suitable only for litter-dominated vegetation communities. However, there are extensive areas of the savanna woodlands in low rainfall regions where grasses dominate the fine fuels. Skilled fire management in these regions can also lead to reduced greenhouse gas emissions; however, quantifying the emissions reductions with the accuracy required for certifiable carbon credits requires the development of new accounting methodologies. A critical component of the process is evaluating the parameters of the accounting algorithm. This chapter reviews the procedures for assessing emission factors and the assumptions and pitfalls involved in assigning accurate values to a parameter set to produce project-based inventory methodologies suitable for emission accounting of low rainfall savanna fires.

Introduction

This chapter aims to recommend a set of savanna burning emission factors appropriate for the low rainfall savanna region of northern Australia. It extends the study in the high rainfall savanna in western Arnhem Land (Meyer *et al.* 2012) which underpins the high rainfall savanna accounting methodology (Russell-Smith *et al.* 2009; Commonwealth of Australia 2013).

Biomass combustion products are major determinants of atmospheric chemistry, and therefore understanding the processes and rates of their production has occupied the attention of atmospheric scientists for more than 40 years. The generic description of the production rate is:

$$\text{Emission} = \text{Fuel burned} \times \text{Emission Factor} \quad (1)$$

The mass of fuel burned is determined by the fuel loads, combustion completeness, patchiness, and fire severity. The fuel load is the net result of vegetation dynamics and biological decomposition. These components are discussed at length in Russell-Smith *et al.* (2009) in the context of the high rainfall Australian savannas, and in other chapters of this book in the context of the low rainfall savannas. The emission factor summarises the combustion process itself; this single term represents all the aspects of combustion chemistry and the factors that control it. When developed and published by Seiler and Crutzen (1980), the purpose of Eqn 1 was to estimate average regional production rates of the trace gases and particulates important for the global and regional atmospheric budgets. To define the scale of the regional and global issues, they focused on aggregate emissions and on large-area averages rather than on the detailed combustion chemistry of the fuel constituents and the processes that control them. The importance of smoke from fires for regional pollution and greenhouse gas budgets is now well established and is reported at length in the international literature and, in the case of greenhouse gases, in the IPCC assessment reports. The focus has now moved to management and mitigation of the emissions following discussion of the issues by Cook (2003), and for that we need to look more deeply into the nature of emission factors and, particularly, the assumptions implicit in them.

There are two seminal reviews of emission factors published in the global literature: the review of Andreae and Merlet (2001) that comprehensively summarised the data to the date of publication, and the review of Akagi *et al.* (2011), which covers emissions close to the source mostly published since 2001. Both reviews are aimed at global emission determination and therefore reduce the land surface to six broad classes: savanna, forest, tropical forest, extra-tropical forest (boreal forest and temperate forest), crop residue and peat, supplemented with several other industrial and domestic biomass combustion categories. Much of the new work in the last decade has addressed the trace volatile organic species, organic and inorganic nitrogen species and particulates with, in some cases, significant revision to emission factors. Nitrous oxide has mostly been neglected.

The recommended emission factors for the greenhouse gases, CO_2, CH_4 and N_2O range, respectively, from 1489 to 1686, 1.9 to 6.8, and 0.07 to 0.41 g (kg fuel consumed)$^{-1}$, (Table 9.1) or expressed as fractions of fuel elemental content, ~81–91% fuel C as CO_2, 0.29–1.02% fuel C as CH_4 and 0.89–5.5% fuel nitrogen as N_2O. These reviews remain the primary source of emission factors for inventory accounting, but due to their high level of aggregation should be applied with caution to regional and project accounting. To explain why, we first need to review how combustion progresses in the field from ignition to extinction, and the methods used to measure emission factors.

Biomass combustion mechanics

In their review of biomass combustion processes that occur in the field, Sullivan and Ball (2012) emphasise that biomass mostly comprises cellulose, with some additives such as lignin, other carbohydrates, terpenes, protein and lipids. For the most part, the emitted

Table 9.1. Comparison between recommended emission factors from two recent reviews of the global literature, Andreae and Merlet (2001) and Akagi *et al.* (2011)

		Emission factor (g (kg fuel^{-1}))					
	Trace species	Savanna	Tropical forest	Extratropical forest	Temperate forest	Boreal forest	Agricultural residues
Andreae and Merlet (2001)	CO_2	1613	1580	1569	–	–	1515
	CO	65	104	107	–	–	92
	CH_4	2.3	6.8	4.7	–	–	2.7
	NMHC[1]	3.4	8.1	5.7	–	–	7
	N_2O	0.21	0.2	0.26	–	–	0.07
	PM2.5[2]	5.4	8.1	13	–	–	3.9
Akagi *et al.* (2011)	CO_2	1686	1643	1509	1637	1489	1585
	CO	63	93	122	89	127	102
	CH_4	1.94	5.07	5.68	3.92	5.68	5.83
	NMHC	2.54	3.26	–	3	6.08	3.81
	NMOC[3]	24.7	51.9	54	23.7	58.7	51.4
	N_2O	–	–	0.38	0.16	0.41	–
	PM2.5	7.17	9.1	15	12.7	15.3	6.26

1 Non-methane hydrocarbons.
2 Particulate matter less than 2.5 μm diameter.
3 Non-methane organic compounds.

organic products are from cellulose combustion. Hence, much of the variation in combustion products between fires is likely to be due to the way in which the fuel burns rather than the composition of the fuel.

To paraphrase Byram (1959) in Sullivan and Ball (2012), open burning of biomass has three distinct phases: (1) pre-ignition when thermal decomposition starts to produce volatile products once fuels are heated and dried; (2) flaming combustion of the volatilised gases; and (3) smouldering combustion of the char residues. Preheating is caused by heat transfer from adjacent burning fuel, by both radiation and flame contact.

Thermal decomposition proceeds through two competing processes: (1) volatilisation of the cellulose, largely via levoglucosan formation, followed by chemical cracking of the volatiles to yield small molecular weight gas-phase fuels, and (2), a wide range of other chemical reactions including chemical dehydration, decarboxylation and decarbonylation, to produce low molecular weight gases and solid char (Sullivan and Ball 2012). The former process is endothermic and the latter is exothermic. The balance between the two processes, plus the heat transfer from adjacent burning fuels, determines whether the thermal decomposition is sustainable and hence whether combustion proceeds.

There are also two classes of combustion reactions: gas phase oxidation by air of the gas-phase products of thermal decomposition, and solid phase oxidation by air of char. Both processes are exothermic and both occur during the flaming and smouldering combustion phases but to different extents: flaming combustion is sustained by the turbulent convection at the flame front which ensures sufficient oxygen for the gas-phase oxidation reactions. After this diminishes, the solid-phase oxidation continues to dominate the smouldering combustion phase.

A combination of physical and chemical properties determines the nature and extent of the combustion reaction. The size of the fuel elements and the structure of the fuel bed influence the rate of heat transfer during preheating; the production rate of gas-phase decomposition products coupled with the entrainment of air determines the rate and extent of flaming combustion. The net effect of the processes is well known: dry grasses burn fiercely because their open structure and thin leaf blades favour efficient heat transfer from burning to unburned fuel elements and allow rapid mixing of air into the burning matrix. Large dead branches and logs burn by flaming combustion only as long as there is a sufficient heat input from adjacent burning fuels or coal beds to sustain significant production of volatile gases. As this diminishes, smouldering combustion begins to dominate until the balance between the two pathways of thermal decomposition shifts to the endothermic reactions and combustion ceases to be sustainable. Combustion of peat and duff is limited by the air supply, rather than by heat transfer, and consequently these fuels tend to smoulder for very long periods.

Wind direction relative to the progression of the fire front controls preheating and therefore is a well known determinant of fire intensity. In controlled environments, (particularly in wind-tunnels), it is possible to create, selectively, heading, flanking and backing fires; however, in the field all three classes occur simultaneously. The combustion chemistry of these fire classes is distinct, and, while not conclusively demonstrated to date, probably leads to differences in emission factors.

Therefore the nature of the combustion is possibly more important than the chemical composition of the fuel, and combustion characteristics are critical to how we interpret emission factor measurements.

Measuring emission factors

Emission factors are almost invariably derived from emission ratios of combustion gases. The differences between the studies are the sampling techniques. Measurement programs include (1) laboratory studies where fuels collected from the field are burned under controlled conditions and the combustion gases are, where possible, analysed in real time; (2) aircraft sampling of smoke plumes from either prescribed fires lit for ecosystem management objectives or uncontrolled wildfires; and (3) ground-level sampling. Laboratory-based sampling allows the deployment of complex analysers under optimal operating conditions; however, it is a challenge to recreate both a fuel bed of similar composition and structure and a fire behaviour with properties similar to a field burn. There are relatively few facilities where this has been achieved with success; these include the US Fire Service facilities at Missoula (e.g. Christian *et al.* 2004), burn rooms and huts, which were usually developed to study toxic emissions (e.g. dioxins, Gullett and Touati 2003), and wind-tunnels (e.g. the CSIRO Pyrotron, Sullivan *et al.* 2013).

Ground-based sampling has ranged from (1) flask or grab sampling of smoke either from flaming or smouldering sources, or from smoke eddies mixed by turbulence from the rising plume to ground level (e.g. Hurst *et al.* 1994a, 1994b); (2) continuous smoke sampling from masts 6 m or longer to monitor the progression of the fire (e.g. Ward and Radke 1993; Meyer *et al.* 2004, 2009); (3) sampling from targeted sources, either from specific fuel elements or the flaming fire front (Meyer *et al.* 2012) or by open-path spectrometry (Paton-Walsh *et al.* 2014; Smith *et al.* 2014). Mostly the physical smoke samples are transported to a laboratory for subsequent analysis, but a few parameters can be measured with sufficient precision using portable monitors to provide direct measures of smoke properties at the fire ground. These include CO_2, CO, and sometimes particulate matter (PM). A range of organic species can be measured continuously with Fourier transform infrared spectrometry (FTIR). This approach is relatively simple logistically, and bridges the gap between the detailed control that is possible in the laboratory and the uncontrolled complexities of fire in the field. However, in the field, the closer you are to the source then the greater is the heterogeneity of the smoke.

To avoid the problems posed by heterogeneity of the smoke, large-scale sampling from aircraft of the rising smoke plume has been the preferred method for measuring the smoke composition (e.g. Hurst *et al.* 1994a, 1996). Sampling from aircraft also has its limits; it is expensive, complex and, because of the high speed of the aircraft, requires extensive plumes, and hence relatively large fires. Smaller experimental and prescribed burns pose a challenge. Also, because the plume is generated mostly at the fire front, smoke emitted from slow combustion of logs, which is a large source of some smoke constituents, is not measured.

Laboratory methods allow detailed studies of the chemistry, but rely on the assumption that the fire properties in the laboratory represent combustion properties in the field. Field measurements, because they sample real fires, avoid this assumption, but ground sampling must address the issue of heterogeneity in the smoke, while aircraft can only sample the mixed plumes emitted from the fire front and also need to take into account for the chemical ageing of the smoke that occurs between emission and collection.

So how consistent are the results of the three methods? The most comprehensive analysis is by Yokelson *et al.* (2013), which compares the results of large-scale laboratory burns with ground and airborne field measurements in pine understorey and semi-arid shrubland fires. While laboratory and field measured CO_2 and CO emission factors, and therefore the modified combustion efficiency (MCE, see next section for the definition) were similar, the laboratory method underestimated CH_4 in the arid shrubland fuels, underestimated other organic species by 2.7-fold on average, and overestimated oxidised nitrogen (as nitric oxide) between two- and three-fold. This suggests that the laboratory fires favoured oxidative combustion chemistry rather more than normally occurs in the field. Ground-based emission factors tend to be weighted towards smouldering combustion sources (Burling *et al.* 2011), although ground-based sampling of targeted flaming sources tend to underestimate the degree of smouldering sources compared to airborne samples (Hurst *et al.* 1994b).

The purpose of measuring emission factors is their application in emissions accounting, either for inputs into dispersion models, which are mostly used to analyse or forecast patterns of smoke dispersion and impacts on regional populations, or for emissions inventories. Global emission inventories can be calculated using combustion and emissions parameters aggregated to broad vegetation classes. However, more detailed approaches are needed for dispersion modelling, for quantifying the impacts of land management on emissions, and, particularly, the accounting of carbon mitigation for carbon trading. Both the model Consume (Prichard *et al.* 2006), which is applied widely in North America, and the methodology applied in the Carbon Farming Initiative (CFI) projects in Australia (Russell-Smith *et al.* 2009; Commonwealth of Australia 2013) require emission factors for each fuel element class and each vegetation class. The database of emission factors, although rapidly expanding, remains sparse, and the variation in reported emission factors between individual fires, and between vegetation classes, remains large, particularly for the reactive organic smoke constituents but perhaps a little less for more stable species.

The number of strata that require specific emission factors in the emissions models is much larger than the number of measurements in the database, and therefore the selection of emission factors for each inventory methodology is underpinned by many assumptions, most of which are unstated. The key assumption is that the fire characteristics (and associated emission factor) for each stratum of the emissions methodology are consistent, on average, and that this average can be estimated from replicate measurements of measured fires. At high levels of aggregation there are usually adequate numbers of measurements to enable a mean and some estimate of variance to be derived, with the caveat that, due to the small database, large variability both within and between strata, and uneven statistical sampling of the spatial domain of each stratum, these estimates are likely to be biased to the substrata with the highest degrees of sampling. At high levels of stratification a deal of expert judgement is required in the allocation of a small number of measured emission factors to a greater number of inventory strata.

If combustion processes can be expressed reliably in terms of some explanatory variables that can be readily measured, then the task of extrapolating from the database to the emission model classes becomes simpler. A potential explanatory variable is combustion efficiency, or its simplification, modified combustion efficiency (MCE), which describes the extent to which smoke constituents are fully oxidised. First used in the early 1990s (Hao and Ward 1993; Ward *et al.* 1996), its explanatory power continues to be explored for a wide range of both oxidised and reduced compounds. Recently, Meyer *et al.* (2012), Smith *et al.* (2014) and Paton-Walsh *et al.* (2014) have shown that for CH_4 at least it explains much of the variation observed in the emission factors. In turn, MCE for the fine fuels and grass can be predicted from fuel moisture content (Hoffa *et al.* 1999) and the fine fuel composition (Korontzi 2005). Although the data underpinning the relationships are sparse, the potential that such explanatory relationships offer for improving the extrapolation from the current emission factor database to the inventory methodology strata is considerable and justifies active exploration of the approach.

The aim of this chapter is to describe how we have applied this pathway to our measurements of emission factors for CH_4, N_2O, CO_2 and CO in northern Australia to establish emission factor values for application in an emission accounting methodology for the low rainfall (600 mm–1000 mm) savanna region.

Definitions: emission factors and MCE

First we need to define some terms that are used throughout this chapter. For application to both emissions inventories and dispersion models, researchers are increasingly turning to emission models in which both fuel load and emission factor terms are disaggregated to the levels required to account for the effects of fire intensity, severity, the relationship between the size of fuel elements and their burning completeness, and vegetation type.

The emission ratios and emission factors for CH_4, N_2O, and the modified combustion efficiency (MCE), can be calculated following Meyer *et al.* (2012). The emission ratios (ER_i) are defined in this chapter as:

$$ER_i = \Delta C_i / (\Delta CO_2 + \Delta CO + \Delta CH_4)$$

where $\Delta C_i = [i]_{smoke} - [i]_{amb}$, i.e. the difference between the molar concentration (or mole fraction) of trace species i in the smoke sample and its concentration in ambient air upwind of the combustion source. For example $\Delta CO_2 = [CO_2]_{smoke} - [CO_2]_{amb}$ is the difference between the molar concentration (or mixing ratio) of $CO_2 - C$ in the smoke sample and in the upwind air, etc.

Modified combustion efficiency (MCE) is defined as:

$$MCE = \Delta CO_2/(\Delta CO + \Delta CO_2)$$

Emission factors are commonly expressed relative to either the mass of fuel consumed in combustion (Andreae and Merlet 2001) or to its elemental content (Hurst *et al.* 1994b).

For emission factors that are defined relative to fuel mass:

$$EF_i = ER_i/CC$$

where CC = the carbon content of the burned fuel.

In this chapter we follow Hurst *et al.* (1994b) and define emission factors (EF_i) relative to the elemental content of the combusted fuel rather than relative to combusted fuel mass. For the carbon species, the emission factor is equivalent to the emission ratio. Hence:

$$EF_{CH4} = ER_{CH4}$$

For the nitrogen species, the reference mass is fuel nitrogen, which is related to emitted carbon by the nitrogen to carbon ratio (NC). By convention, we use the N:C ratio of the fuel before combustion:

$$EF_{N2O} = ER_{N2O}/NC$$

There are good theoretical arguments that NC should be the N:C ratio of the burning fuel. However, there are major practical difficulties in routinely estimating this value in the field, and therefore in this chapter we have persisted with the current convention.

Emissions from low rainfall savannas in northern Australia

Good practice guidelines for greenhouse gas inventories require accounting for methane (CH_4) and nitrous oxide (N_2O). The emissions of these gases from savanna burning make up between 2–4% of accountable gases from Australia (DoE 2012). Meyer *et al.* (2012) established that savanna emissions of CH_4 and N_2O measured in a high rainfall region near Kulgnuki in the Northern Territory of Australia (12°38′S, 133°55′E) do not show a seasonal dependence (in contrast to African studies such as Hoffa *et al.* (1999), Korontzi *et al.* (2003), Korontzi *et al.* (2004), Korontzi (2005)) but instead CH_4 emission factors are dependent on fuel type (Meyer *et al.* 2012).

The fuel classes defined in accounting methodologies are fine fuels (<6 mm: grass and tree litter), coarse woody debris (CWD: ≥6 mm–<50 mm coarse, ≥50 mm heavy) and shrubs. The high rainfall savannas are dominated by two broad vegetation classes: open woodlands and forests with either tussock or hummock grass understory; and sandstone heaths dominated by hummock grasses and shrubs. In the open woodlands and forest the fine fuel is composed, substantially, of tree litter, while the fine fuel in the sandstone heath has a substantial grass component. Meyer *et al.* (2012) showed that the difference in EF_{CH4} between vegetation classes was mostly due to the difference between grass and tree litter combustion: the well-aerated grass tussocks and hummocks burn fiercely and efficiently producing smoke with a very high MCE, in contrast to tree litter, which burns more slowly and produces smoke with lower MCE. The N_2O emission factors were not related to MCE and showed variation across vegetation and fuel size class. From the known temperature dependence of the formation and destruction reactions for N_2O, Meyer *et al.* (2012) suggested the variability observed in EF_{N2O} might be associated with local variability in fire intensity, and hence flame length.

In the low rainfall savanna woodlands the main vegetation classes are open woodlands dominated by tussock or hummock grasses or mixtures of the two, and grass forms a substantial fraction of the fine fuel component. This suggests that fires in this region might be more efficient, with higher MCE and fire intensity for given fuel loads than in the high rainfall savannas, and therefore that that EF_{CH4}, and possibly EF_{N2O}, might also be substantially different from the high rainfall savannas. Because grasses in this region are often burned before they are fully cured, seasonality in emission factors could also be significant. To test whether this was the case, a series of field campaigns were planned to determine EF_{N2O} and EF_{CH4} in low rainfall savanna vegetation classes, and specifically:

1. to assess the dependence of MCE and EF_{CH4} on the grass content of the fine fuels and on the moisture content of grass
2. to investigate further the dependence of EF_{N2O} on MCE, and whether this relationship is dependent on fire intensity as indicated by flame length.

Emission factors of specific fuels can be measured either by selectively targeting the specific fuel elements during a normal prescribed burn when the fire behaviour is unconstrained, or by measured experimental burning of pure samples of each fuel/size class, in effect a laboratory burn conducted in the field. The advantage of the latter method is

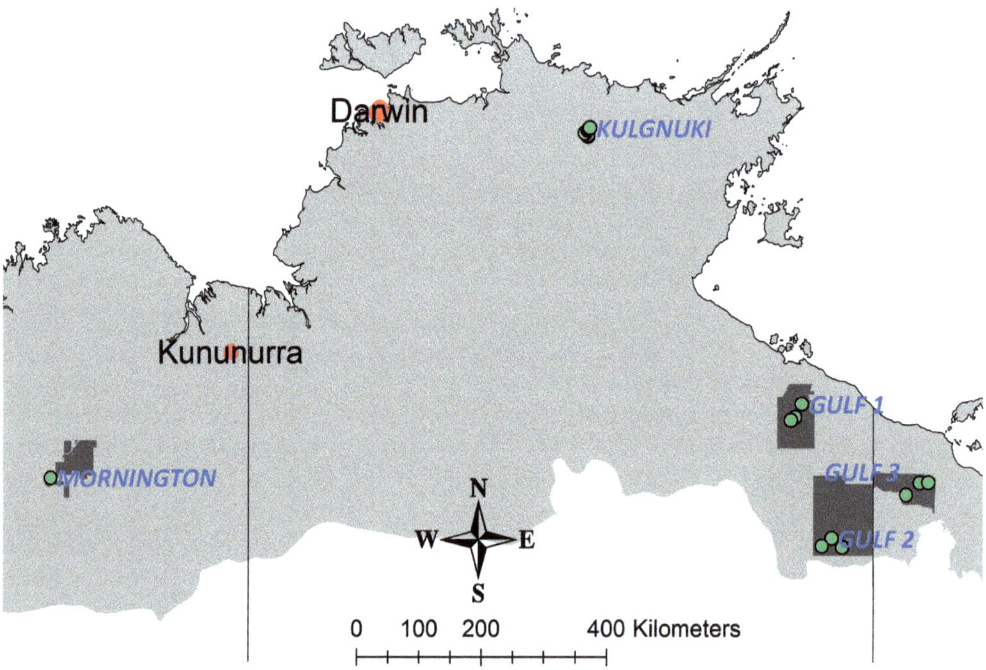

Figure 9.1. Campaign site locations. The savanna woodlands are shaded light grey. Regional cities are shown in red, campaign regions are shown in dark grey. Campaign sites are shown as green points.

that the smoke sample is uncontaminated by other fuel size elements, and for some fuels it allows sampling when an unconstrained burn would pose a safety hazard. The risk is that the combustion conditions differ from those of a normal, unconstrained fire. The experimental tests can be conducted with the fuel *in situ*, by manually igniting individual hummocks or tussocks; this ensures that the fuel structure is not disturbed. Alternatively the fuel can be burned in piles, allowing fire intensity to be controlled by the initial fuel mass and by the rate of fuel supply. A combination of both approaches is usually the best option; it provides estimates from a wide range of fuels and combustion states, and tests whether the small experimental burns adequately reproduce the conditions found in field burns.

Measurements were carried out in the low rainfall savanna woodlands in two campaigns. The first campaign was conducted in the Gulf of Carpentaria region of the Northern Territory and Queensland (Fig. 9.1) in September 2011. The ground-based smoke sampling method developed and successfully applied at Kulgnuki (Meyer *et al.* 2012; Fig. 9.2a) was also applied in this campaign due to isolation of the field sites. The sampling

Figure 9.2. Smoke sampling techniques: (a) sampling flaming combustion of fine fuel at the fire front; (b) sampling a high intensity pile burn using the Woozle, a portable, high-volume smoke sampler.

device was named 'the backpack'. The collected smoke was analysed at a field laboratory instead of being returned to Darwin. CO_2 and CH_4 concentrations were measured with a cavity ring-down spectrometer (Picarro, USA), a high precision and extremely robust instrument.

In addition to the general objective of measuring emission factors from fires in undisturbed fuel beds, this campaign had the objective of investigating the dependence of EF_{CH4} on the grass content of the fine fuels and on the moisture content of grass. Three classes of burning experiments were conducted:

1. flaming combustion of fine fuels, i.e. leaf litter and grass hummocks and tussocks
2. the role of overlying shrub layer on emissions
3. smouldering combustion of heavy fuels (logs >50 mm diameter).

Hence, this campaign focused on sampling the available parameter space rather than evenly sampling the majority vegetation. The Gulf campaign was conducted during the late dry season when fuels are fully cured, therefore sampling a useful range of fuel moisture contents presented a challenge. Nevertheless, although rare, there were a few small locations where soils and drainage combined to support green grass throughout the season. We found a few such locations within or adjacent to the experimental burn sites which differed from the majority of the site area only by the presence of live grass. These locations were selectively targeted for smoke sampling. Table 9.2 lists the experimental burns together with their fuel loads and the grass moisture content.

The second field campaign was conducted at Mornington Sanctuary in the Kimberley region of Western Australia in August 2013. This region has extensive open woodlands dominated by either tussock grasses or mixtures of hummock and tussock grasses. The campaign focused on the relationship between EF_{N2O} and MCE and the influence of flame length. It relied mostly on sampling of pure fuel types in pile burns using a high volume smoke sampler referred to in the international literature as 'the Woozle' (Fig. 9.2b; see Meyer *et al.* 2009 for a detailed description). As with the Gulf campaign, sample gases were analysed for, CO_2, CH_4 and N_2O concentration at a field laboratory using cavity ring-down spectrometers (Los Gatos Research, USA), and emission factors were calculated as described by Meyer *et al.* (2012).

Several classes of burning experiments were carried out at in order to sample from a wide range of fuels, combustion efficiencies and fire intensities. These included:

1. pure fuel types including spinifex, tussock grasses, tree leaf litter and logs under flaming and smouldering conditions
2. flaming combustion of tussock and hummock grasses at two intensities: long flames (~1.5–2 m long) and short flames (0.3–1.2 m)
3. the time course of emissions from tussock grass from flaming through to smouldering char combustion
4. field burns of tussock and hummock grass swards.

The objective of this campaign was to investigate combustion processes, rather than to measure emissions from intact fuel beds during a normal fire. The first three series of measurements were conducted using pile burns and were sampled using the Woozle; the fourth series of burns was conducted in the field on intact fuel beds. The pile burns were commenced with a small initial fuel load, and fuel was progressively added to maintain the

Table 9.2. Experimental burns sampled during the Gulf of Carpentaria campaign and their fuel loads and grass moisture content

Burn	Fuel type	Fine fuel (t ha^{-1})				CWD (t ha^{-1})	Fine fuel (% grass)	Moisture content (% dry weight)
		Grass	Litter	Twigs	Total			
G1	Mixed litter/ tussock grass	2.11	2.46	0.84	5.42	0.67	44	18.5
G2	Mixed litter/ tussock grass	2.87	2.62	0.49	5.98	0.00	52	39.2
G3	Mixed litter/ tussock grass	4.10	1.05	0.10	5.25	1.04	77	26.8
G4	Acacia shrubs over hummock grass	5.63	1.78	1.56	8.97	0.06	64	19.0
G5	Acacia shrubs over hummock grass	2.79	0.62	0.20	3.61	0.46	81	18.0
G6	Acacia shrubs over hummock grass	4.35	0.85	0.07	5.27	0.52	80	17.6
G7	Acacia shrubs over hummock grass	3.43	0.96	0.07	4.47	0.03	75	22.2
G8	Mixed tussock grass grass/ hummock grass	3.05	2.35	0.39	5.78	0.16	94	18.7
G9	Mixed tussock grass/ hummock grass	1.72	2.05	0.07	3.83	0.29	51	12.6
G10	Mixed tussock grass/ hummock grass	3.14	0.72	0.02	3.88	0.67	81	39
G11	Mixed tussock grass/ hummock grass	3.14	0.72	0.02	3.88	0.67	81	16.7
G12	Pure hummock grass	–	–	–	–	–	100	20.7

required degree of flaming and flame length while avoiding excessive accumulation of burning char. The field burns were not unconstrained fires; instead they were conducted in late afternoon in still conditions and the fire progressed as a series of small patches radiating from point ignitions, or ignitions of individual grass hummocks. This is similar to methods used for sampling pure grass fuels in undisturbed swards. The experimental burns and their average emission factors are presented in Table 9.3. The elemental contents of all fuels sampled in the Mornington campaign were also measured (Table 9.4).

Table 9.3. Experimental burns sampled during the Mornington Sanctuary campaign

Date	Burn number	Sampling method	Fuel type	Treatment	Site
2/08/2013	M1	Backpack	Tussock grass	Savanna fire	Basalt
3/08/3013	M2	Woozle	Spinifex	Long flame	Sandstone
4/08/2013	M3	Woozle	Spinifex	Short flame	Sandstone
4/08/2013	M4	Woozle	Spinifex	Long flame	Sandstone
4/08/2013	M5	Woozle	Spinifex	Smouldering	Sandstone
5/08/2013	M6	Woozle	Tussock grass	Short flame	Sandstone
5/08/2013	M7	Woozle	Tussock grass	Long flame	Sandstone
5/08/2013	M8	Woozle	Tussock grass	Smouldering	Sandstone
6/08/2013	M9	Woozle	Weathered litter	Smouldering	Basalt
6/08/2013	M10	Woozle	Fresh litter	Flaming	Basalt
7/08/2013	M11	Woozle	CWD	Flaming	Sandstone
7/08/2013	M12	Backpack	Tussock grass	Savanna fire	Basalt
8/08/2013	M13	Woozle	CWD	Smouldering	Sandstone
8/08/2013	M14	Backpack	Spinifex	Savanna fire	Basalt
9/08/2013	M15	Woozle	Tussock grass	Burnout	Basalt
9/08/2013	M16	Backpack	Spinifex	Savanna fire	Sandstone

The average MCE, and the CH_4 and N_2O emission factors for each burn, are summarised in Tables 9.5 and 9.6.

Methane emission factors

The low rainfall woodlands in northern Australia tend to be dominated by grasses, and therefore the fine fuels contain a much higher proportion of grass and a lower proportion of tree litter than in the high rainfall regions. This is reflected in higher MCE values, 0.92 to 0.98, and lower EF_{CH4} values, 0.05–0.35, than in the denser woodlands of the north. However the relationship between EF_{CH4} and MCE (Fig. 9.3) was essentially unchanged from that found at Kulgnuki (Meyer *et al.* 2012), and the empirical explanatory function, developed for the available Australian and global data applied in the Kulgnuki study, the exponential Stirling curve (Eqn 2), provides a good fit and explains most of the variance in

Table 9.4. Elemental content of fuels sampled at Mornington Sanctuary

Fuel type	Substrate	Carbon (%)	Nitrogen (%)	N:C ratio
Coarse woody debris	n/a	48.2	0.19	0.0039
Spinifex	Sandstone	40.2	0.35	0.0086
	Basalt	43.6	0.31	0.0071
Tussock grass	Basalt	43	0.28	0.0066
Tree leaf and twig litter (fresh)	Basalt	48.5	0.73	0.015
Tree leaf and twig litter (decomposed for >1 yr)	Basalt	39.2	0.56	0.014

Table 9.5. Mean MCE and EF_{CH4} (% emitted C) measured at Gulf of Carpentaria burns

Flaming combustion was measured at the fire front and sampled emissions from the fine fuels and shrubs (when present). Emission from burning logs included smouldering, flaming and mixed emission. Emissions from cow pats were smouldering combustion only.

	Fine fuels – Flaming		Logs – Smouldering		Cow pats – Smouldering	
Fire	**MCE**	**EF_{CH4}**	**MCE**	**EF_{CH4}**	**MCE**	**EF_{CH4}**
G1	0.959 (0.006)	0.143 (0.016)	0.80 (0.05)	1.80 (0.50)	–	–
G2	0.927 (0.010)	0.342 (0.024)	0.85 (0.03)	1.50 (0.32)	–	–
G3	0.934 (0.007)	0.227 (0.013)	0.86 (0.02)	1.57 (0.39)	–	–
G4	0.927 (0.018)	0.318 (0.031)	0.83 (0.01)	1.04 (0.11)	–	–
G5	0.957 (0.010)	0.166 (0.017)	0.88 (0.01)	1.07 (0.09)	–	–
G6	0.962 (0.012)	0.137 (0.016)	0.83 (0.02)	1.87 (0.39)	–	–
G7	0.944 (0.010)	0.231 (0.022)	0.87 (0.02)	1.37 (0.42)	–	–
G8	0.939 (0.009)	0.177 (0.012)	0.86 (0.04)	2.70 (1.37)	–	–
G9	0.939 (0.012)	0.178 (0.014)	0.87 (0.03)	1.31 (0.32)	0.76	3.34
G10	0.945 (0.007)	0.223 (0.012)	–	–	–	-
G11	0.949 (0.010)	0.175 (0.013)	0.84 (0.01)	1.60 (0.33)	0.78 (0.015)	3.01 (0.66)
G12	0.975 (0.006)	0.063 (0.004)	–	–	–	–

the data. The Stirling function is an exponential curve that becomes approximately linear at MCE >0.9.

$$EF_{CH4} = 0.239(e^{(10.6(1-MCE)}-1)\ (r^2 = 0.74) \quad (2)$$

However, indications are emerging that the function has a limited domain. The range of combustion processes that can be supported by open burning of an intact fuel beds is likely to be a subset of the potential range of combustion possible for the fuel type. Weather conditions, fuel continuity, fuel condition and fuel structure determine whether a fire is sustainable or self-extinguishing and, in turn, limit the range of MCE and related emission factors that are observed in the field. Large fuel elements tend to burn more slowly than small fuel elements. Detailed laboratory studies found that the rate of weight loss for round wood fuels declined exponentially with diameter of the element (Burrows 2001): jarrah (*Eucalyptus marginata*) twigs 3 mm in diameter were consumed at the rate of 80% per minute compared to 25 mm branches, which burned at 11% per minute, and hence the smaller the fuel element the greater the amount of gas-phase combustion. Flaming combustion of larger sticks and logs only occurs when there is sufficient preheating from surrounding burning fuel or an ash bed, and hence burning logs rarely have MCE values >0.9. In cases where high MCEs are observed from large fuel elements the associated EF_{CH4} usually diverges from Eqn 2 (Fig. 9.3b).

When MCE was introduced as a potential explanatory variable it soon became apparent that a single regression might explain emission properties of a wide range of fuel and combustion classes. Hao and Ward (1993) also came to that conclusion from their analysis of the SAFARI campaign. The conclusion later lost traction, with much of the work in the last decade focusing on the variability of the regressions rather than their uniformity. Meyer *et al.* (2012) again raised the suggestion that a single function might be sufficient to

Table 9.6. Mean values of MCE, EF_{CH4} and EF_{N2O} for each of the burns at Mornington Sanctuary

Figures in parentheses indicate standard error.

Burn	Sampling method	Fuel type	Site/fire type	MCE	EF_{CH4} (% fuel C)	EF_{N2O} (% fuel N)	Number of samples
M1	Field	Tussock grass	Basalt	0.972 (0.002)	0.12 (0.017)	2.02 (0.116)	13
M12	Field	Tussock grass	Basalt	0.971 (0.003)	0.14 (0.018)	1.81 (0.165)	19
M14	Field	Spinifex	Basalt	0.980 (0.002)	0.05 (0.007)	1.01 (0.158)	13
M16	Field	Spinifex	Sandstone	0.980 (0.003)	0.16 (0.023)	0.55 (0.050)	18
M2	Woozle	Spinifex	Long flame	0.991 (0.001)	0.05 (0.014)	0.36 (0.031)	9
M3	Woozle	Spinifex	Short flame	0.985 (0.001)	0.07 (0.009)	0.37 (0.020)	12
M4	Woozle	Spinifex	Long flame	0.992 (0.001)	0.05 (0.011)	0.34 (0.033)	6
M5	Woozle	Spinifex	Smouldering	0.858 (0.018)	0.93 (0.199)	1.11 (0.118)	9
M6	Woozle	Tussock grass	Short flame	0.987 (0.002)	0.06 (0.010)	0.79 (0.082)	14
M7	Woozle	Tussock grass	Long flame	0.988 (0.001)	0.05 (0.004)	0.93 (0.081)	18
M8	Woozle	Tussock grass	Smouldering	0.843 (0.025)	0.73 (0.128)	0.78 (0.099)	6
M9	Woozle	Litter	Smouldering	0.933 (0.005)	0.62 (0.078)	1.35 (0.050)	13
M10	Woozle	Litter	Flaming	0.982 (0.002)	0.12 (0.015)	0.64 (0.033)	13
M11	Woozle	CWD	Flaming	0.988 (0.001)	0.08 (0.014)	0.91 (0.050)	15
M13	Woozle	CWD	Smouldering	0.894 (0.008)	1.11 (0.104)	1.46 (0.070)	12
M15	Woozle	Tussock grass	Burnout	0.892 (0.021)	0.60 (0.136)	1.40 (0.124)	12

describe most of the variation in EF_{CH4}; however, the Mornington campaign showed that this is not always the case: clearly, the pile burns conducted in the Mornington campaigns were not described by Eqn 2, and although the form of the relationship remained the Stirling function, EF_{CH4} values were substantially greater than predicted by the function that described the field measurements (Fig. 9.4).

Before addressing the possible reasons for this, it should be remembered that MCE and EF_{CH4} can be measured in two ways: as a time course throughout the combustion of a single fuel mass, and as and average from a bulked sample of a fuel type or fire area.

Grab samples can be taken as combustion of a single fuel mass proceeds though the combustion phases of preheating, thermal decomposition and volatilisation, gas phase combustion of mostly volatile organics, and finally glowing char combustion.

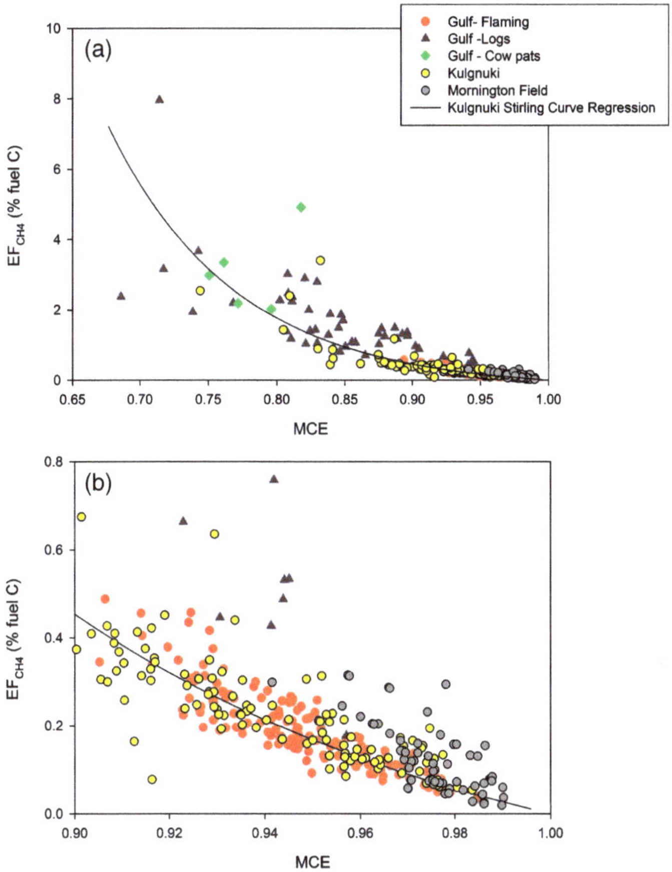

Figure 9.3. The relationship between EF_{CH4} and MCE for all field measurements in northern Australia: (a) Full MCE domain; (b) Detail of the high MCE domain.

Instantaneous MCE and EF_{CH4} can be calculated from the concentration of the emitted gases collected in this way, but the time course of EFs relates to the progression of combustion chemistry rather than the total CO_2, CO and CH_4 emissions from the fuel type. Examples of this are shown in many publications describing the open burning of fine fuels (e.g. Lobert *et al.* 1990) and in wood combustion in domestic stoves and heaters (e.g. Gras *et al.* 2002).

The alternative is to collect gases from a defined fuel type or fire area throughout the combustion process and calculate average MCE and EF_{CH4} from the bulked sample. This yields MCE and EF_{CH4} related to fuel or vegetation class and to fire behaviour; these are the EF estimates required for inventory accounting. In most cases, gases collected on the fire ground or from aircraft and open-path measurements include mixtures of smoke from all combustion stages. The objective is to collect samples that are either unbiased to any single combustion phase, or that capture the phases that account for most of the emissions. In most Australian ecosystems, fine fuels burn completely within a few minutes (e.g. Burrows 2001), forming a combustion zone that is mostly only a few metres wide. The sampling technique used by Meyer *et al.* (2012) collects aggregated samples over 5–10 min from within this zone; aircraft sampling relies on turbulence in the smoke plume to

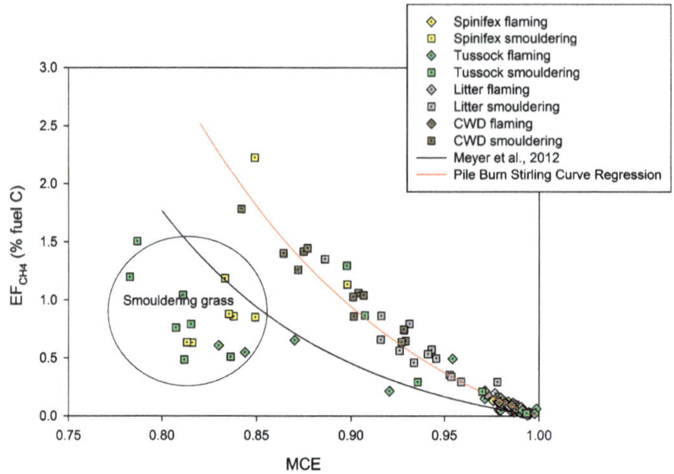

Figure 9.4. The relationship between EF_{CH4} and MCE for the pile burns.

produce unbiased mixed samples of flaming combustion. With larger fuel elements combustion is more protracted, but nevertheless progresses though the sequence described above. The exception is the final stages of glowing char combustion, which occurs in ash beds and tussocks as combustion proceeds to extinction. The emission products of charcoal combustion are almost entirely CO and CO_2, and therefore both MCE and EF_{CH4} are low. Emission factors from char combustion of grasses are evident in Fig. 9.4.

Small experimental burns, in the laboratory or in fireplaces in the field, pose a challenge for sampling average smoke concentrations unbiased by combustion phase; and in most cases unbiased sampling can only be achieved using dilution tunnels and combustion hoods that collect the complete smoke plume. This is not possible in the field, and therefore the Woozle tests at Mornington Sanctuary were aimed explicitly at analysing emissions by combustion phase rather than total emissions. To maintain flaming

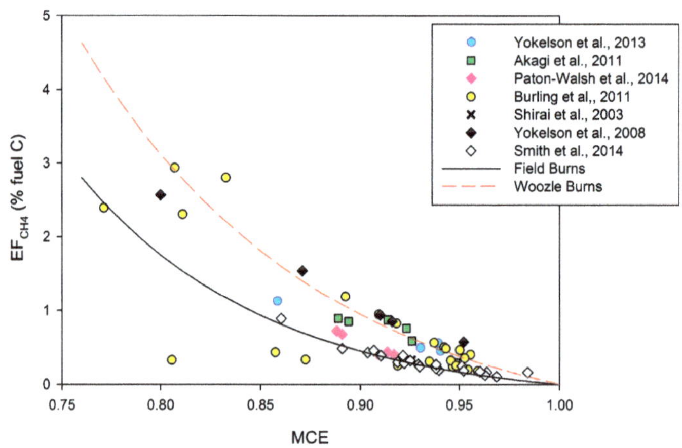

Figure 9.5. Relationship between EF_{CH4} and MCE reported for North American forests and fuels (Yokelson *et al.* 2008; Akagi *et al.* 2011; Burling *et al.* 2011; Yokelson *et al.* 2013) and Australian savanna woodlands (Shirai *et al.* 2003; Smith *et al.* 2014) and Australian temperate forests (Paton-Walsh *et al.* 2014).

combustion of the required intensity, fuel was fed continuously into the fire at a single location. This approach produces a fuel bed increasing in depth as the fire progresses. Air access is limited and therefore flaming combustion is suppressed and hence emissions proceed by charring and char combustion. Deep fuel beds are uncommon in most Australian ecosystems and are rare in savannas; however, they are common in many coniferous and deciduous forests in North America and northern Europe, which develop thick duff layers. Fig. 9.5 presents a comparison between some recently published airborne and open path measurements of EF_{CH4} from North America and Australia. While the Australian data (Paton-Walsh *et al.* 2014; Smith *et al.* 2014) are consistent with the Kulgnuki data (Meyer *et al.* 2012), the North American data are closer to the pile burns. This leads us to conclude that open combustion can be described by a family of EF_{CH4}/MCE curves with the gradient increasing with the extent of charring and char combustion, rather than by a single curve. The emissions from South African savannas (Ward *et al.* 1996) and Australian savanna and eucalypt forests lie on the lower boundary, while fires in the temperate and boreal forests in the northern hemisphere in which duff combustion contributes emissions lie closer to the upper boundary. Hence, while this work confirms that MCE is a useful descriptor of EF_{CH4} for inventory applications, care is required to select measurements only from well-defined or undisturbed fuel beds and vegetation types that are characteristic of the relevant ecosystems.

Explanatory relationships for MCE

Noting the caveats discussed above, EF_{CH4} can be explained largely by MCE, and MCE appears to depend on the structure of the fine fuels (Meyer *et al.* 2012). Direct measurements of emission factors are complex and expensive tasks, and therefore the measurements are relatively few and spatially sparse. However, there are now extensive databases of fuel loads and structure with comprehensive spatial coverage, and therefore if explanatory relationships can be found to quantitatively describe MCE in terms of fuel condition and fine fuel structure this will simplify the task of determining EF_{CH4} for each vegetation and fuel stratum in the inventory methodology, and the uncertainties and biases that result from extrapolating a small set of emission factor measurements to a wide range of vegetation classes will be reduced.

We know from observation that fuel structure and fuel moisture also influence combustion properties and hence EF_{CH4}. Meyer *et al.* (2012) concluded that in the high rainfall savanna fires mostly occurred when tree litter was dry and the grasses were mostly cured. Even when curing is not completed by the start of the fire season, tussock grasses typically are a small component of the fine fuel and their higher fuel moisture is unlikely to dominate the fine fuel emission factors.

This is not the case for the lower rainfall savannas, where grass comprises a substantial fraction of the fine fuels; therefore, in the Gulf campaign we explicitly addressed the effect of fuel moisture on MCE. The fuel moisture content of relevance here is the moisture content of the tussock components that actually burn – the senescing and dry mature leaves – rather than the average moisture content of the complete tussock, which is dominated by the green base. The relationship between moisture content of the combustible grass components and fine fuel EF_{CH4} is presented in Fig. 9.6. It should be noted that grass comprises approximately 50% of the fine fuel in the Australian low rainfall savanna; the remainder is tree and shrub litter, which is uniformly dry. Although EF_{CH4} increases substantially with grass fuel moisture content, the response is only half that reported by Hoffa *et al.* (1999) for African savanna grasses. Therefore, although seasonal variation in EF_{CH4}

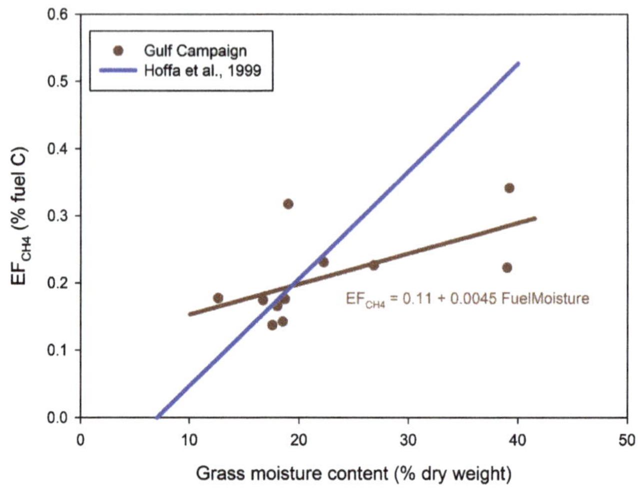

Figure 9.6. The relationship between EF_{CH4} from fine fuel and grass moisture content observed in the Gulf region of the Australian low rainfall savanna (brown). The relationship reported by Hoffa *et al.* (1999) for African grasslands is shown in purple.

due to grass curing in the early dry season is likely to occur in these lower rainfall open woodland, its magnitude is likely to be much smaller than the seasonal changes reported for African savannas (Korontzi 2005).

However, variation in fine fuel composition is a significant determinant of MCE and hence EF_{CH4}. Korontzi (2005) reported that MCE could be described as a function of the relative proportions of grass, dead grass, leaf litter and twigs in the fine fuel in African savannas (Eqn 3). (There was an error in the published equation due to double counting of the total fuel mass; Eqn 3 is the corrected version of Korontzi (2005) Eqn 7).

$$\text{MCE} = 0.938P_{greengrass} + 0.963P_{deadgrass} + 0.94P_{leaflitter} + 0.860P_{twiglitter} \quad (3)$$

Using similar logic, we find that MCE observed in the Gulf fires can be described as:

$$\text{MCE} = 0.9328 + 0.0427 * P_{grass} - 0.0007 * \text{M} \ (r^2 = 0.43) \quad (4)$$

where:

P is the fraction of grass in the fine fuel, and
M is the grass moisture content (% dry weight).

Table 9.7. Statistics of the grass fraction to MCE regression

Factor	Value	SE	t	P
Intercept	0.9328	0.0174	53	0
Moisture	–0.0007	0.0004	–1.6	0.134
Fraction grass	0.0427	0.021	2.0348	0.072
Significance of regression				0.047

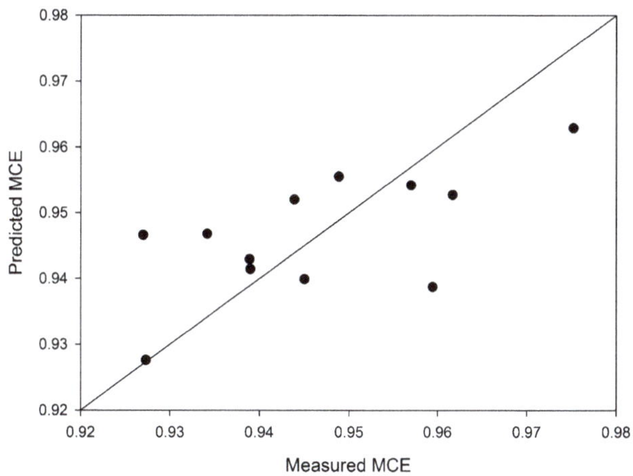

Figure 9.7. Comparison of predicted and measured MCE, where predicted MCE is determined from the regression coefficients listed in Table 9.7.

Table 9.8. Summary of Australian and global EF_{N2O} values

Source	Vegetation		EF
Australia			
Hurst *et al.* (1994a)	Hummock grass open woodland	0.0076	±0.00381
Meyer *et al.* (2012)	Tussock grass open woodland – litter	0.0077	±0.0021
Meyer *et al.* (2012)	Hummock grass open woodland – litter	0.0066	±0.0005
Meyer *et al.* (2012)	Sorghum – grass sward	0.0064	±0.0031
Meyer *et al.* (2012)	Spinifex – grass sward	0.0053	±0.0003
Meyer *et al.* (2012)	CWD	0.0036	±0.0005
Volkova *et al.* (2014)	Temperate eucalypt forest (Vic) – litter	0.0109	±0.0080
Volkova *et al.* (2014)	Temperate eucalypt forest – CWD	0.0076	±0.0009
Paton-Walsh *et al.* (2014)	Temperate eucalypt forest (NSW), Gulgar Nature Reserve	0.027	
Paton-Walsh *et al.* (2014)	Temperate eucalypt forest (NSW), Alfords Point	0.015	
Other			
Cofer *et al.* 1989		0.0391	(0.027–0.078)
Lobert *et al.* 1990		0.0077	0.0021
Cofer *et al.* 1990a		0.0487	(0.034–0.078)
Cofer *et al.* 1990b		0.1420	(0.087–0.24)
Cofer *et al.* 1991		0.0370	(0.024–0.073)
Nance *et al.* 1993		0.0089	
Laursen *et al.* 1992		0.0440	(0.004–0.165)
Delmas *et al.* 1995		0.0350	
Cofer *et al.* 1996		0.0204	(0.003–0.05)

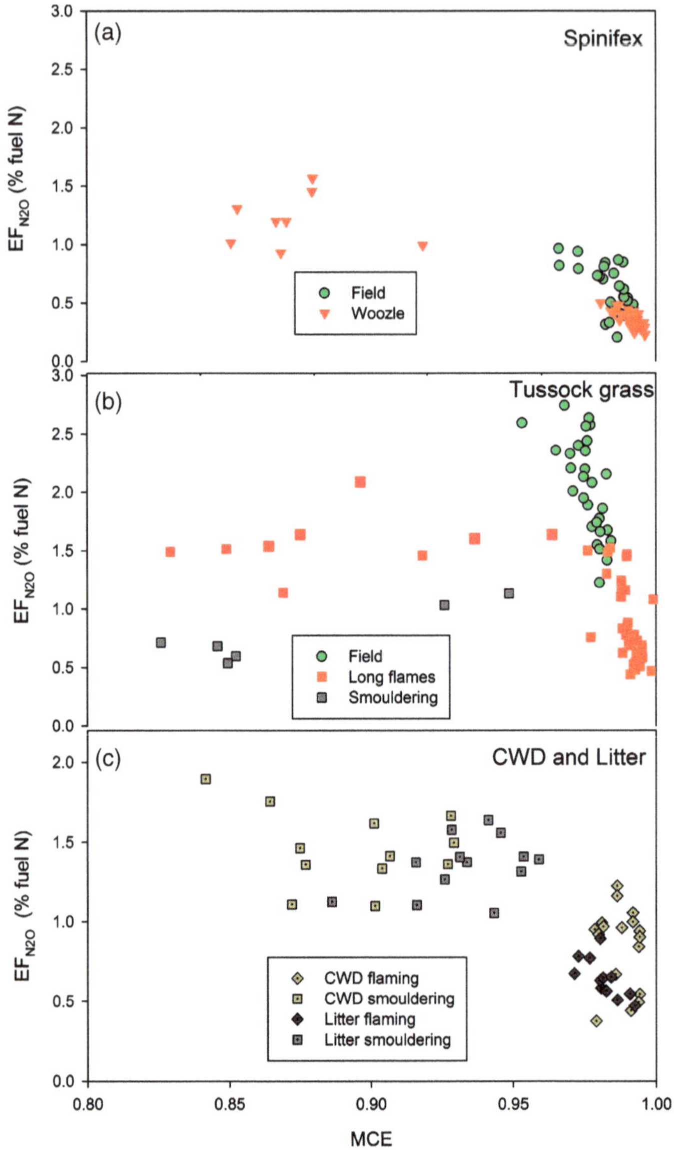

Figure 9.8. Relationship between EF_{N2O} and MCE for flaming and smouldering combustion of (a) spinifex; (b) tussock grass; and (c) litter and coarse woody debris.

The regression described by Eqn 4 is statistically significant at $P<0.05$ (Table 9.7). Fig. 9.7 shows the goodness of fit.

To summarise, MCE can be estimated from fuel load and fuel moisture, and, EF_{CH4} can be estimated from MCE (Eqn 2). The measurements of fuel load and fuel moisture are simpler and quicker than measurements of emission factors. The current fuels database is large with relatively uniform coverage across the much of the Australian savannas, and is much more extensive than the emissions measurements. There are also well-established models available to estimate moisture content of fine fuels with reasonable reliability. Therefore, assuming that Eqns 2 and 4 apply generally across the Australian savannas, they offer a better alternative for estimating EF_{CH4} than averages of a small number of direct measurements, with less risk of unrecognised biases, and provide a sound statistical basis for assessing uncertainties in regional emission estimates.

Nitrous oxide

The principal objective of the Mornington campaign was to explore the sources of variation in published EF_{N2O}. Variability was a major feature of the measurements at Mornington Sanctuary (Table 9.6). EF_{N2O} values ranged from 0.34% burned fuel N to 2.02% and, overall, were higher than was observed in the earlier studies in the Australian savanna (Table 9.8); the EF_{N2O} values measured from hummock grass at Mornington Sanctuary were similar to measurements at Kapalga (Hurst *et al.* 1994a) and Arnhem land, NT (Meyer *et al.* 2012) but the measurements from CWD and tussock grass were two- to three-fold higher.

EF_{N2O} was also independent from MCE for most of the MCE domain. Figure 9.8 shows the relationships between EF_{N2O} and MCE for spinifex (Fig. 9.8a), tussock grass (Fig. 9.8b) and the fuels derived from shrubs and trees (leaf litter and CWD, Fig. 9.8c) for

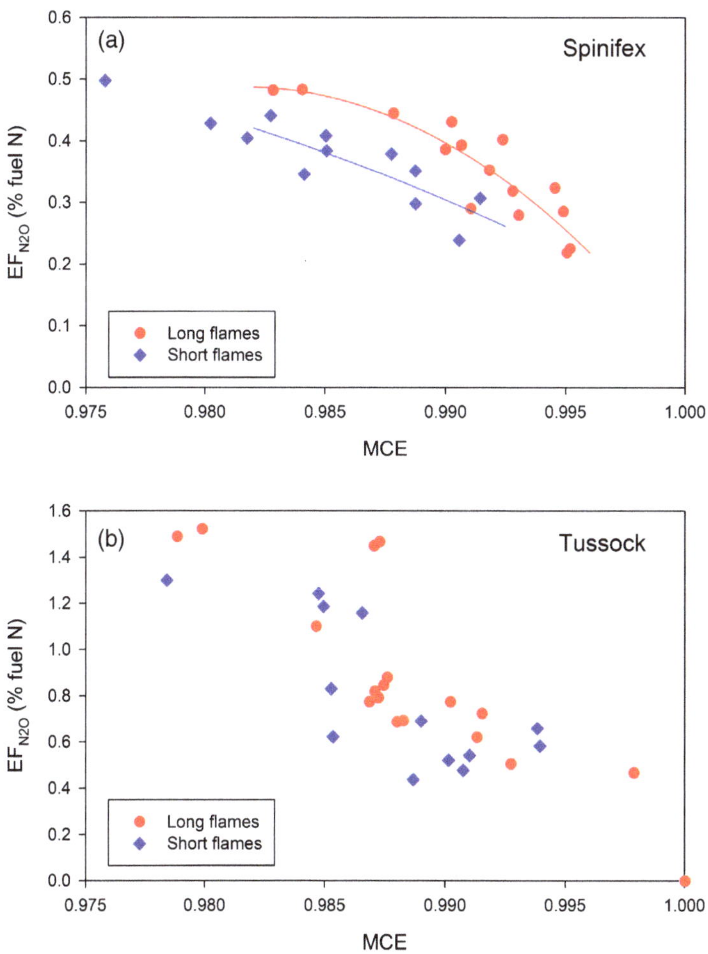

Figure 9.9. Effect of fire intensity on EF_{N2O} of (a) spinifex; and (b) tussock grass.

smouldering and flaming burns. In most cases EF_{N2O} increases sharply as MCE declines from 1 to 0.97, after which it stabilises. The EF_{N2O} from the spinifex field burns aligned reasonably with the pile burns but is approximately double in value at the same MCE (Fig. 9.8a), and a similar result was observed with tussock grass field burns (Fig. 9.8b), which indicates, as with EF_{CH4}, that combustion characteristics are as important as fuel composition for determining EF_{N2O}.

Fire intensity is one of the combustion characteristics of importance. Flame temperature is related to flame length (Wotton *et al.* 2012), and although accurate control of flame length is not possible in a semi-field experiment, by varying the rate at which fuel was added to the piles it was possible to produce qualitatively different tests with either short flames (<1.5 m) or long flames (>1.5 m). There was no influence of flame length on EF_{N2O} at MCE less than 0.97, but beyond this value of MCE, EF_{N2O} tended to decline as MCE increased. With spinifex, a species that retains its growth structure when harvested, burns with long flames produced 20% higher EF_{N2O} values than burns with short flame (Fig. 9.9a). With tussock grass, maintaining a well-aerated burn in a pile proved to be more difficult; although on average EF_{N2O} produced by long flame burns were 12% higher than EF_{N2O} from short flame burns, larger variability observed in both MCE and EF_{N2O} masked any statistical difference (Fig. 9.9b).

In contrast to most other smoke components, there are, surprisingly, very few measurements of EF_{N2O}. To the best of our knowledge, apart from Meyer *et al.* 2012, there are almost none published since 2000, and most were from measurements between 1980 and 1996. Some of the data were affected by a sampling artefact that caused gross overestimates, as was discovered by Muzio and Kramlich (1988): when combustion gas samples are stored in stainless steel containers for extended periods, the N_2O is synthesised by surface reactions involving SO_2. Most of the N_2O emission factors in the global database date to between 1988 and 1996; most are based on samples that were stored in stainless steel flasks and, although the majority were analysed within a few days, some were stored for weeks before analysis. The issue is discussed by Cofer *et al.* (1991) and identifies the boreal

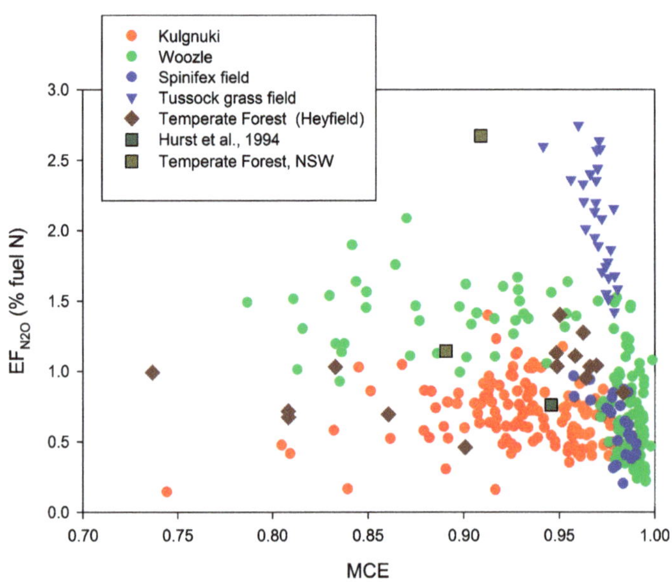

Figure 9.10. The complete dataset of EF_{N2O} from Australian fires.

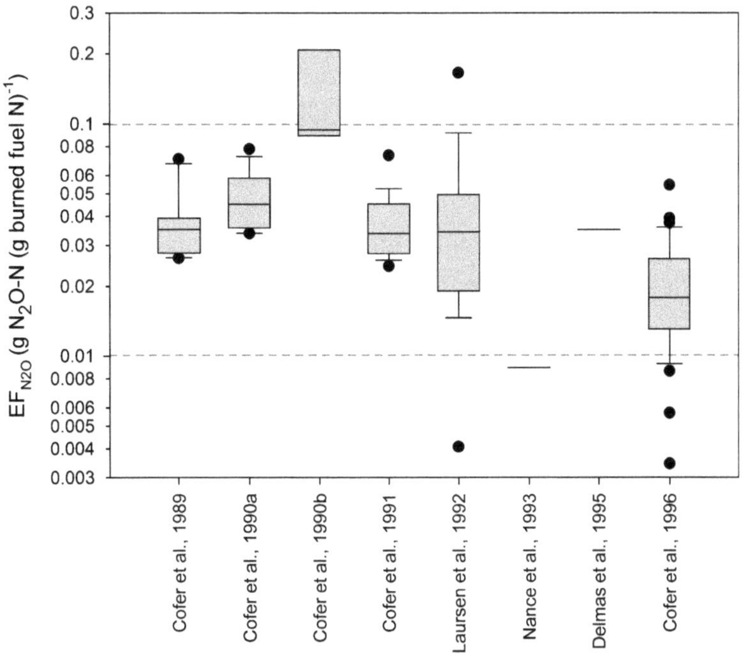

Figure 9.11. Mean and range of EF$_{N2O}$ from North American forests (Cofer *et al.* 1989, 1990a, 1990b, 1991, 1996; Delmas *et al.* 1995; Laursen *et al.* 1992; Nance *et al.* 1993).

measurements published in 1990 (Cofer *et al.* 1990b) as being substantially affected by the problem. The only Australian study before Meyer *et al.* (2012) was that of Hurst *et al.* (1994a), who collected smoke samples in glass flasks and so produced results that were free from the artefact. The only other Australian measurements (Volkova *et al.* 2014) were measured during a prescribed burn in a temperate eucalypt forest near Heyfield in Victoria. The complete dataset is presented in Fig. 9.10. The values cover a wide range (0.1% to 2.8% fuel N emitted as N_2O); Kapalga (Hurst *et al.* 1994a) and Kulgnuki (Meyer *et al.* 2012) measurements are similar; Heyfield litter EF$_{N2O}$ values are 50% higher than Kulgnuki but similar to the Mornington measurements but the EF$_{N2O}$ values from tussock grass at Mornington Sanctuary (1%–2.8% fuel N) are the highest recorded to date in Australia. These values are two-fold to 10-fold less than most EF$_{N2O}$ estimates for tropical forests, northern hemisphere temperate forests and boreal forests (Fig. 9.11, Table 9.8).

Of all the combustion gas species that are relevant to atmospheric chemistry, nitrous oxide remains the least well documented. The current database remains extremely sparse and variable. Fortunately, all the recent the recent data is Australian, mostly savanna, and therefore is directly applicable to Australian inventory accounting methods. Most of the remainder is pre 1996, some using sampling methods that can be subject to artefacts and is of less direct relevance to a carbon mitigation project in Australia. However, we can draw a few conclusions from the data that can guide us in applying the available observations to accounting methodologies.

The nitrous oxide emission factor:

- depends on MCE but only at MCE values >0.97. Across the MCE range encountered in the field EF$_{N2O}$ is independent of the combustion state

- increases with flame length, as postulated, but not sufficiently to explain a significant fraction of the observed variability within a fire or fuel class; but
- varies both with fuel type and with combustion state. This is evident in a comparison between EF_{N2O} from an undisturbed sward of tussock grass burnt in the field and the same fuel burned in a pile, in which the differences in EF_{N2O} can only be caused by the difference in combustion conditions.

In contrast to EF_{CH4}, we conclude that there are no easily measured combustion or fuel properties that explain the observed variability in EF_{N2O}.

Emission factors and the CFI

The final task to address in this chapter is the method for applying the observed EF_{CH4} and EF_{N2O} to a low rainfall region CFI emissions inventory methodology. The constraints are that the methodology must be consistent with the project accounting methodology developed for the high rainfall savanna woodlands (Commonwealth of Australia 2013) and with the Australian National Methodology (DoE 2012), and must also follow IPCC Good Practice (IPCC 2000). Commencing in 2015, all National Communications will be required to implement the 2006 IPCC Guidelines (IPCC 2006) and its revision (IPCC 2014), and therefore so must this low rainfall savanna CFI methodology.

IPCC (2000) and IPCC (2006) apply the standard algorithm (Eqn 1) to estimate greenhouse gas emissions from biomass combustion. The emission factor for Tier 1 (default) methods are the biome-averaged estimates sourced from Andreae and Merlet (2001), while Tier 2 methods apply country-specific emission factors sourced from direct measurements in the nationally defined vegetation and fuel classes. IPCC (2000) also recommends estimating EF_{CH4} from direct measurements of MCE using a specified linear EF_{CH4} vs. MCE function to account for seasonal variation of EF_{CH4}: however, this has not been implemented in any national or regional methodology. Model-based algorithms such as Consume (Prichard *et al.* 2006), FEPS (Anderson *et al.* 2004) and FINN (Wiedinmyer *et al.* 2011), which are used extensively for air quality modelling in some jurisdictions, are Tier 3 methodologies. IPCC (2000), IPCC (2006), and IPCC (2014) do not discuss the approaches explicitly; however, in principle, they are consistent with good practice. Both Consume and FEPS are spatially and temporally explicit process models that calculate hourly emissions and require detailed temporal and spatial inputs of vegetation and weather data. In these models hourly values of EF_{CH4} and a range of other combustion products (e.g. CO_2, CO, PM2.5) are calculated from the relative contributions of smouldering and flaming; in FEPS through a version of MCE, and in Consume by weighting flaming and smouldering emissions factors by the fraction of smouldering combustion. In contrast, FINN is a classification model with parameter variables defined explicitly for each class. Significantly, none of the models estimates N_2O emissions.

The current Australian national methodology for savanna burning (DoE 2012) and the high rainfall CFI methodology (Commonwealth of Australia 2013) apply Tier 2 approaches. Tiers 1 and 2 are classification-based methods and therefore, in common with all classifications, assume that the parameters of the emissions algorithm are time-invariant properties of the classes. This is in contrast to Tier 3 approaches, which usually model emissions as continuous processes with parameters that vary in space and time. If the classification is effective, such that within-class variance is small and between-class variance is large, then the Tier 3 approach can be aggregated to Tier 2 by averaging the parameter values across the time and space domains of the classes. However, in the absence of process models, Tier 2 methodologies commonly source their parameter values from measurement campaigns targeted at the vegetation groups that comprise the inventory classification. This introduces the risk of bias to the emission estimates through sampling errors in the parameter space. If within-class variability is small, then bias errors will also be small, but if the classification is not so robust and within-class variability is significant, then so is the potential for bias errors. Directly testing for sampling bias is complex and rare, and although approaches that use independent top-down methods to constrain or verify emission estimates are possible and encouraged by IPCC, they lack sufficient precision to accurately quantify bias errors in inventory parameters. However, occasionally the required parameters can be related to variables with better spatial coverage and hence less risk of spatial bias. This is the approach that we use here to estimate parameter values for EF_{CH4} for the inventory classes of the low rainfall savanna woodland. Where there are no explanatory relationships or process models available, as is the case for N_2O emissions, we are reliant on direct measurements of EF_{N2O} and expert knowledge to select locations and times that are representative of the inventory classes. The proposed low rainfall CFI methodology implements the IPCC (2006) algorithm as a Tier 2 methodology stratified by vegetation class (*i*) fuel size class (*j*) and season (*k*). For methane:

$$E = \Sigma_k(\Sigma_i A_{ik} P_k \Sigma_j (F_{ij} BEF_{jk}\ CC_{ij}\ EF_{ij})) \tag{5}$$

and for N_2O:

$$E = \Sigma_k(\Sigma_i A_{ik} P_k \Sigma_j F_{ij} BEF_{jk}\ CC_{ij}\ NC_{ij} EF_{ij}) \tag{6}$$

where A is area burned, P is patchiness, F is fuel load, BEF is burning efficiency, CC is fuel carbon content, NC is fuel nitrogen to carbon ratio, and EF is emission factor (EF_{CH4} or EF_{N2O}).

Consistent with the national and high rainfall CFI, EF_{CH4}, EF_{N2O}, CC and NC vary only with vegetation class (woodland with hummock grass understorey, WHu; woodland with mixed hummock and tussock grass understorey, WM; woodland with tussock grass understorey, WTu; open woodland with hummock grass understorey, OWHu; open woodland with tussock grass understorey, OWTu; and heathland with hummock grass understorey, HHu) and fuel size class (grass and leaf litter, <6 mm; coarse fuel, 6 mm–50 mm; live shrubs and heavy fuel, >50 mm).

Table 9.9. Fuel load and fractional composition proposed for Low rainfall CFI (from Chapter 6)

Fuel type	Fuel load (t ha^{-1})	Grass (%)	Litter (%)	Coarse (%)	Heavy (%)	Shrub (%)	Fraction of fine fuel	
							Grass	Litter
WHu	9.17	24	24	20	12	20	50	50
WM	5.9	18	25	12	34	11	41	59
WTu	7.3	27	34	19	17	4	44	56
OWHu	5.87	38	17	13	14	19	69	31
OWTu	5.68	34	26	12	14	13	57	43
HHu	5.05	41	27	14	3	18	63	42

Table 9.10. Elemental carbon content (% fuel dry weight) for low rainfall savanna vegetation and fuel size classes

	Fuel size class			
Vegetation type	**Fine <6 mm**	**Coarse 6–50 mm**	**Shrub**	**Heavy >50 mm**
WHu	39.7	48.2	48.5	48.2
WM	39.7	48.2	48.5	48.2
WTu	41.0	48.2	48.5	48.2
OWHu	39.9	48.2	48.5	48.2
OWTu	41.5	48.2	48.5	48.2
HHu	39.8	48.2	48.5	48.2

The Gulf and the Mornington Sanctuary campaigns, coupled with the mean fuel loads for each fuel class described in chapter 6, provide the data from which we estimate the emission factors and elemental content. The fuel elemental content is presented in Table 9.4 and fuel composition is presented in Table 9.9. From these, the CC and NC values applicable for the low rainfall methodology are derived (Tables 9.10 and 9.11) as follows: the inventory strata, coarse and heavy fuels are CWD of different sizes but similar chemical composition; we assume that live shrub elemental content is similar to fresh litter; and for fine fuel we assume that the elemental content is the average of the appropriate grass (hummock or tussock) and leaf litter for each vegetation class. We assume that mixed grass understorey is mostly hummock grass (i.e. spinifex) from sandstone soils. This average is weighted by the relative contribution of grass and litter to burned fine fuel biomass assuming that the BEF for grass is 1 and BEF for litter is 0.9.

The methane emission factor for fine fuel is calculated from MCE using Eqn 2, and MCE is calculated using Eqn 4 from the fraction of grass in fine fuel and assuming a mean grass fuel moisture content of 14%. This value is the mean minimum of the range observed in the field, and hence this step removes the effect of fuel moisture on EF_{CH4} by normalising Eqn 4 to moisture content of dry fuel. EF_{CH4} for the heavy fuels are the mean EF_{CH4} for CWD from the smouldering CWD combustion. For the vegetation classes W-H, W-T and W-M we assign the mean EF_{CH4} for smouldering CWD observed in the Gulf campaign (0.0158, Table 9.5) and for the OWHu, OWTu and HHu we assign the EF_{CH4} for smouldering CWD observed at Mornington Sanctuary (0.0111, Table 9.6). Shrubs rarely burn in the absence of flaming combustion from an underlying fine fuel layer; hence, in the absence of

Table 9.11. Nitrogen to carbon ratio for low rainfall savanna vegetation and fuel size classes

	Fuel size class			
Vegetation type	**Fine <6 mm**	**Coarse 6–50 mm**	**Shrub**	**Heavy >50 mm**
WHu	0.0113	0.00389	0.01497	0.00389
WM	0.0118	0.00389	0.01497	0.00389
WTu	0.0105	0.00389	0.01497	0.00389
OWHu	0.0102	0.00389	0.01497	0.00389
OWTu	0.0096	0.00389	0.01497	0.00389
HHu	0.0107	0.00389	0.01497	0.00389

Table 9.12. Methane emission factors for low rainfall savanna

Fuel type	EF_{CH4} (% burned fuel C)			
	Fine <6 mm	Coarse 6–50 mm	Shrub	Heavy >50 mm
WHu	0.15	0.15	0.15	1.58
WM	0.17	0.17	0.17	1.58
WTu	0.16	0.16	0.16	1.58
OWHu	0.12	0.12	0.12	1.11
OWTu	0.14	0.14	0.14	1.11
HHu	0.13	0.13	0.13	1.11

any reliable measurements of EF_{CH4} from pure shrub fuels, we will assume that EF_{CH4} for shrub fuels are equal to EF_{CH4} from fine fuels. Coarse fuels (6 mm–50 mm) are more complex. This fuel size class covers a wide range of combustion rates; at the lower end (<25 mm) are twigs and small sticks that burn quickly, usually within a few minutes (Burrows 2001), and completely. Recent observations of test burns in an experimental wind tunnel (Meyer, unpublished data) largely confirmed the general rule-of-thumb that fuels less than 25 mm tend to sustain combustion until the fuel element is completely consumed, while the fuel elements larger than 25 mm diameter tend to self-extinguish and only burn partially. The coarse fuel size class therefore includes both fully combusting and self-extinguishing fuel elements, with most of the consumption likely to occur in the same period as the fine fuel combustion. Our EF_{CH4} measurements of smoke emission at the fire front includes the flaming coarse fuels, hence we assign EF_{CH4} for the fine fuel to the coarse fuel size class. Emission factors for EF_{CH4} for the low rainfall savanna are presented in Table 9.12.

Nitrous oxide poses more of a challenge, because currently there are no explanatory relationships for EF_{N2O} values. The Mornington campaign provides the only EF_{N2O} vales for the Australian low rainfall savanna (Table 9.6). While these data are larger on average then EF_{N2O} values observed in high rainfall savanna woodland, with the exception of the OW-Tu vegetation class, which does not occur in the high rainfall zone, EF_{N2O} values for all other classes are broadly similar. However the data are measures of single fuel classes, not measures of mixed plumes, and therefore we will need to derive EF_{N2O} for some of the inventory strata from combinations of the Mornington burn results. We apply the following logic:

- For some of the inventory strata the Woozle tests cover MCE ranges and combustion states that might not occur in the field, therefore the field burn samples provide the more reliable measures of EF_{N2O} from combustion of pure fuels in undisturbed fuel beds.
- We estimate EF_{N2O} for the fine fuels to be a weighted average of the mean EF_{N2O} for litter from the Woozle tests (burn M9, Table 9.6) and either (1) EF_{N2O} for hummock grass on sandstone-derived soil (burn M16, Table 9.5) or (2) tussock grass on basalt soil (burns M1 and M12, Table 9.6). The weightings were the same as those applied to EF_{CH4}. In the high rainfall savanna there was little difference between EF_{N2O} from hummock and tussock grasses, both of which are confined to sandstone-derived soils, therefore we assume that EF_{N2O} for W-M is the same as W-Hu.

Table 9.13. Nitrous oxide emission factors for low rainfall savanna

	EF_{N2O} (% burned fuel N)			
Fuel type	**Fine <6 mm**	**Coarse 6–50 mm**	**Shrub**	**Heavy >50 mm**
WHu	0.60	0.60	0.60	1.46
WM	0.60	0.60	0.60	1.46
WTu	1.20	1.20	1.20	1.46
OWHu	0.60	0.60	0.60	1.46
OWTu	1.36	1.36	1.36	1.46
HHu	0.59	0.59	0.59	1.46

- EF_{N2O} for smouldering CWD (burn M13, Table 9.6) is assigned to the heavy fuel class while EF_{N2O} for coarse and shrub fuels are assigned the fine fuel EF_{N2O} values using the logic described above for EF_{CH4}.

Emission factors for EF_{N2O} for the low rainfall savanna are presented in Table 9.13.

For the most part the EF_{CH4} and EF_{N2O} values recommended for the low rainfall CFI are consistent with our knowledge of the vegetation structure and, where direct comparisons are relevant, are relatively minor changes from the factors currently applied in the national methodology and the high rainfall CFI methodology (Table 9.14). The EF_{CH4} values for the fine fuels are a factor of three higher in the open forest, open woodland and sandstone woodland vegetation classes of the high rainfall zone, where fuels are predominantly (>70%) leaf litter, and 50% higher in the sandstone, also due to the a larger contribution of leaf litter to the fine fuels. EF_{CH4} for CWD is similar in both zones, reflecting the higher variance of this parameter rather than a significant difference between means. EF_{N2O} values for fine fuels are similar in both high and low rainfall zones on hummock and mixed hummock/tussock woodland, but higher by a factor of two in the tussock grass woodlands. Tussock grass woodlands are confined to the low rainfall region and have not been measured previously. The EF_{N2O} for CWD is ~fourfold higher than recorded in the high rainfall zone; however, when combined with an N:C value that is half the value of

Table 9.14. Emission factors and fuel elemental content applied in the current Australian national and high rainfall Carbon Farming Initiative savanna burning methodologies

	Fine fuel			CWD		
Vegetation class	**MCE**	**EF_{CH4}**	**EF_{N2O}**	**MCE**	**EF_{CH4}**	**EF_{N2O}**
Open forest	0.92	0.31	0.66	0.87	1.01	0.36
Open woodland	0.92	0.31	0.75	0.87	1.01	0.36
Sandstone woodland	0.92	0.31	0.75	0.87	1.01	0.36
Sandstone heath	0.96	0.15	0.66	0.87	1.01	0.36
Fuel size class	**C (%fuel dwt)**			**N:C**		
Grass	44			0.0087		
Fine	49			0.0101		
Shrub	50			0.0093		
Coarse	50			0.0081		
Heavy	50			0.0081		

high rainfall savanna CWD, the overall difference in N_2O emissions from the heavy fuel strata will be less pronounced. Therefore, the differences in EF_{CH4} and EF_{N2O} between high and low rainfall zones are consistent with known differences between the regions.

Conclusions

To summarise, as the spatial resolution of inventory methodologies increases, so too does the requirement for accurate parameter values for each of the inventory strata. Regional or national inventories can achieve accuracy by spatial averaging. However, project accounting methodologies cannot rely on aggregation to avoid addressing the sources of variability, and therefore assigning parameter values requires greater attention to emission processes, and greater awareness of the underlying assumptions. The use of explanatory functions, such as MCE, fuel moisture and fuel composition, reduces the risks of introducing bias in emission factor estimates; however, that approach was not possible for all strata of the methodology and we will conclude this chapter by highlighting two caveats.

1. Nitrous oxide, in contrast to most other combustion products, has been very poorly studied, probably because, until recently, its chemical analysis was difficult. The result is that the global database is small and therefore our knowledge of the variation in EF_{N2O} across ecosystems is poor in comparison to most other combustion products. The Kulgnuki and Mornington campaigns have substantially improved this situation for Australia, and we are confident that our recommendations for EF_{N2O} values for the low rainfall savanna woodland are robust; nevertheless, only a small number of fuels and fire types have been analysed to date. It is now clear that EF_{N2O} does vary with fuel and with combustion type, and it is probable that our current best estimates of EF_{N2O} will be revised with further measurement campaigns covering a wider range of fires and vegetation. On the positive side, it is also now clear that EF_{N2O} is not dependent on MCE for most of the observed range, and therefore it is highly unlikely to be influenced by factors such as fuel moisture content that may vary seasonally.
2. EF_{CH4} for fine fuels can be estimated with confidence; however, EF_{CH4} from smouldering combustion is less well characterised. We have confirmed that EF_{CH4} can be reliably estimated from MCE, but that the relationship depends on the form of combustion. Volkova *et al.* (2014) report that in southern eucalypt forests fire intensity profoundly affects CWD fuel dynamics, affecting burning efficiency, char production and the combustion to the organic soil surface layer (sometimes called duff). It is likely that mean EF_{CH4} for CWD elements would also vary under these conditions. Whether increased fire intensity leads to an extended flaming phase for CWD and lower EF_{CH4}, or persistent smouldering combustion and larger EF_{CH4}, can only be speculated. Hence, our current estimates of EF_{CH4} for CWD strata must remain indicative rather than definitive.

Determining the significance of these caveats for the accuracy of regional accounts must wait for the outcome of future experimental field studies.

Acknowledgements

This chapter is the result of the efforts of many people. In particular we thank Fabienne Reisen, John Schatz, Adam Liedloff and Dick Williams of CSIRO, who ensured the success of the field campaigns. We thank the Bushfires NT team (now CDU) who planned and coordinated the Gulf campaign, and the Indigenous ranger groups from Robinson River, Wangalingi and Burketown who conducted the experimental burns with skill and safety. The Mornington Sanctuary campaign would not have been possible without the support of the Australian Wildlife Conservancy and we are very grateful for their generous support. The work was commissioned and funded by the North Australian Indigenous Land and Sea Management Alliance.

References

Akagi SK, Yokelson RJ, Wiedinmyer C, Alvarado MJ, Reid JS, Karl T, Crounse JD, Wennberg PO (2011) Emission factors for open and domestic biomass burning for use in atmospheric models. *Atmospheric Chemistry and Physics* **11**, 4039–4072. doi:10.5194/acp-11-4039-2011

Anderson GK, Sandberg DV, Norheim RA (2004) 'Fire Emission Production Simulator (FEPS) User's Guide.' Version 1.0. USDA Forest Service, Pacific Northwest Research Station: Seattle, WA.

Andreae MO, Merlet P (2001) Emission of trace gases and aerosols from biomass burning. *Global Biogeochemical Cycles* **15**, 955–966. doi:10.1029/2000GB001382

Burling IR, Yokelson RJ, Akagi SK, Urbanski SP, Wold CE, Griffith DWT, Johnson TJ, Reardon J, Weise DR (2011) Airborne and ground-based measurements of the trace gases and particles emitted by prescribed fires in the United States. *Atmospheric Chemistry and Physics* **11**, 12197–12216. doi:10.5194/acp-11-12197-2011

Burrows ND (2001) Flame residence times and rates of weight loss of eucalypt forest fuel particles. *International Journal of Wildland Fire* **10**, 137–143. doi:10.1071/WF01005

Byram GM (1959) Combustion of forest fuels. In *Forest Fire Control and Use.* (Ed. KP Davies) pp. 61–89. McGraw-Hill: New York.

Christian TJ, Kleiss B, Yokelson RJ, Holzinger R, Crutzen PJ, Hao WM, Shirai T, Blake DR (2004) Comprehensive laboratory measurements of biomass-burning emissions: 2. First intercomparison of open-path FTIR, PTRMS, and GC-MS/FID/ECD. *Journal of Geophysical Research* **109**, D02311. doi:10.1029/2003JD003874

Cofer WR III, Levine JS, Sebacher DI, Winstead EL, Riggan PJ, Stocks BJ, Brass JA, Ambrosia VG, Boston PJ (1989) Trace gas emissions from chaparral and boreal forest fires. *Journal of Geophysical Research* **94**, 2255–2259. doi:10.1029/JD094iD02p02255

Cofer WR III, Levine JS, Winstead EL, LeBel PJ, Koller AM Jr, Hinkle CR (1990a) Trace gas emissions from burning Florida wetlands. *Journal of Geophysical Research* **95**, 1865–1870. doi:10.1029/JD095iD02p01865

Cofer WR III, Levine JS, Winstead EL, Stocks BJ (1990b) Gaseous emissions from Canadian boreal forest fires. *Atmospheric Environment* **24**, 1653–1659. doi:10.1016/0960-1686(90)90499-D

Cofer WR III, Levine JS, Winstead EL, Stocks BJ (1991) New estimates of nitrous oxide emissions from biomass burning. *Nature* **349**, 689–691. doi:10.1038/349689a0

Cofer WR III, Levine JS, Winstead EL, Cahoon DR, Sebacher DI, Pinto JP, Stocks BJ (1996) Source compositions of trace gases released during African savanna fires. *Journal of Geophysical Research* **101**, 23597–23602. doi:10.1029/95JD02048

Commonwealth of Australia (2013) Carbon Credits (Carbon Farming Initiative) Reduction of Greenhouse Gas Emissions through Early Dry Season Savanna Burning—1.1 Methodology Determination 2013. www.comlaw.gov.au/Series/F2013L01165.

Cook GD (2003) Fuel dynamics, nutrients, and atmospheric chemistry. In *Fire in Tropical Savannas: The Kapalga Experiment.* (Eds AN Andersen, GD Cook, RJ Williams) pp. 47–58. Springer-Verlag: New York.

Delmas R, Lacaux JP, Menaut JC, Abbadie L, Le Roux X, Helas G, Lobert J (1995) Nitrogen compound emission from biomass burning in tropical African savanna FOS/DECAFE 1991 experiment (Lamto, Ivory Coast). *Journal of Atmospheric Chemistry* **22**, 175–193. doi:10.1007/BF00708188

DoE (Australian Government Department of the Environment) (2012) 'National Inventory Report 2012.' Volume 1. The Australian Government Submission to the United Nations Framework Convention on Climate Change Australian National Greenhouse Accounts. Commonwealth of Australia: Canberra.

Gras JL, Meyer CP, Weeks IA, Gillett RW, Galbally IE, Todd J, Carnovale F, Joynt RC, Hinwood A, Berko H, Brown S (2002) 'Emissions from domestic solid fuel burning appliances [wood-heaters, open fireplaces].' Technical report no. 5. Environment Australia: Canberra.

Gullett B, Touati A (2003) PCDD/F emissions from burning wheat and rice field residue. *Atmospheric Environment* **37**, 4893–4899. doi:10.1016/j.atmosenv.2003.08.011

Hao WM, Ward DE (1993) Methane production from global biomass burning. *Journal of Geophysical Research* **98**, 20657–20661. doi:10.1029/93JD01908

Hoffa EA, Ward DE, Hao WM, Susott RA, Wakimoto RH (1999) Seasonality of carbon emissions from biomass burning in a Zambian savanna. *Journal of Geophysical Research* **104**, 13841–13853. doi:10.1029/1999JD900091

Hurst DF, Griffith DWT, Carras JN, Williams DJ, Fraser PJ (1994a) Measurements of trace gases emitted by Australian savanna fires during the 1990 dry season. *Journal of Atmospheric Chemistry* **18**, 33–56. doi:10.1007/BF00694373

Hurst DF, Griffith DWT, Cook GD (1994b) Trace gas emissions from biomass burning in tropical Australian savannas. *Journal of Geophysical Research* **99**, 16441–16456. doi:10.1029/94JD00670

Hurst DF, Griffith DWT, Cook GD (1996) Trace gas emissions from biomass burning in Australia. In *Biomass Burning and Global Change Volume 2: The Tropical and Temperate Zones.* (Ed. JS Levine) pp. 787–792. MIT Press: Cambridge, MA.

IPCC (2000) 'Good Practice Guidance and Uncertainty Management in National Greenhouse Gas Inventories.' IPCC National Greenhouse Gas Inventories Programme, Technical Support Unit, Kanagawa, Japan.

IPCC (2006) Agriculture, forestry and other land use. In *Guidelines for National Greenhouse Gas Inventories.* (Eds S Eggleston, L Buendia, K Miwa, T Ngara, K Tanabe) Vol. 4. Institute for Global Environmental Strategies (IGES) for the Intergovernmental Panel on Climate Change: Hayama, Japan.

IPCC (2014) '2013 Revised Supplementary Methods and Good Practice Guidance Arising from the Kyoto Protocol.' (Eds T Hiraishi, T Krug, K Tanabe, N Srivastava, J Baasansuren, M Fukuda, TG Troxler). IPCC: Switzerland.

Korontzi S (2005) Seasonal patterns in biomass burning emissions from southern African vegetation fires for the year 2000. *Global Change Biology* **11**, 1680–1700. doi:10.1111/j.1365-2486.2005.001024.x

Korontzi S, Ward DE, Susott RA, Yokelson RJ, Justice CO, Hobbs PV, Smithwick EAH, Hao WM (2003) Seasonal variation and ecosystem dependence of emission factors for selected trace gases and PM2. 5 for southern African savanna fires. *Journal of Geophysical Research* **108**, 8406. doi:10.1029/2003JD003730

Korontzi S, Roy DP, Justice CO, Ward DE (2004) Modeling and sensitivity analysis of fire emissions in southern Africa during SAFARI 2000. *Remote Sensing of Environment* **92**, 255–275. doi:10.1016/j.rse.2004.06.010

Laursen KK, Hobbs PV, Radke LF, Rasmussen RA (1992) Some trace gas emissions from North American biomass fires with an assessment of regional and global fluxes from biomass burning. *Journal of Geophysical Research* **97**, 20687–20701. doi:10.1029/92JD02168

Lobert JM, Scharffe DH, Hao WM, Crutzen PJ (1990) Importance of burning in the atmospheric budgets of nitrogen-containing gases. *Nature* **346**, 552–554.

Meyer CP, Beer T, Müller J, Gillett R, Weeks I, Powell J, Tolhurst K, McCaw L, Cook GD, Marney D (2004) 'Dioxins Emissions from Bushfires in Australia.' Technical Report No. 1. Environment Australia: Canberra.

Meyer CP, Mueller J, Symons RK (2009) 'Determination of the source and level of dioxin emissions from bushfires in Australia, National Dioxins Program. Final report to the Department of Water, Heritage and the Arts.' CSIRO Marine and Atmospheric Research: Aspendale.

Meyer CP, Cook GD, Reisen F, Smith TEL, Tattaris M, Russell-Smith J, Maier SW, Yates C, Wooster MJ (2012) Direct measurements of the seasonality of emission factors from savanna fires in northern Australia. *Journal of Geophysical Research* **117**, D20305. doi:10.1029/2012JD017671

Muzio LJ, Kramlich JC (1988) An artifact in the measurement of N_2O from combustion sources. *Geophysical Research Letters* **15**, 1369–1372.

Nance JD, Hobbs PV, Radke LF, Ward DE (1993) Airborne measurements of gases and particles from an Alaskan wildfire. *Journal of Geophysical Research* **98**, 14873–14882. doi:10.1029/93JD01196

Paton-Walsh C, Smith TEL, Young EL, Griffith DWT, Guérette É-A (2014) New emission factors for Australian vegetation fires measured using open-path Fourier transform infrared spectroscopy. Part 1: methods and Australian temperate forest fires. *Atmospheric Chemistry and Physics Discussion* **14**, 4327–4381. doi:10.5194/acpd-14-4327-2014

Prichard SJ, Ottmar RD, Anderson GK (2006) Consume 3.0 User's Guide. USDA Forest Service. Pacific Northwest Research Station, Seattle, WA. www.fs.fed.us/pnw/fera/research/smoke/consume/index.shtml

Russell-Smith J, Murphy BP, Meyer CP, Cook GD, Maier S, Edwards AC, Schatz J, Brocklehurst P (2009) Improving estimates of savanna burning emissions for greenhouse accounting in northern Australia: limitations, challenges, applications. *International Journal of Wildland Fire* **18**, 1–18. doi:10.1071/WF08009

Seiler W, Crutzen P (1980) Estimates of gross and net fluxes of carbon between the biosphere and the atmosphere from biomass burning. *Climatic Change* **2**, 207–247. doi:10.1007/BF00137988

Shirai T, Blake DR, Meinardi S, Rowland FS, RussellSmith J, Edwards A, Kondo Y, Koike M, Kita K, Machida T, Takegawa N, Nishi N, Kawakami S, Ogawa T (2003) Emission estimates of selected volatile organic compounds from tropical savanna burning in northern Australia. *Journal of Geophysical Research* **108**, 8406. doi:10.1029/2001JD000841

Smith TEL, Paton-Walsh C, Meyer CP, Cook GD, Maier SW, Russell-Smith J, Wooster MJ, Yates CP (2014) New emission factors for Australian vegetation fires measured using open-path Fourier transform infrared spectroscopy. Part 2: Australian tropical savanna fires. *Atmospheric Chemistry and Physics Discussion* **14**, 6311–6360. doi:10.5194/acpd-14-6311-2014

Sullivan AL, Ball R (2012) Thermal decomposition and combustion chemistry of cellulosic biomass. *Atmospheric Environment* **47**, 133–141. doi:10.1016/j.atmosenv.2011.11.022

Sullivan AL, Knight IK, Hurley RJ, Webber C (2013) A contractionless, low-turbulence wind tunnel for the study of free-burning fires. *Experimental Thermal and Fluid Science* **44**, 264–274. doi:10.1016/j.expthermflusci.2012.06.018

Volkova L, Meyer CP, Murphy S, Fairman T, Reisen F, Weston C (2014) Fuel reduction burning mitigates wildfire effects on forest carbon and greenhouse gas emission. *International Journal of Wildland Fire***23**(6), 771–780. doi:10.1071/WF14009

Ward DE, Radke LF (1993) Emissions measurements from vegetation fires: a comparative evaluation of methods and results. In *Fire in the Environment: The Ecological, Atmospheric, and Climatic Importance of Vegetation Fires*. (Eds PJ Crutzen, JG Goldammer) pp. 53–76. John Wiley and Sons: New York.

Ward DE, Hao WM, Susott RA, Babbitt RE, Shea RW, Kauffman JB, Justice CO (1996) Effect of fuel composition on combustion efficiency and emission factors for African savanna ecosystems. *Journal of Geophysical Research* **101**, 23569–23576. doi:10.1029/95JD02595

Wiedinmyer C, Akagi SK, Yokelson RJ, Emmons LK, Al-Saadi JA, Orlando JJ, Soja AJ (2011) FINN The Fire INventory from NCAR (FINN): a high resolution global model to estimate the emissions from open burning. *Geoscientific Model Development* **4**, 625–641. doi:10.5194/gmd-4-625-2011

Wotton BM, Gould JS, McCaw WL, Cheney NP, Taylor WS (2012) Flame temperature and residence time of fires in dry eucalypt forest. *International Journal of Wildland Fire* **21**, 270–281.

Yokelson RJ, Burling IR, Gilman JB, Warneke C, Stockwell CE, de Gouw J, Akagi SP, Urbanski SK, Veres P, Roberts JM, Kuster WC, Reardon J, Griffith DWT, Johnson TJ, Hosseini S, Miller JW, Cocker DRIIIJung H, Weise DR (2013) Coupling field and laboratory measurements to estimate the emission factors of identified and unidentified trace gases for prescribed fires. *Atmospheric Chemistry and Physics* **13**, 89–116. doi:10.5194/acp-13-89-2013

Yokelson RJ, Christian TJ, Karl TG, Guenther A (2008) The tropical forest and fire emissions experiment: laboratory fire measurements and synthesis of campaign data. *Atmospheric Chemistry and Physics* **8**, 3509–3527.

10

Application of a lower rainfall savanna burning emissions abatement methodology

Jeremy Russell-Smith, Cameron P Yates, Jay Evans, CP (Mick) Meyer and Andrew C Edwards

Summary

This chapter sets out the recommended calculation framework for lower rainfall savanna burning projects as a formal methodology for inclusion in Australia's national agricultural emissions offset program, the Carbon Farming Initiative. With reference to three project-scale worked examples, the chapter illustrates that three factors have a significant bearing on achievable emissions abatement: the size of the project area, the proportional extent of respective fuel types, and the baseline fire frequency.

Introduction

Based on data and discussion presented in earlier chapters, here we set out a recommended Carbon Farming Initiative methodology for calculating accountable (CH_4 and N_2O) greenhouse gas emissions from savanna burning activities under lower rainfall (600–1000 mm p.a.) conditions. For practical purposes the calculation framework adopted in this methodology follows closely that of the approved 'higher rainfall' methodology (Commonwealth of Australia 2013), with noted exceptions. Key parameter tables are provided in the Appendix.

In the final section we present worked examples of savanna burning accountable emissions, and realistically achievable emissions abatement, for three project areas in the Northern Territory's Gulf region.

Fire frequency in the lower rainfall savanna burning abatement regional envelope

Over the 14-year period 2000–2013 for which fire mapping derived from MODIS imagery is available (Fisher and Edwards, Chapter 3), mean fire frequency in the regional envelope describing the lower rainfall savanna burning abatement methodology was 28%, with 7% occurring in the early dry season (EDS) and 21% occurring in the late dry season (LDS) (Table 10.1). Mean annual fire extent increases progressively from 19% in the 600–699 mm isohyet class to 35% in the 900–999 mm class. LDS fires dominate in all isohyet classes

Table 10.1. Seasonal distribution (mean ± s.e.m) of fire over the period 2000–2013 in respective isohyet classes for the lower rainfall savanna burning abatement envelope

Fire mapping derived from MODIS imagery.

Rainfall isohyet envelope (mm)	Area (km^2)	Mean annual proportion burnt (%)		
		Early dry season	Late dry season	Annual
600–699	211 308	4 ± 0.75	16 ± 2.35	19 ± 2.83
700–799	180 984	6 ± 1.3	17 ± 2.65	23 ± 3.48
800–899	192 357	9 ± 1.07	20 ± 2.55	30 ± 2.99
900–999	120 449	12 ± 0.85	23 ± 2.41	35 ± 2.44
Total	**705 098**	**7 ± 0.92**	**21 ± 2.37**	**28 ± 2.89**

(Table 10.1). The geographic patterning of fire frequency for the entire lower rainfall region is given in Fig. 10.1. Based on experience with higher rainfall (>1000 mm p.a.) savanna burning projects, these data indicate that lower rainfall savanna burning projects should be feasible, especially under relatively high rainfall LDS fire frequency conditions.

Outline of the lower rainfall savanna burning abatement methodology

The formulae for calculating methane and nitrous oxide emissions from savanna burning are provided in Eqns 1 and 2 below (Russell-Smith *et al.* 2009; Commonwealth of Australia 2013).

For methane, CH_4:

$$E_{oc} = M_o \sum_{pk} \left(A_{pk} P_k \sum_{l} \left(EF_{pl} FL_{knpl} CC_l \sum_{m} (S_m BEF_{klm}) \right) \right) \quad (1)$$

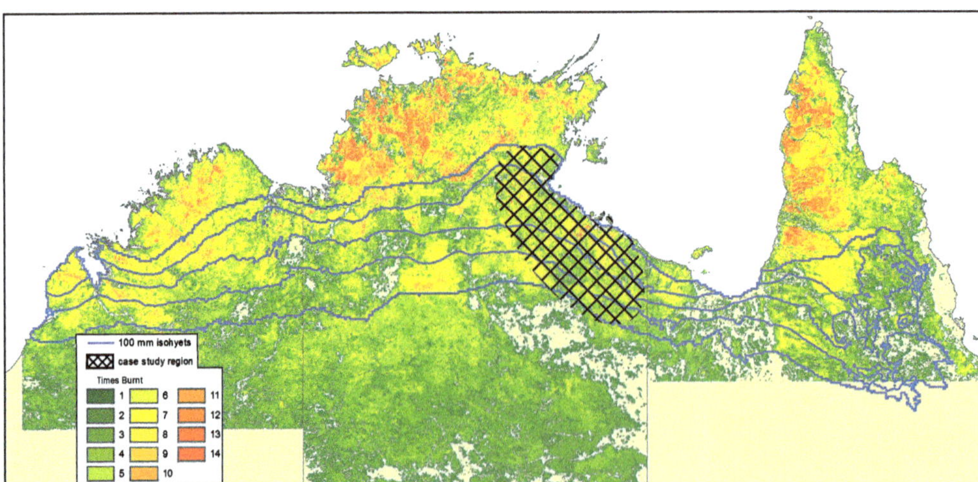

Figure 10.1. Fire frequency (2000–2013) for the lower rainfall savanna burning methodology envelope with respect to rainfall isohyet classes. Fire mapping derived from MODIS imagery. Applying the methodology outlined in this chapter, savanna burning emissions calculations are presented for three project examples included within the hatched area.

For nitrous oxide, N_2O:

$$E_{on} = M_o \sum_{pk} \left(A_{pk} P_k \sum_{l} \left(EF_{pl} FL_{knpl} CC_l NC_l \sum_{m} (S_m BEF_{klm}) \right) \right) \quad (2)$$

Descriptions of all parameters and sub-classes in Eqns 1 and 2 are given in Table 10.2. Both equations are the same as those for the higher rainfall methodology (Russell-Smith *et al.* 2009; Commonwealth of Australia 2013), with the exception that this lower rainfall methodology includes an additional seasonal term for fuel load (i.e. FL_k) reflecting greater leaf litter inputs in the late dry season period as described in Chapters 6 and 7. Five parameters in this lower rainfall methodology warrant further explanation. Additionally, we consider further the issue of setting the baseline (pre-project) period discussed previously in Chapter 2.

Fire season

Prior to the development of the earlier higher rainfall savanna burning methodology (Commonwealth of Australia 2013), substantial effort was given to see if fire seasons could be regionalised (i.e. 'early vs. late') across northern Australia on the anecdotal evidence that, in a typical year, late dry season (LDS) conditions (i.e. fully cured fine fuels at landscape scales fuelling extensive wildfires) appear to develop earlier in the west (e.g. West Kimberley) and progressively develop a month or two later in the east (e.g. Cape York Peninsula). A prospective method for determining a regionalised approach was illustrated in Maier and Russell-Smith (2012), where the onset of late dry season conditions was described as the month where the proportion of night-time:day-time 'hotspots' detected from satellite imagery (refer Jacklyn, Chapter 5) reached a certain threshold – in other words, when it appeared that savanna fires typically start burning through the night. The method was shown to work for western and northern Australia, but apparently not for Cape York, and in consequence the approach was abandoned in favour of a consistent seasonal cut-off to be applied across northern Australia – 31 July being prescribed as the end of the early dry season (EDS) period.

Although seemingly arbitrary, and evidently warranting further investigation, we have maintained this convention in the present methodology. For practical purposes it is evident from long-term modelled rainfall data that if we simply define the wet season period as being those months which consistently receive 50 mm rainfall p.a., the dry season period effectively begins in May across most of the lower rainfall savanna region, and ends mostly in November or else December (Fig. 10.2). As illustrated also in Table 10.1, currently 75% of fire extent across the lower rainfall region occurs typically as wildfire/unmanaged fire in the above defined late dry season period.

Fire severity class

We have adopted essentially the same plot-based assessment procedure for determining fire severity class (parameter *m* – refer Table 10.2) as given in Russell-Smith *et al.* (2009). For the higher rainfall abatement methodology we utilised published fire severity records for sandstone vegetation derived from long-term monitoring plots in Kakadu and Nitmiluk national parks (Russell-Smith and Edwards 2006). Here we have utilised fire severity assessments for 392 fires recorded from 267 long-term plots in the lower rainfall region over the period 2009–2013, as described in Chapter 6.

Although substantial work is ongoing to develop a remotely sensed fire severity mapping product (refer Chapter 8), further development is required especially under lower

Table 10.2. Descriptions of parameters (capital letters) and classes (non-capitals) relating to methodology Eqns 1 and 2

Parameters and classes	Description	Classes	References
A	Fire affected (burnt) area (ha), estimated for respective fire season *k*		Commonwealth of Australia (2013), Fisher and Edwards, Chapter 3
BEF	Burning efficiency factor – an estimate of the proportion of the fuel affected by fire which is emitted (pyrolised), estimated for respective fire season *k*		Russell-Smith *et al.* (2009); Yates *et al.*, Chapter 6
CC	Carbon content of fuel (gram of carbon per gram of dry fuel)		Commonwealth of Australia (2013)
EF	Emission factor (% of fuel elemental content released in fire)		Meyer *et al.*, Chapter 9
E_{oc}	Emission (Gg) of CH_4		Commonwealth of Australia (2013)
E_{on}	Emission (Gg) of N_2O		Commonwealth of Australia (2013)
FL	Fuel load (t dry matter ha^{-1})		Yates *et al.*, Chapter 6; Cuff and Brocklehurst, Chapter 7
k	*fire season*	• Early dry season (EDS), before 1 August • Late dry season (LDS) from 1 August	Russell-Smith *et al.* (2009); Commonwealth of Australia (2013)
l	*fuel size class*	• Fine fuels (litter, grass) ≤0.6 mm diameter • Coarse woody fuels (sticks) >0.6–50 mm diameter • Heavy woody fuels (branches, logs) >50 mm diameter • Shrubs (woody plants) <50 mm diameter at breast height	Russell-Smith *et al.* (2009); Yates *et al.*, Chapter 6
M	Ratio of molecular mass to the elemental mass: CH_4 = 1.3333, N_2O = 1.5714		Commonwealth of Australia (2013)
m	*fire severity class*	• Low severity (vegetation scorched to 2 m; often patchy fires) • moderate severity (tree canopy layer unscorched) • High (tree canopy layer scorched)	Russell-Smith *et al.* (2009); Edwards and Russell-Smith, Chapter 8

Table 10.2. (Continued)

Parameters and classes	Description	Classes	References
n	*number of years since the patch of vegetation was last burned*	derived from fire mapping, where each mapped pixel (250 × 250 m MODIS imagery) is attributed as being 1, 2, 3, 4, 5, or more than 5 years unburnt	Commonwealth of Australia (2013)
NC	Elemental nitrogen to carbon ratio		Commonwealth of Australia (2013)
o	*greenhouse gas species*	*oc* = CH_4; *on* = N_2O	Commonwealth of Australia (2013)
P	Patchiness – an estimate of the mean proportion of mapped burnt pixels affected by fire in respective fire season *k*		Yates *et al.*, Chapter 6
p	*vegetation fuel class*	• WHu – Woodland with hummock grasses • WMi – Woodland with mixed hummock and tussock grasses • WTu – Woodland with tussock grasses • OWM – Open woodland with mixed grasses • SHH – Shrubland heath with hummock grasses	Lynch *et al.*, Chapter 4; Yates *et al.*, Chapter 6; Meyer *et al.*, Chapter 9
S	Severity class – fraction of fires of severity class *m* in fire season *k*		Edwards and Russell-Smith, Chapter 8

rainfall, relatively more open-canopied savanna conditions before it can be used in emissions accounting methodology applications. However, once the procedure is deemed sufficiently reliable it will remove the necessity for considering fire seasonality at all, and instead emissions calculations could be based directly on remotely sensed observations of fire severity.

Vegetation fuel class

Four vegetation fuel classes (eucalypt open forest, eucalypt woodland, sandstone woodland, sandstone heath) are included in the higher rainfall savanna burning methodology (Commonwealth of Australia 2013). In Chapter 4, Lynch *et al.* differentiate nine fuel types that occur to some extent across the lower rainfall region. Of these, Lynch *et al.* consider that four fuel types need to be either amalgamated or omitted as follows:

1. The eucalypt open forest type is of such restricted extent across the region that it should be amalgamated with woodland classes where it occurs.
2. Lancewood (*Acacia shirleyi*-dominated) communities should be excluded given that internal patches are effectively non-flammable due to very sparse leaf litter fuels.

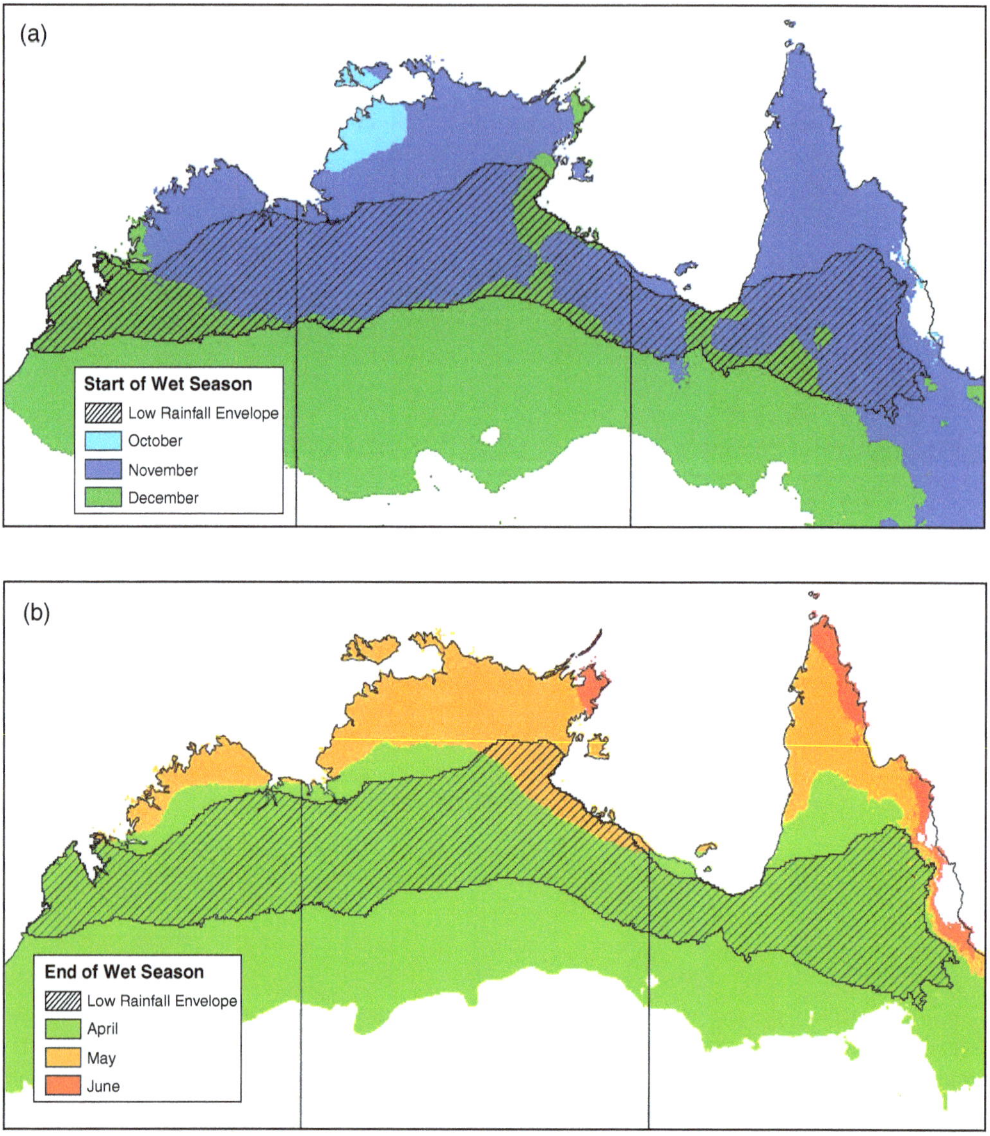

Figure 10.2. 'Start' (a) and 'end' (b) of wet season period, as defined by the mean month per 0.05 × 0.05° grid cell over the period 1970–2012: (a) first receiving 50 mm rainfall, (b) last receiving 50 mm rainfall, over respective annual wet season periods. Rainfall data were acquired from the Bureau of Meteorology as monthly rainfall grids, each of 0.05 × 0.05°, generated using the Barnes successive correction technique (Jones and Weymouth 1997; Jones *et al.* 2006).

3 and 4. Fertile open woodlands with tussock grasses and open tussock grasslands are effectively indistinguishable, in many situations, differing only to the extent that the latter type may support 'sparse trees' whereas the open woodland type may have up to 10% tree foliage projective cover. Omitting both fuel types is also consistent with arguments advanced in a recent assessment of analogous fuel types occurring under higher rainfall conditions (Russell-Smith *et al.* 2014), namely

> that these relatively productive fuel types comprise a major cattle industry pasture resource and are consequently seldom if ever burnt, and that application of a savanna burning methodology focusing on imposition of relatively low intensity patchy fires may result in the perverse outcome of facilitating woody plant invasion and thickening.

As an indication of the relative distribution of the five accountable fuel classes, being woodland with tussock grasses (WTu), woodland with hummock grasses (WHu), woodland with mixed tussock and hummock grasses (WMi), open woodland with mixed grasses (OWM) and heath shrublands with hummock grasses (SHH) across the lower rainfall region, a coarse resolution fuel class map is given in Fig. 10.3. This map is derived from the 1:1 000 000 'Vegetation of the Australian tropical savannas' (Fox *et al.* 2001b), comprising 249 map units representing the natural vegetation of the 22 biogeographical regions (Environment Australia 2000) covering the Australian tropical savannas. For present purposes, we have re-classified the map units (Table 10.3) within the lower rainfall region, using detailed floristic, structural and ground-storey descriptions as given in Fox *et al.* (2001a). We note that the scale of these data is too coarse for the requirements of describing savanna burning project areas under a lower rainfall methodology.

These indicative mapping data suggest that the major eligible fuel types conducive for undertaking early dry season savanna burning emissions abatement activities comprise open woodland with mixed grasses (26% of the region) and woodland with tussock grasses (23%). A further 25% of the region comprises ineligible fuel types, especially open woodland tussock grasslands.

Fuel load

As described by Yates *et al.* (Chapter 6), fuel load sampling was undertaken under relatively unmodified savanna conditions, avoiding very heavily grazed, weedy, and cleared situations. Although it is estimated that as much as 90% of the northern savannas are ostensibly used for pastoral (beef cattle) productions (NLAW 2009), the reality is that only

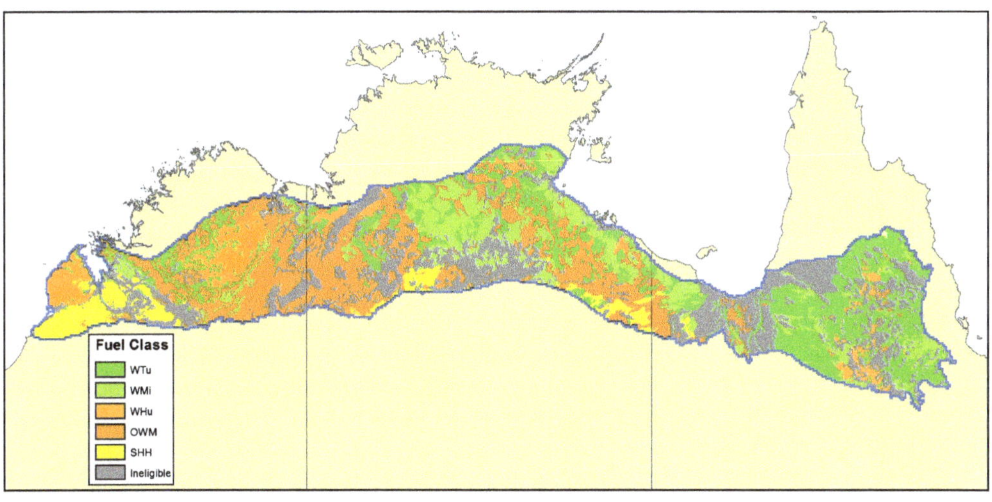

Figure 10.3. Indicative distribution of the lower rainfall fuel classes. Mapping is derived from 1:1 000 000 digital vegetation mapping of the Australian tropical savannas (Fox *et al.* 2001b).

Table 10.3. The reclassification of the 'Vegetation of the Australian tropical savannas' 1:1 000 000 classes (Fox *et al.* 2001a) for the lower rainfall classification scheme

CFI class	Vegetation of the Australian tropical savannas (1:1 000 000) classes
WTu	C3, 4, 12–17, 19; D3, 8, 10, 12, 14, 20, 22, 25, 35, 54; E2; F4–6; G1; H5–6, 10, 12–13; J3, 5–6, 8–11; K4, 10, 14–15
WMi	D7, 9, 11, 13, 15–17, 23, 26, 32; J13, 17; K13, 22
WHu	D21, 28; H9, 11, 18; J4
OWM	C27; D27, 30, 33, 34, 44–47, 49, 51, 53; E4; F9, 14; G3, 4, 14; H3, 7, 14, 17, 20–28; J14, 15; K17–19, 24–26
SHH	D42, 43, 50, 55, 57–58, 63, 65, 68; G12

small fertile areas of that vast estate are utilised relatively intensively and economically for pastoral production (Dyer *et al.* 2001; McCosker *et al.* 2010; Gleeson *et al.* 2012). As for the higher rainfall methodology, we have thus estimated fuel load values, and allowed for extensive grazing in the methodology itself, under ambient low-to-moderate grazing utilisation conditions.

As described in Chapter 6, there is a substantial literature comparing the utility of stocking rates vs. annual utilisation rates (AURs) and, applying available pastoral industry standards, it can be argued that, as a general rule applicable to all soil types, heavily grazed paddocks with average AURs exceeding 20% should be excluded (Dr Dionne Walsh, pers. comm.). However, the requirements of project proponents to demonstrate AURs at paddock scale would be both very onerous and presumably reliant on sophisticated modelling (e.g. Walsh and Cowley 2011). For these pragmatic reasons, and in keeping with the original higher rainfall methodology, we consider it appropriate that projects should be approved, and credits granted, on the basis that while cattle may be present on the project area they should not be introduced, nor should cattle stocking rates be materially increased, to reduce fuel loads and emissions from fire under this methodology (Commonwealth of Australia 2013, p. 10).

As noted also in Chapter 6, for establishing values for fine fuel accumulation with time we have utilised Olson curve (Olson 1963) as opposed to linear natural log-transformed relationships, even though the latter afford greater predictive power. This methodological choice was determined by the precedent set in national greenhouse accounts (DoE 2014) and Carbon Farming Initiative determinations (Commonwealth of Australia 2013) concerning calculation of fuel loads under higher rainfall savanna burning conditions. Olson curves implicitly address biological processes describing accumulation modulated by decay. Conversely, linear relationships provided in Chapter 6 make no such assumptions although arguably better reflect inherent measurement variability associated with time-for-space sampling and different grazing and environmental site conditions.

Emission factors

Full details of the measurement and assignment of CH_4 and N_2O emission factors to the vegetation and fuel class strata for the lower rainfall methodology are presented by Meyer and Cook, Chapter 9. Here we summarise the key points.

The CH_4 and N_2O emission factors assigned to the vegetation and fuel class strata of the higher rainfall CFI were sourced from the study of Meyer *et al.* (2012). There are close similarities between these vegetation fuel classes and many of the low rainfall vegetation types, hence the high rainfall methodology has informed the analysis and assignment of emission factors for the low rainfall methodologies. Clearly there are also differences. The high rainfall open forest and woodland classes have a dominant tree overstorey, and hence most of the fine fuel category consists of tree leaves and small twigs. Grass litter contributed from the mixed hummock/tussock understorey comprised less than 25% of the fine fuel mass. Leaf litter and grass litter support contrasting modes of combustion. Due to the open structure of grass hummocks and tussocks, grass fuels burn efficiently; tree leaf litter lies flat on the ground and therefore is less efficiently aerated, and burns at a lower temperature because some of the heat of combustion is lost to the ground. Hence tree leaf litter burns at lower modified combustion efficiency (MCE: refer Chapter 9 for definition) and therefore has a higher EF_{CH4} than grass litter. In general, the vegetation of the lower rainfall savanna is substantially grassier than the higher rainfall open mixed hummock/tussock grass woodlands; tussock grass open woodlands are confined to the low rainfall region. The new measurements presented in Chapter 9 specifically addressed these issues, and the differences between vegetation types are reflected in the new emission factors.

For the lower rainfall methodology:

1. the EF_{CH4} values for fine, coarse woody, and shrub fuels are about half those of comparable open forest and woodland fuel types in the higher rainfall methodology; this is a consequence of the substantially higher grass content of the fine fuel, and therefore the more efficient combustion and higher MCE values. Thus, EF_{CH4} values for all vegetation classes in the lower rainfall zone are now similar to the EF_{CH4} of hummock grass-dominated shrubland in the higher rainfall zone. Combustion of heavy fuels is similar in both high and low rainfall zones, and therefore the EF_{CH4} for heavy fuels is effectively the same in both methodologies.
2. globally, N_2O emission factor measurements are sparse and variable; however, in contrast to EF_{CH4}, the source of the variability remains unknown. The dataset produced by Meyer *et al.* (2012) was one of the more extensive contributions; the new data presented in Chapter 9 substantially increase the size of the database, and consequently there have been revisions to EF_{N2O} values, partly due to a larger sample base and partly due to sampling new vegetation classes. EF_{N2O} for fine, coarse woody, and shrub fuels are slightly greater in the lower rainfall region, 0.0075 g N_2O N (g emitted N)$^{-1}$ compared to 0.006 g N_2O N (g emitted N)$^{-1}$ for most fuel classes and fuel types, save WTu. For WTu, EF_{N2O} is 0.0120, i.e. approximately double EF_{N2O} for the other fuels, and is the mean of tree litter and pure tussock grass weighted by their respective contributions to the fine fuel mass. This value is well within the non-Australian, global database range (0.009–0.052 g N_2O N (g emitted N)$^{-1}$; Andreae and Merlet 2001; Akagi *et al.* 2011), but substantially higher than EF_{N2O} values measured to date for other Australian fuel and vegetation classes. The EF_{N2O} for heavy fuel, 0.0146, is also larger than the high rainfall value (0.0036 g N_2O N (g emitted N)$^{-1}$). The two sets of values are statistically different, reflecting differences in fuel/combustion properties.

Setting the baseline pre-project period

For the existing higher rainfall methodology, emissions baselines are set as the annual mean estimated over 10 years, with an additional 5 years of fire mapping required to establish fuel loads at the start of the emissions baseline. In Chapter 2, Whitehead *et al.* (2014) suggest at least two ways in which baselines in the lower rainfall methodology could achieve equivalent precision to Fish River and WALFA projects in the higher rainfall region: careful choice of project sites where fire has been relatively less variable; and extending baselines to include more annual observations to reduce error in estimation of the mean. Taking a random set of 40 lower rainfall 30 000 km^2 sites, Whitehead *et al.* (2014) found that 22.5% were at least as precise as the WALFA precision estimate of 13.3% when the baseline period was calculated over 15 years. Given that project sites will never be selected randomly but, rather, comprise savanna vegetation frequently burnt, particularly under LDS conditions, we conclude that 15-year baseline periods can effectively provide equivalent emissions estimate precision as that observed for higher rainfall abatement projects, and as such are to be applied to the lower rainfall abatement methodology.

Applying the lower rainfall abatement methodology to project examples

Key input parameter tables required for the lower rainfall methodology are set out in the Appendix. For convenience, these are presented in the same general order and format as set out in the earlier higher rainfall abatement methodology (Commonwealth of Australia 2013). For the purposes of setting pre-project baselines, we have adopted the recommendation of Whitehead *et al.* (Chapter 2) that a 15-year baseline calculation period is required to address increased variability in inter-annual fire extent in the lower rainfall region. Although most parameter values differ from those used in the earlier published higher rainfall methodology (Commonwealth of Australia 2013), all other calculation procedures are similar with the previously noted exception that here we have adopted separate fuel load (*FL*) estimates per fire season (*k*).

For illustrative purposes we have applied the lower rainfall savanna burning abatement methodology to three realistic example projects occurring in the Northern Territory's Gulf region (Fig. 10.1). Fuel type mapping is available for this region (Lynch *et al.*, Chapter 4). Taking into account excluded fuel types, net project areas for Projects 1 and 2 are of the order of 15 000 km^2, and 9000 km^2 for Project 3 (Table 10.4). Projects 1 and 3 comprise a majority of woodland fuel types, whereas Project 2 comprises mostly open woodland and shrubland heath (Table 10.4). Mean annual fire frequency over the assessed 15-year baseline period ranges from 19% for Project 1 to 43% for Project 3; LDS fire extent predominates in all project areas (Table 10.5).

Resultant accountable baseline annual emissions range from 73 602 t CO_2-e for Project 1 to 33 963 t CO_2-e for Project 2 (Table 10.5). Assuming that implementation of more

Table 10.4. Proportion of accountable fuel types* in three example project areas

Note that accountable fuel types add to only a proportion of respective total project areas. Refer Table 10.2 for full spelling of fuel types.

Project	Total area (km^2)	Net area* (km^2)	Accountable vegetation fuel type as a proportion of total project area				
			WTu	WMi	WHu	OWM	SHH
1	21 772	14 236	30	10	5	16	4
2	19 955	15 224	11	2	0	44	19
3	10 000	8696	18	23	29	14	3

* excluding non-flammable vegetation (lancewood, coastal plains, coastal flats, marine plains), and excluding also open woodland with tussock grasses, and tussock grassland fuel types on fertile soils.

Table 10.5. Proportion of example project areas burnt in EDS and LDS periods, resultant accountable baseline emissions, and projected abatement assuming potential 38% reduction emissions annually achievable*

Project	Proportion burnt (%)			Baseline emissions (t CO_2-e)			Projected abatement* (t CO_2-e)	Projected abatement* per net project area (t CO_2-e km^{-2})
	EDS	LDS	Annual	EDS	LDS	Annual		
1	5	13	19	12 268	61 334	73 602	27 969	1.96
2	5	20	25	4860	29 103	33 963	11 059	0.73
3	16	27	43	12 554	35 532	48 086	18 273	2.10

* based on example of Western Arnhem Land Fire Abatement project, 2005–2011 (Russell-Smith *et al.* 2013).

strategic EDS fire management could deliver annual emissions abatement of at least 38% (based on Western Arnhem Land Fire Abatement (WALFA) experience – refer Russell-Smith *et al.* 2013), resultant annual accountable emissions abatement would conservatively range from 27 969 t CO_2-e for Project 1 to 11 059 t CO_2-e for Project 2 (Table 10.5). Expressed on a per unit area basis, both Projects 1 and 3 could realistically annually deliver ~2.0 t CO_2-e km^{-2} whereas Project 2 would deliver less than half this (Table 10.5). By contrast, the equivalent emissions abatement per unit area achieved by WALFA over its first 7 years was 4.18 t CO_2-e km^{-2} (Russell-Smith *et al.* 2013). As outlined by Yates *et al.* (Chapter 6), the lesser rate of return achievable under the lower rainfall savanna burning abatement methodology is attributable to lower fuel loads, and especially to markedly lesser seasonal differentiation in BEF parameters.

As illustrated in Table 10.6, the disproportionately larger emissions abatement achievable in Projects 1 and 3 reflects the much larger emissions resulting from woodland fuel types. As noted in Chapter 6, fine fuel accumulation in WTu and WHu fuel types is substantially greater than for other fuel types and, as noted in Chapter 9, N_2O emission factors for fine, coarse and shrub fuels in WTu are double those than of all other fuel types. It follows from the above that, when assessing the viability of lower rainfall savanna burning project areas, at least three factors addressed here have a significant bearing: the size of the accountable project area, the proportional extent of respective fuel types, and the baseline fire frequency.

The accounting and reporting process for developing and implementing lower rainfall savanna burning abatement projects is summarised in Fig. 10.4. As noted in the caption, this flow diagram is current at the time of chapter finalisation (August 2014), but is likely

Table 10.6. Relative accountable emissions per unit area achieved from burning respective fuel types in EDS and LDS periods applying the lower rainfall savanna burning abatement methodology

Calculation based on burning an entire 100 ha site where 1/6th (16.66%) of fuels have accumulated over 1 year since-last-burnt, 1/6th of fuels have accumulated over 2 years since-last-burnt ... and 1/6th of fuels were last burnt more than 5 years previously.

Fire season	Accountable emissions achievable per vegetation fuel type (t CO_2–e/100 ha)				
	WTu	WMi	WHu	OWM	SHH
EDS	16.64	8.66	9.99	7.05	6.07
LDS	24.81	14.97	15.61	10.47	8.76

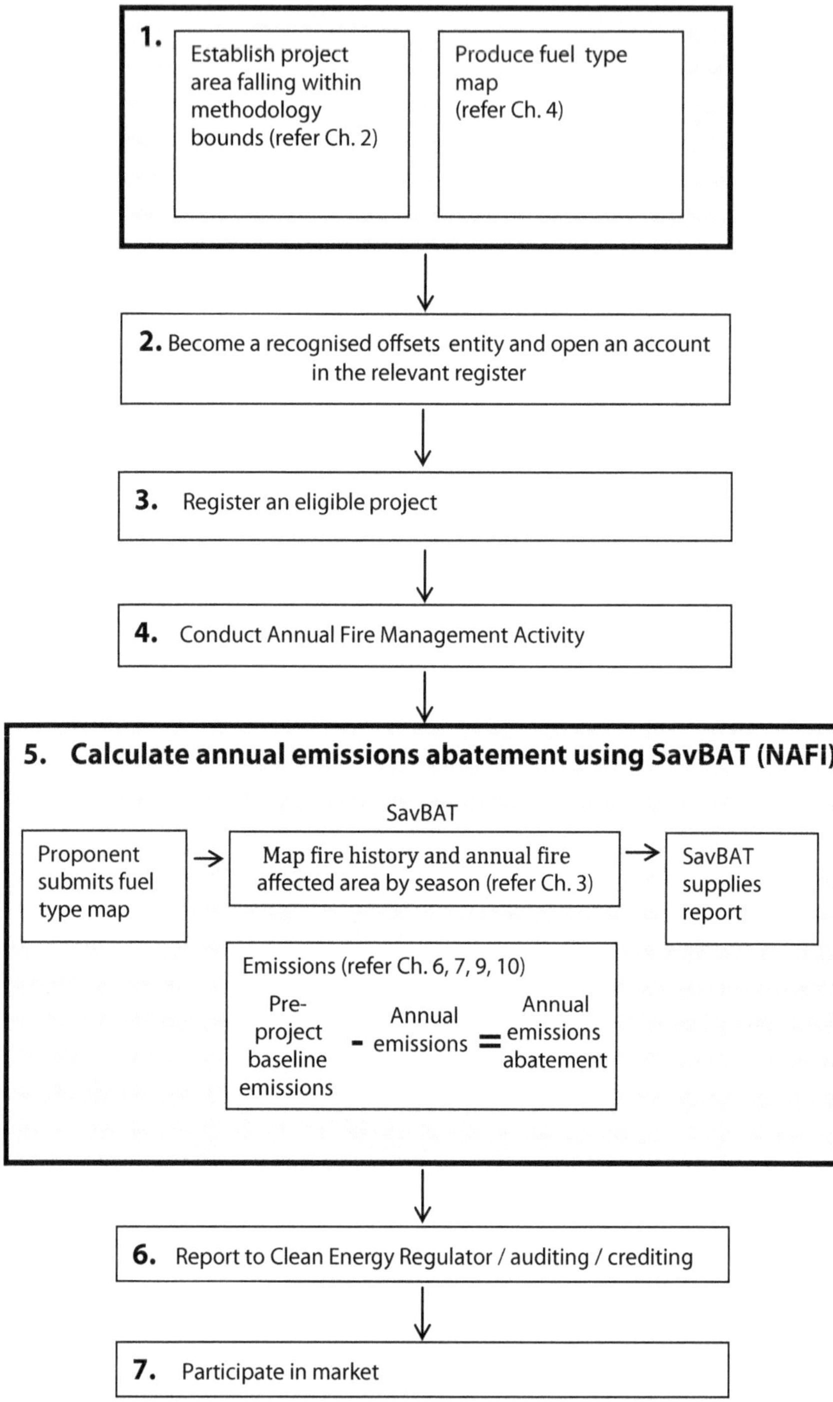

Figure 10.4. Accounting and reporting process for lower rainfall savanna burning emissions abatement projects. Note that procedures for steps 2, 3, 6, and 7 were current at time of writing but may be subject to change (www.environment.gov.au/climate-change/emissions-reduction-fund).

to be subject to change given ongoing changes to Australian GHG emissions accounting policy and procedures.

References

Akagi SK, Yokelson RJ, Wiedinmyer C, Alvarado MJ, Reid SJ, Karl T, Crouse JD, Wennberg PO (2011) Emission factors for open and domestic biomass burning for use in atmospheric models. *Atmospheric Chemistry and Physics* **11**, 4039–4072. doi:10.5194/acp-11-4039-2011

Andreae MO, Merlet P (2001) Emission of trace gases and aerosols from biomass burning. *Global Biogeochemical Cycles* **15**, 955–966. doi:10.1029/2000GB001382

Commonwealth of Australia (2013) Carbon Credits (Carbon Farming Initiative) Reduction of Greenhouse Gas Emissions through Early Dry Season Savanna Burning—1.1) Methodology Determination 2013. www.comlaw.gov.au/Series/F2013L01165

DoE (Australian Government Department of Environment) (2014) National Inventory Report 2012. Vol. 2. www.environment.gov.au/climate-change/greenhouse-gas-measurement/publications/national-inventory-report-2012

Dyer R, Jacklyn P, Partridge I, Russell-Smith J, Williams RJ (Eds) (2001) 'Savanna burning: understanding and using fire in northern Australia.' Tropical Savannas Cooperative Research Centre: Darwin.

Environment Australia (2000) 'Revision of the Interim Biogeographic Regionalisation of Australia (IBRA) and the Development of Version 5.1.' Summary Report. Australian Government Department of Environment and Heritage: Canberra.

Fox ID, Nelder VJ, Wilson GW, Bannink PJ (2001a) *The Vegetation of the Australian Tropical Savannas*. Environmental Protection Agency: Brisbane.

Fox ID, Neldner VJ, Wilson GW, Bannink PJ, Wilson BA, Brocklehurst PS, Clark MJ, Dickinson KJM, Beard JS, Hopkins AJM, Beeston GR, Harvey JM, Thompson EJ, Ryan TS, Thompson SL, Butler DW, Cartan H, Addicott EP, Bailey LP, Cumming RJ, Johnson DC, Schmeider M, Stephens KM, Bean AR (2001b) *The Vegetation of the Australian Tropical Savannas*. Queensland Herbarium, Environmental Protection Agency and the Cooperative Research Centre for the Sustainable Development of Tropical Savannas: Brisbane and Darwin.

Gleeson T, Martin P, Mifsud C (2012) 'Northern Australian beef industry: Assessment of risks and opportunities.' ABARES report to client prepared for the Northern Australia Ministerial Forum, Canberra, May, CC BY 3.0.

Jones DA, Weymouth G (1997) 'An Australian monthly rainfall data set.' Technical Report No. 70. Bureau of Meteorology: Melbourne.

Jones DA, Wang W, Fawcett R, Grant I (2006) 'The generation and delivery of Level-1 historical climate data sets.' Australian Water Availability Project Milestone Report. Bureau of Meteorology: Melbourne.

Maier SW, Russell-Smith J (2012) Measuring and monitoring of contemporary fire regimes in Australia using satellite remote sensing. In *Flammable Australia: Fire Regimes, Biodiversity and Ecosystems in a Changing World*. (Eds RA Bradstock, AM Gill, RW Williams) pp. 79–95. CSIRO Publishing, Melbourne.

McCosker T, McLean D, Holmes P (2010) *Northern beef situation analysis 2009*. Meat & Livestock Australia: Sydney.

Meyer CP, Cook GD, Reisen F, Smith TEL, Tattaris M, Russell-Smith J, Maier SW, Yates CP, Wooster MJ (2012) Direct measurements of the seasonality of emission factors from savanna fires in northern Australia. *Journal of Geophysical Research, D, Atmospheres* **117**, D20305. doi:10.1029/2012JD017671

NLAW (Northern Australia Land and Water Taskforce) (2009) 'Sustainable development of northern Australia.' Australian Government Department of Infrastructure, Transport, Regional Development and Local Government: Canberra.

Olson JS (1963) Energy storage and the balance of producers and decomposers in ecological systems. *Ecology* **44**, 322–331. doi:10.2307/1932179

Russell-Smith J, Edwards A (2006) Seasonality and fire severity in savanna landscapes of monsoonal northern Australia. *International Journal of Wildland Fire* **15**, 541–550. doi:10.1071/WF05111

Russell-Smith J, Murphy BP, Meyer CP, Cook GD, Maier S, Edwards AC, Schatz J, Brocklehurst P (2009) Improving estimates of savanna burning emissions for greenhouse accounting in northern Australia: limitations, challenges, applications. *International Journal of Wildland Fire* **18**, 1–18. doi:10.1071/WF08009

Russell-Smith J, Cook GD, Cooke PM, Edwards AC, Lendrum M, Meyer CP, Whitehead PJ (2013) Managing fire regimes in north Australian savannas: applying customary Aboriginal approaches to contemporary global problems. *Frontiers in Ecology and the Environment* **11**, e55–e63. doi:10.1890/120251

Russell-Smith J, Yates CP, Evans J, Desailly M (2014) Developing a savanna burning emissions abatement methodology for tussock grasslands in high rainfall regions of northern Australia. *Tropical Grasslands – Forrajes Tropicales* **2**, 154–166.

Walsh D, Cowley R (2011) Looking back in time: can safe pasture utilisation rates be determined using commercial paddock data in the Northern Territory? *The Rangeland Journal* **33**, 131–142. doi:10.1071/RJ11003

Whitehead PJ, Russell-Smith J, Yates CP (2014) Fire patterns in north Australian savannas: extending the reach of incentives for savanna fire emissions abatement. *The Rangeland Journal* **36**, 371–388. doi:10.1071/RJ13129

Appendix: Parameter tables for the lower rainfall savanna burning abatement methodology following formatting conventions as set out in Commonwealth of Australia (2013)

Section 3.7. Patchiness (P)

Fire season	Patchiness
EDS	0.771
LDS	0.948

Table 1. Burning efficiency (BEF)

Fire season	Fine	Coarse	Heavy	Shrub
EDS	0.866	0.217	0.145	0.146
LDS	0.877	0.225	0.149	0.147

Table 2. Non-fine fuel loads (t ha^{-1})

Class	Coarse	Heavy	Shrub
WHu	1.85	1.15	1.84
WMi	0.69	2.02	0.66
WTu	1.39	1.25	0.27
OWM	0.76	0.8	1.13
SHH	0.73	0.17	0.87

Table 3. Fine fuel accumulation in Year Since Burnt (t ha^{-1})

Early dry season (EDS)						
Class	**1**	**2**	**3**	**4**	**5**	**>5**
WHu	1.75	3.14	4.12	4.80	5.27	5.60
WMi	2.00	2.69	2.89	2.95	2.97	2.98
WTu	2.71	4.48	5.52	6.12	6.47	6.68
OWM	2.79	3.66	3.91	3.98	4.01	4.01
SHH	1.00	2.11	3.08	3.93	4.68	5.32
Late dry season (LDS)						
Class	**1**	**2**	**3**	**4**	**5**	**>5**
WHu	2.01	3.40	4.38	5.06	5.53	5.86
WMi	2.28	2.97	3.17	3.23	3.25	3.26
WTu	3.05	4.82	5.86	6.46	6.81	7.02
OWM	2.97	3.84	4.09	4.16	4.19	4.19
SHH	1.28	2.39	3.36	4.21	4.96	5.60

Table 4. Emissions factors – CH_4

Class	Fine	Coarse	Heavy	Shrub
WHu	0.0015	0.0015	0.0158	0.0015
WMi	0.0017	0.0017	0.0158	0.0017
WTu	0.0016	0.0016	0.0158	0.0016
OWM	0.0012	0.0012	0.0111	0.0012
SHH	0.0013	0.0013	0.0111	0.0013

Table 5. Emissions factors – N_2O

Class	Fine	Coarse	Heavy	Shrub
WHu	0.006	0.006	0.0146	0.006
WMi	0.006	0.006	0.0146	0.006
WTu	0.012	0.012	0.0146	0.012
OWM	0.006	0.006	0.0146	0.006
SHH	0.0059	0.0059	0.0146	0.0059

Table 6. Carbon content

Class	Fine	Coarse	Heavy	Shrub
WHu	0.397	0.482	0.482	0.485
WMi	0.397	0.482	0.482	0.485
WTu	0.41	0.482	0.482	0.485
OWM	0.399	0.482	0.482	0.485
SHH	0.398	0.482	0.482	0.485

Table 7. Nitrogen to carbon ratio

Class	Fine	Coarse	heavy	Shrub
WHu	0.0113	0.00389	0.01497	0.00389
WMi	0.0118	0.00389	0.01497	0.00389
WTu	0.0105	0.00389	0.01497	0.00389
OWM	0.0102	0.00389	0.01497	0.00389
SHH	0.0107	0.00389	0.01497	0.00389

Table 8. Molecular to elemental mass conversion factors

CH_4	N_2O
1.3333	1.5714

Table 9. Month representing start of LDS

Year	Month
All	August

Table 23. Total annual emissions – GWP from NGER regulations

Gas	GWP
CH_4	21
N_2O	310

11

Reconciling 'bottom-up' and 'top-down' greenhouse gas accounting methodologies for savanna burning in northern Australia

CP (Mick) Meyer, Martin Cope, Ross Mitchell, Stuart Young, Sunhee Lee, Ashok Luhar and Susan Campbell

Summary

This chapter presents an assessment of the potential for using 'top-down' accounting methodologies to constrain or replace 'bottom-up' accounts. The increasing availability of remotely sensed spatial datasets of fire activity and atmospheric concentrations of the emission products from biomass combustion presents an opportunity for greenhouse gas accounting methodologies that are independent of the current bottom-up methods. Three potential applications for top-down accounting are identified: (1) independent assessment of the accuracy of bottom-up accounts; (2) a method for constraining bottom-up emission estimates to match the observed atmospheric concentrations of emitted compounds; and (3) a fully independent method for determining regional emissions.

Assessment of consistency between independent datasets used as input to bottom-up accounts confirmed that current Australian practice and method yield emission estimates that are reliable. However, global data products appear to significantly underestimate fire area in the Australian savanna woodlands and rangelands, and overestimate fuel loads. Regional monitoring and analysis is therefore essential for regional accounts.

Atmospheric dispersion models of proven accuracy in the Australian region, coupled with a chemical transport model, are also found to perform reasonably reliably in Australian applications. This confirms that current emissions estimates produce atmospheric concentrations of smoke that are consistent with observed concentrations. At three observing sites (Lake Argyle, Darwin and Jabiru), there is good correspondence between the magnitude and seasonal variation in aerosol optical depth (AOD) – the loading of particulate matter observed though the atmospheric column. This could only be achieved by accounting for all sources of aerosol (i.e. marine, wind-blown dust, secondary aerosol produced from biogenic sources of volatile organic compounds, and both fresh and aged smoke). Model predictions of the smoke contribution to total AOD were slightly smaller than indicated by the surface measurements. This work is consistent with previous studies, but indicates that the empirical methods used in previous studies may have overestimated the significance of smoke relative to other aerosol sources. Clearly, some components of the chemical transport model require further refinement to fully describe the known aerosol chemical and physical transformations in the atmosphere.

There is also good agreement between AOD predicted from emissions estimates and AOD predicted from atmospheric transport with observations of smoke plumes derived from lidar backscattering profiles from the Cloud-Aerosol Lidar Infrared Pathfinder Satellite Observations (CALIPSO) satellite. However, the comparison was limited by the mismatch between the spatial resolution of the model and that of the CALIPSO observations.

To develop the top-down technique to the point where it can be used to constrain and even replace bottom-up accounts will require further refinement of model components and development of inverse modelling software. While both are currently in planning, it appears unlikely that top-down accounting can be applied in the near future for totally independent accounting due to the limited precision and temporal coverage of the satellite observations.

Introduction

Carbon trading and offsets mechanisms require, at a minimum, the consistent application of a scientifically valid methodology, with the implicit assumption that the methodology is accurate. While this is a challenge for all national greenhouse gas reporting sectors, it is especially problematic for diffuse emissions sources in the agricultural and land use sectors, where both activity data and emission process data are sparse. Currently, all standard national and Intergovernmental Panel on Climate Change (IPCC) methodologies use 'bottom-up' approaches in which the individual emission sources are identified and quantified, then aggregated to regional total emissions. Because they are unconstrained, the accuracy of the inventories is entirely dependent on the accuracy of both the algorithms and the underlying datasets.

The net change in smoke concentration over a region (after correction for advection, deposition and chemical transformation processes) must equal the sum of the net surface emissions. This mass balance in principle provides the basis for estimating total regional emissions by direct measurement of smoke concentration in the atmosphere. Termed a 'top-down' approach, it contrasts with conventional 'bottom-up' accounting methods, which estimate emissions from consumption of fuel. With bottom-up methods, there is a risk that uncertainties in knowledge of emission processes (i.e. parameters) accumulate, resulting in large uncertainties in the estimates of total emissions. Additionally, if the measured parameters describing fuel emissions are not truly representative of the region, bottom-up estimates may be biased towards those parts of the landscape that are most studied. Quantifying these uncertainties is no small challenge. However, the variety and number of satellite-borne instruments that measure concentrations of trace chemicals in the atmosphere with precision is increasing. Many of these sensors measure concentrations of the relevant greenhouse gas species, or compounds that are emitted at rates directly correlated with the greenhouse gases. This opens up the possibility for top-down accounting. The aim of this study was to assess the potential for top-down accounting for savanna fire emission in the tropical woodlands of northern Australia.

In all regions, a wide range of emission sources contribute to the atmospheric concentrations of each greenhouse species. Some of these sources are anthropogenic, i.e.

associated directly with human activity, and others are natural, forming components of the carbon and other nutrient cycles in the biosphere that occur independently of human presence. In the northern Australian savannas, anthropogenic sources include industrial, transport, domestic, agricultural, land clearing, forestry, and fires of all classes (wildfires and prescribed fires, i.e. fires ignited under prescription for management purposes such as fuel reduction, biodiversity management and silviculture). Non-anthropogenic sources in the savanna landscape include emissions from natural wetlands, net exchanges of carbon from the vegetation, and emissions from termite activity. Bottom-up accounting easily distinguishes between natural and anthropogenic sources; however, top-down methods can measure only the aggregate of all sources. Where natural sources are large, as is often the case, partitioning the total emission into the anthropogenic and natural components is a major challenge. However, most sources emit a range of chemical species, some of which are specific to, or at least characteristic of, that source, so it is usually possible to find a surrogate tracer that is emitted predominantly from the source of interest and which can be measured in the atmosphere at appropriate precision. Aerosol (or particulate matter – PM) is emitted by biomass combustion; in the unpopulated regions of tropical Australia, open burning (i.e. wildfires and prescribed fires) is the dominant source of PM, and emission rates of PM from combustion are well correlated with the emissions rates of CH_4 (Gras *et al.* 2002). Aerosol concentration in the atmosphere is routinely measured at ground stations at high precision by a range of optical and gravimetric instruments, and from space by optical instruments. Therefore PM is the most promising tracer for top-down accounting of smoke emissions from savanna fires. This chapter discusses the potential for using measures of PM concentration for constraining estimates of smoke emissions, and, implicitly, through the correlation between CH_4 and PM emission rates, the constraints on the greenhouse gas emission budgets.

Reconciling bottom-up with top-down accounts requires several components. These are:

- a chemical transport model that accounts for the movement and chemical transformation of emitted trace species through the atmosphere
- spatially disaggregated bottom-up estimates of the emission rates of the relevant chemical species from the major sources. The primary tracer of relevance here is PM; however, because PM can also be formed and transformed or removed by chemical reactions in the atmosphere, an emissions inventory of relevant reactive species is also required. Savanna fires, however, are not the only source of PM in the region; other primary PM emissions sources include marine aerosol (sea salt), wind-blown dust, and emissions from internal combustion engines used in industry and transport. The spatial domain of emission inventory should be sufficiently large to allow for advection across the region of interest.
- a time series of observations of the atmospheric concentrations of the trace species of interest. This includes observations at defined locations of concentrations in surface air, concentrations integrated through the vertical column of the atmosphere from observing stations on the ground, and observations from satellite-borne instruments, giving partial or complete coverage of the region at defined time intervals.

These are combined in a framework in which the PM emissions estimates (which when aggregated constitute the bottom-up account) are transformed into a 3D distribution of concentrations in the atmosphere using the chemical transport model (CTM). This concentration field can then be compared at specific points in time and space with the observations of PM concentration. The framework can be applied in many ways depending on the specific issue being addressed. The three most common options are as follows:

1. Assess the accuracy of the bottom-up accounts by comparing the observed atmospheric concentrations with the concentrations estimated from the model using the spatial field of emission rates. Assuming that the transport model is correct (this can be tested independently), good agreement between observed and predicted concentrations is confirmation that the emissions field is reliable and therefore the emission inventory derived from it is plausible.
2. An extension of the above option uses information from the comparison between observed and predicted concentrations to modify the emission estimates and, if appropriate, the chemical transport model. This can be either manually applied, or implemented through a formal mathematical process termed an inversion.
3. The third option requires a spatially and temporally complete set of atmospheric observations across the region. Using information about air transport provided by the transport model, it should be possible to calculate the complete mass balance of the relevant chemical species in the atmosphere over the region of interest, i.e. quantify advection into and out of the region, changes in concentration due to atmospheric chemistry, and by difference, emissions from the land surface into the atmosphere. This option yields an emission estimate independent of bottom-up accounts.

The first question for this study is whether the current emission estimates are consistent with the atmospheric observations, as previous studies suggest. Using the method developed by Meyer *et al.* (2008a) for estimating spatial emissions inventories of PM from fires, Luhar *et al.* (2008) compared estimated and observed concentrations of PM in Darwin and AOD at Darwin and Jabiru for the 2004 dry season, May to November. Predicted PM concentrations and AOD were calculated using the transport model TAPM assuming that PM is inert and that fires are the only source of primary PM emissions in the region. The agreement between the predictions and observations was good; however, it was also recognised that the study was limited. Although the assumptions were justifiable as a first approximation, the study could not investigate the relative significance of PM from other sources, particularly dust and marine aerosol emitted from within the model domain, or advected into the region from sources well beyond the domain boundaries. Significantly, comparisons between observations and predictions were made at only two locations.

We extend that work by:

- an assessment of the accuracy of the spatial emissions inventory though a comparison of alternative data sources for fire area and fuel load
- including all major primary PM sources
- a full chemical transport model that addresses chemical transformations and production of secondary PM
- a modelling domain that includes all the Australian mainland and coastal waters
- an improved method for estimating AOD from PM concentration
- developing the tools required for comparison of transects though the domain of predicted PM with AOD observed from satellite.

Regional greenhouse gas accounting for the savanna woodlands

In this section we discuss the emissions models, including spatial and regional inventories and accuracy of data sources.

Following the approach first presented by Seiler and Crutzen (1980) and applied subsequently in most regional, national and international pollutant and greenhouse gas inventories (e.g. IPCC 2006; DoE 2012), carbon emitted from fires is calculated as the product of fire area, within-scar patchiness, fuel load and carbon content, and combustion completeness. Usually, the region of interest is stratified into fuel size classes and fire severity, or its surrogate, fire season. Emissions of individual chemical species are then determined as the product of emitted carbon and a species-specific emission factor. The algorithms currently applied in both the Australian National Greenhouse Gas Inventory Methodology and in the accounting methodology of the savanna burning emissions abatement methodology (Commonwealth of Australia 2013) certified by Australia's Carbon Farming Initiative (CFI) legislation are described in detail in Chapter 9. Although the general forms of the algorithms required for inventory accounting and for spatial modelling are similar, atmospheric transport models require hourly surface emissions of each chemical species, and the spatial stratum is not a region but a grid cell of the model domain.

The input variables and parameters of Eqns 9.5 and 9.6 (Chapter 9) can be obtained from a range of sources. Currently fire area is determined from analysis of remotely sensed reflectance from two satellite-borne sensors: MODIS (Moderate Resolution Imaging Spectroradiometer), aboard the Terra and Aqua Earth Observing System satellites of the National Aeronautical and Space Administration (NASA); and AVHRR (Advanced Very High Resolution Radiometers), on the National Oceanic and Atmospheric Administration (NOAA) family of polar-orbiting platforms. Fuel loads are determined from field measurement (in the case of the national and CFI methodologies), or from biogeochemical models including CASA (e.g. Van der Werf *et al.* 2010), VAST version 1.2 (Barrett 2011), BIOS-2 (Harverd *et al.* 2013). Patchiness, burning efficiency, carbon content and emission factors are determined from field and laboratory measurements. However, in the future some of these may also be determined from remotely sensed data, including reflectance or fire radiative power (FRP). The daily emissions inventories can be disaggregated into hourly emissions, either with a fire spread model or by the observed diurnal variation in FRP. A framework for producing the emission inventories is proceeding (Fig. 11.1). With limited data sources, determining the uncertainties in emission estimates has proved problematic, and has been largely based on expert judgment of the likely bias and errors (e.g. Russell-Smith *et al.* 2009). However fire area and fuel load are now available as spatial products from several independent sources that are described below.

Data consistency – burned area

There are currently four independent and publicly available sources of fire area data for the Australian tropical savanna woodlands, and five for the rangelands. They are:

1. The FIREWATCH time series of fire scars produced by Landgate, WA (www.landgate.wa.gov.au/corporate.nsf/web/Fire+Monitoring) from AVHRR imagery at 1 km resolution. The time series currently covers the period 1989 to the present, and is time-resolved to ~10 days. It covers the Australian mainland and Tasmania. The fire scars are mapped by a semi-automated process with manual checking.

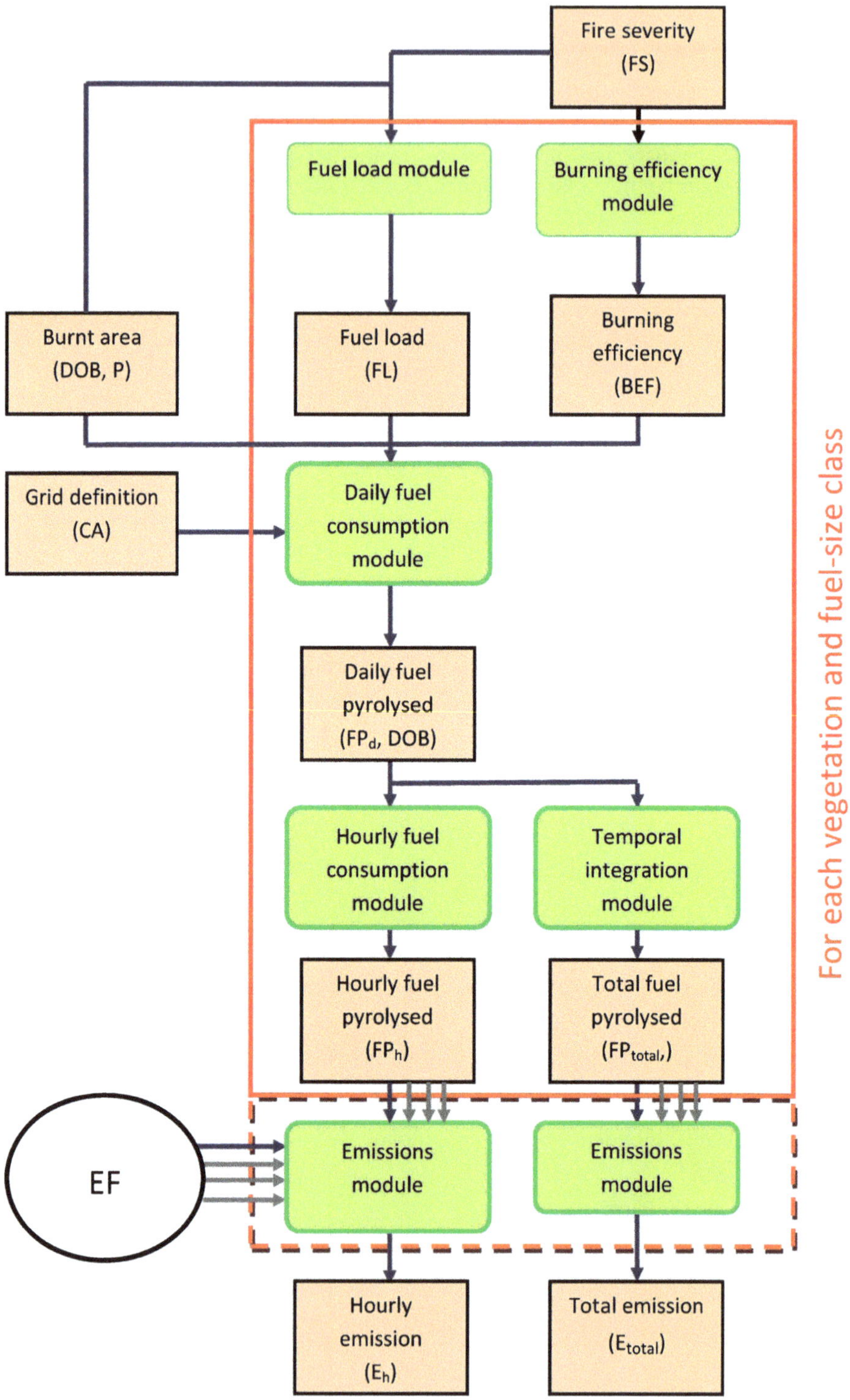

Figure 11.1. Framework of the key data grids and transformation modules required for a spatially explicit emission inventory for chemical transport modelling.

2. Time series of fire scars for the tropical woodlands of northern Australia produced by North Australian Fire Information (NAFI) from MODIS imagery at 250 m resolution. This covers the period from 2005 to the present, time-resolved to ~1 week. Mapping is semi-automated (see Chapter 3).
3. The time series produced for 500 m MODIS imagery by the University of Maryland for NASA (Roy *et al.* 2008) using an automated algorithm. This dataset has global coverage, covers the period 2000 to the present, and is time-resolved to 1 day.
4. The time series of fire scars produced from 500 m MODIS, supplemented by AVHRR using the automated algorithm of Giglio *et al.* (2009). It is available publicly through the GFED3.1 database aggregated to a 0.5 × 0.5° grid with global coverage. This data series is time-resolved to 1 month.

We have supplemented this with an additional dataset produced using the methodology of Meyer *et al.* (2008a). In this, we combine the SENTINEL active fire product (a system developed by CSIRO and currently operated by Geoscience Australia (http://sentinel.ga.gov.au) with the Landgate product to increase the time resolution of the scars to 1 day and to extend the coverage to include fires not detected by the Landgate methodology. This product provides the spatial inventory used for the smoke dispersion modelling in our study.

Fig. 11.2 compares burned area for 2009 estimated using the hotpot-amended Landgate fire scars with Landgate fire scars, NAFI fire scars, the NASA global fire area product, and GFED 3.1. The global burned area products underestimate burned area relative to Landgate over much of northern Australia to a greater extent than the NAFI or uncorrected Landgate product. As discussed in Meyer *et al.* (2008a), fires are occasionally undetected from reflectance data in dissected rocky landscape, in forests with complete canopy cover, if the fires are low intensity or if the burnt area is less than 400 ha. Such fires occur particularly in the rangelands, south-east Queensland and the Arnhem Land escarpment. The NAFI fire scar dataset is mostly similar to Landgate; however, it appears to detect more burned areas in the southern extent of the NT woodland region and in Cape York. In contrast, the NASA and GFED data series detect substantially smaller fire areas in the savanna woodlands and most of the rangelands. There are also grid cells where more fire area occurs in the GFED, NAFI and NASA time series than in the hotspot-amended Landgate time series, but there is no consistency in their location.

We calculated the differences in annual total estimated burnt area from 2002 to 2010. Addition of the SENTINEL hotspot data increases the mean Landgate burnt area in woodland regions by ~20%, while in the rangeland regions the increase is ~70%. For the woodlands, NAFI and Landgate show good agreement except during 2005 and 2010. The global products mostly underestimate burned area relative to Landgate; on average, NASA underestimates by 30% in the woodlands and 50% in the rangelands, and GFED underestimates by 11% in the woodlands and overestimates by 13% in the rangelands. Figure 11.3a compares the distribution of fires in 2009 shown by Landgate fire scars and Sentinel hotspots and demonstrates (in red) the number of fires that the Landgate scars were not able to identify. Comparison with the NASA fire scar data in Fig. 11.3b shows that the global product fails to identify an even greater fraction of fires.

Occasionally, e.g. 2005 and 2010, Landgate mapping appears to substantially underestimate area burnt; however, most of the unmapped scars have potentially been identified by hotspots. The most likely source of this error is scattered cloud coverage on the imagery selected for fire mapping. The issue affects both NAFI and Landgate, but mostly the

(a)

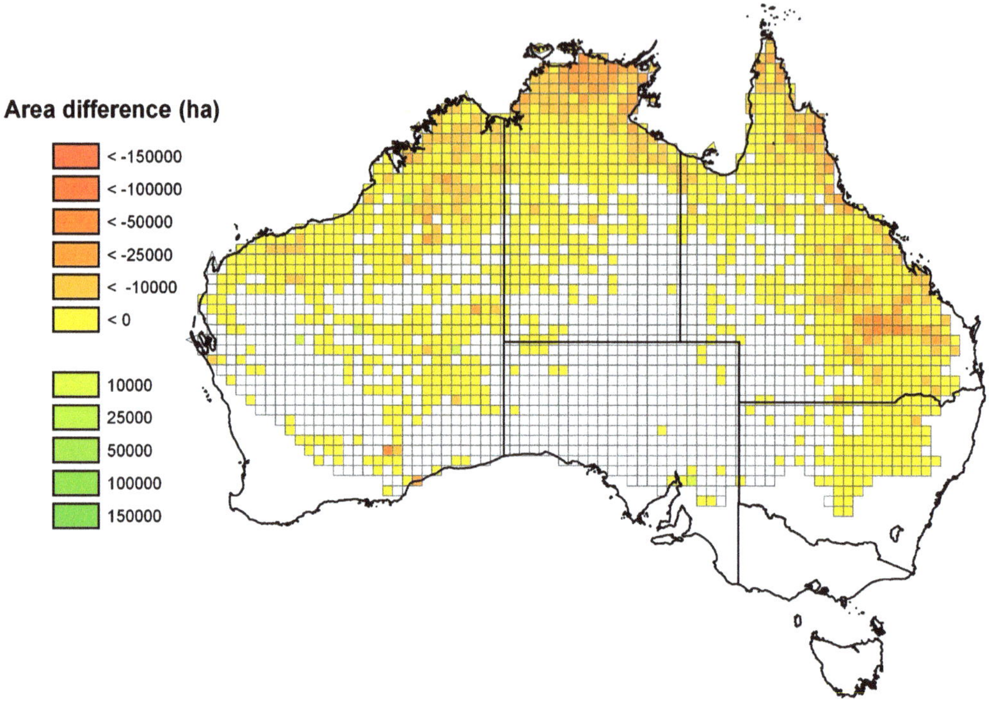

(b)

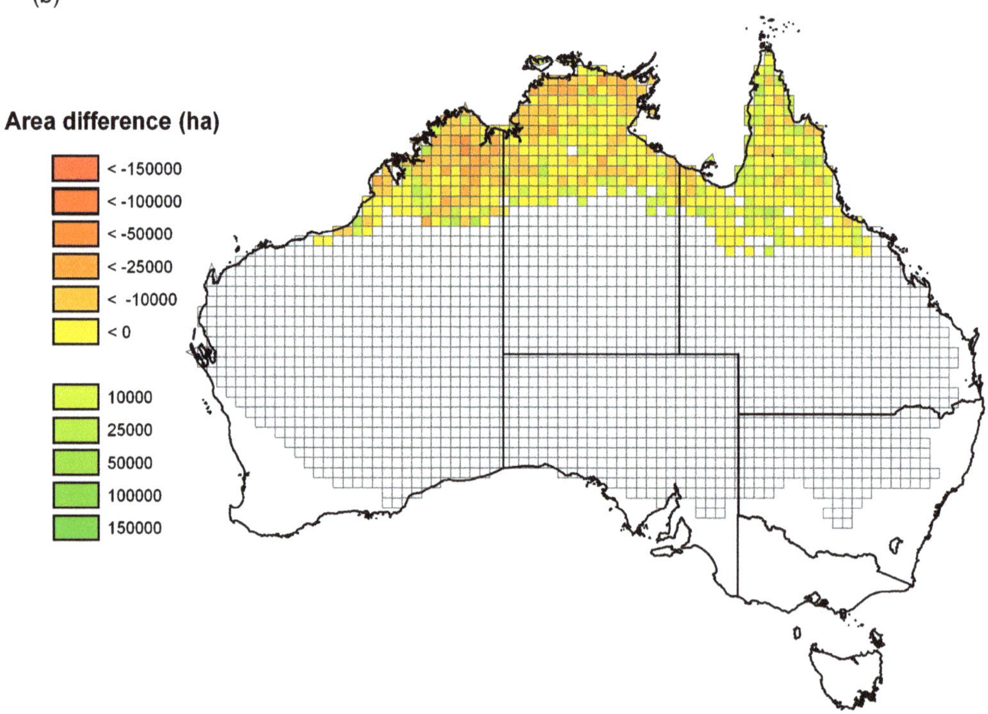

Figure 11.2. Comparison between the Landgate fire scar areas in 2009 amended by the anomalous hotspot area: (a) Landgate scars only; (b) NAFI; (c) NASA global fire area product; (d) GFED 3.1.

(c)

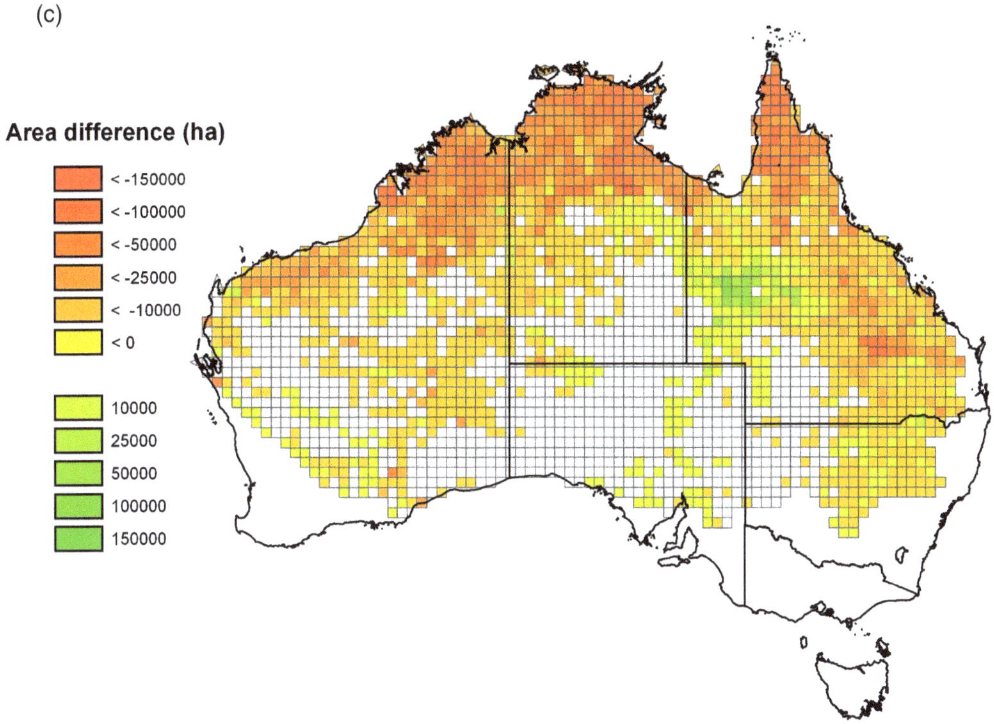

(d)

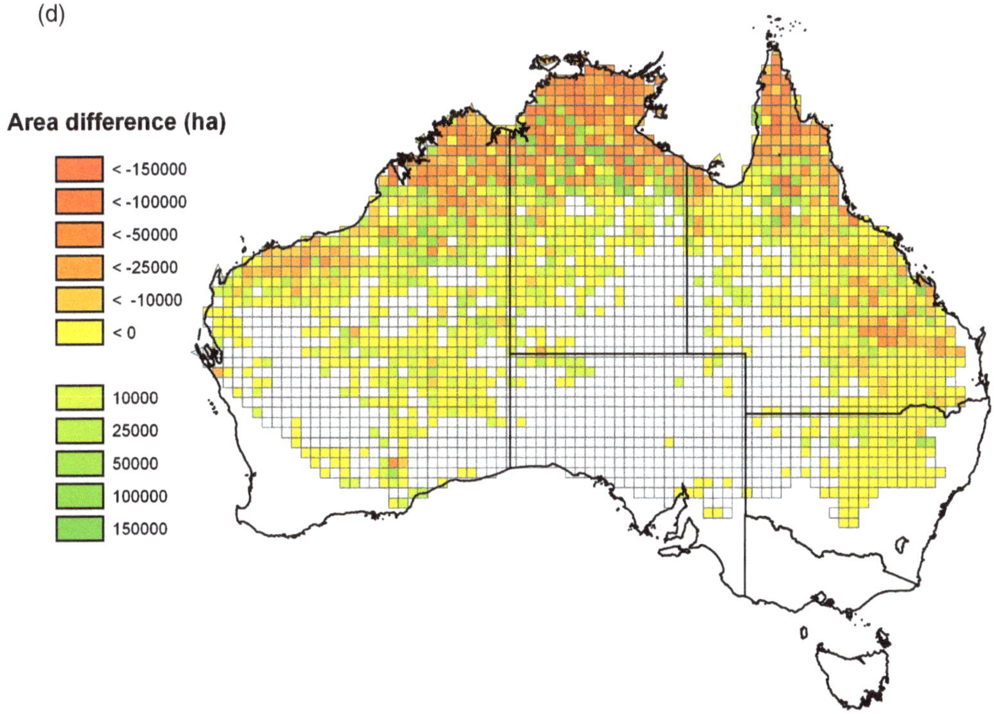

Figure 11.2. (Continued).

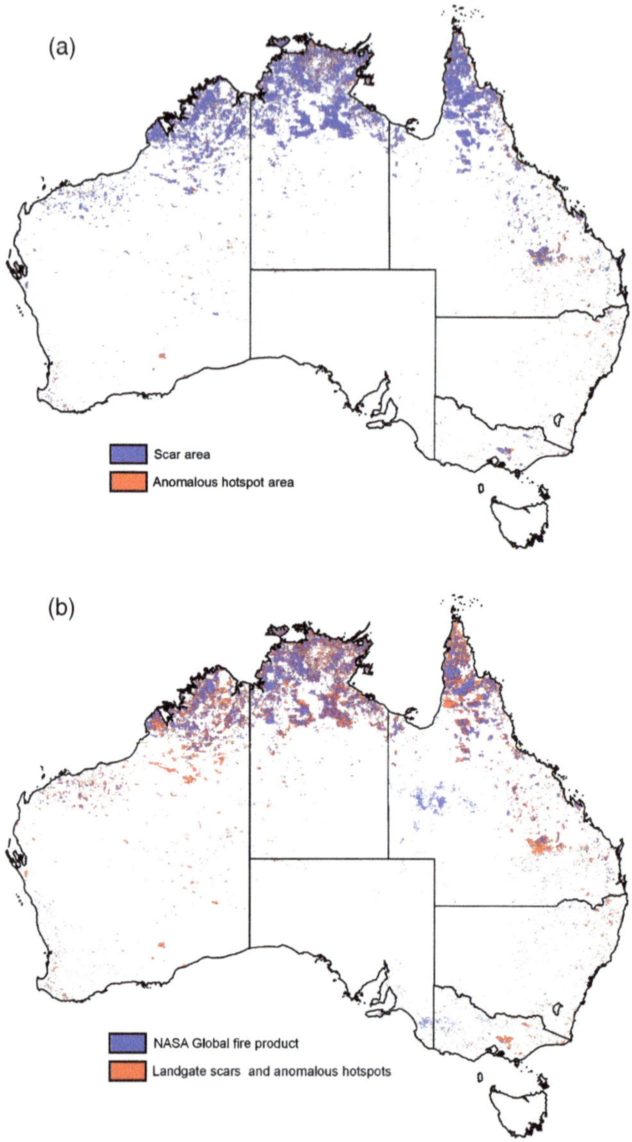

Figure 11.3. Distribution of fires in 2009. (a) Landgate fire scars and anomalous hotspots; (b) Landgate scars and anomalous hotspots overlaid with NASA global fire.

obscured areas are reassessed by supplementing the primary analysis with additional imagery. The NAFI analysis appears to be more sensitive to this issue.

This analysis, therefore, confirms that the approach developed by Meyer *et al.* (2008a) is relatively robust for regional Australia; it agrees well with the meticulous NAFI time series with coverage of both rangelands and woodlands; and with AVHRR and ATSR hotspot data can be extended back to the late 1990s. As described in Chapter 3, detailed assessment of the accuracy of the NAFI fire mapping product was found to be 88%. The publicly accessible global time series appear to be less reliable for Australia, again as demonstrated in Chapter 3. However, regional smoke dispersion modelling requires daily continental emissions fields in order to account for advection from regions beyond the regions of focus. Currently, the NAFI dataset does not extend to national coverage, and the fires are not time-resolved to a day.

Data consistency – fuel loads

A significant amount of effort has gone into mapping fuel loads on the Australian continent. The savanna woodland region of Australia is defined in the national inventory methodology to comprise agroecological zones (AEZ) 1, 2 and 3 of the Interim Biogeographical Regionalisation of Australia (IBRA) version 4.1. The rangelands are defined to comprise AEZs 5, 6 and 11 (Fig. 11.4a). The revision to the national inventory methodology for savanna burning (Meyer 2011) further divided the savanna woodlands into nine fuel regions (Fig. 11.4b). There are currently three readily accessible estimates of fuel loads through these regions:

1. field measurements; these are presented in Russell-Smith *et al.* (2009) and applied in both the current national and CFI methodologies
2. the biogeophysical vegetation model VAST v1.2 (Barrett 2010)
3. the biogeophysical vegetation model BIOS2 (Haverd *et al.* 2013). This model draws on and extends some of the concepts developed in VAST.

Data from field measurement campaigns are classified by the vegetation fuel groups defined in the national and CFI methodologies; CFI compliant fuel type mapping procedures are described in detail in Chapter 4. The contribution of these classes to total emissions depends on both their fuel loads and fire frequencies. The fraction of the total mean savanna fire area (woodlands and rangelands) was calculated for each vegetation/fuel class for the period 1998 to 2010. Most fires occur in the woodlands (27%); mostly *Eucalyptus* woodland in the Northern Territory (16%) and Queensland (7%), and sandstone woodland in Western Australia (8%). A significant fraction of fires also occur in *Melaleuca* woodland in Queensland (4%).

The measured fuel loads reported by Russell-Smith *et al.* (2009) and applied in the national and CFI methodologies are mapped onto the vegetation fuel classes defined in the national and CFI methodologies (Fig. 11.4b) by assuming that open forest is similar to *Eucalyptus* woodland, shrub is similar to sandstone heath, and that the three remaining classes are predominantly grass fuels with the same load as the rangelands (3 t ha^{-1}). Fuel loads in VAST and BIOS2 are assumed to comprise both fine and coarse woody debris pools. These are defined in VAST as two pools (Fig. 11.4c); in BIOS2 they comprise three pools, fine woody, fine grass and coarse woody (Fig. 11.4d). VAST provides estimated mean fuels at a $0.05 \times 0.05°$ grid size averaged for the period 1980–2000. BIOS2 provides monthly estimates from 1990 to 2011, and for the comparisons presented here we use the pools estimated for June 2006, i.e. the fuel loads present in the early dry season. The fuel loads are similar in the savanna woodland regions of Western Australia and Cape York, with VAST predicting slightly higher fuel loads in the southern NT woodland region and in the rangelands.

Overall consistency between the modelled and measured fuel loads lends confidence to the accuracy of the spatial fuel load maps. Table 11.1 compares the mean fuel loads for the three datasets of the vegetation/fuel classes in the savanna woodland regions of Western Australia, Northern Territory and Queensland. The fuel loads estimated by both VAST

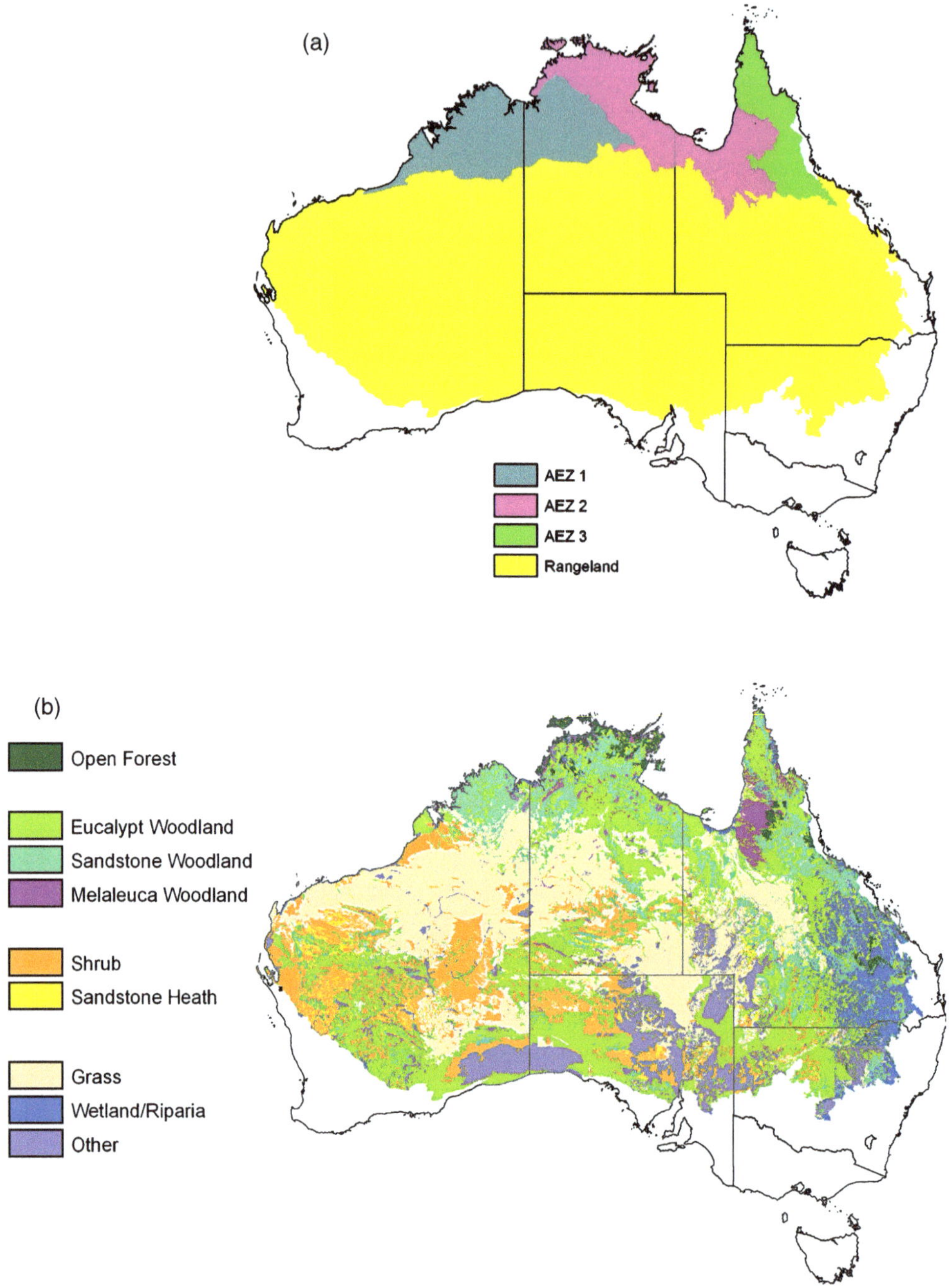

Figure 11.4. Fuel loads: (a) the savanna region – AEZ 1,2 and 3 are savanna woodland; (b) fuel classification; (c) average coarse and fine fuel load predicted by VAST1.2; (d) fine woody, fine grass and coarse woody debris predicted by BIOS2 for early dry season (June) 2006.

(c)

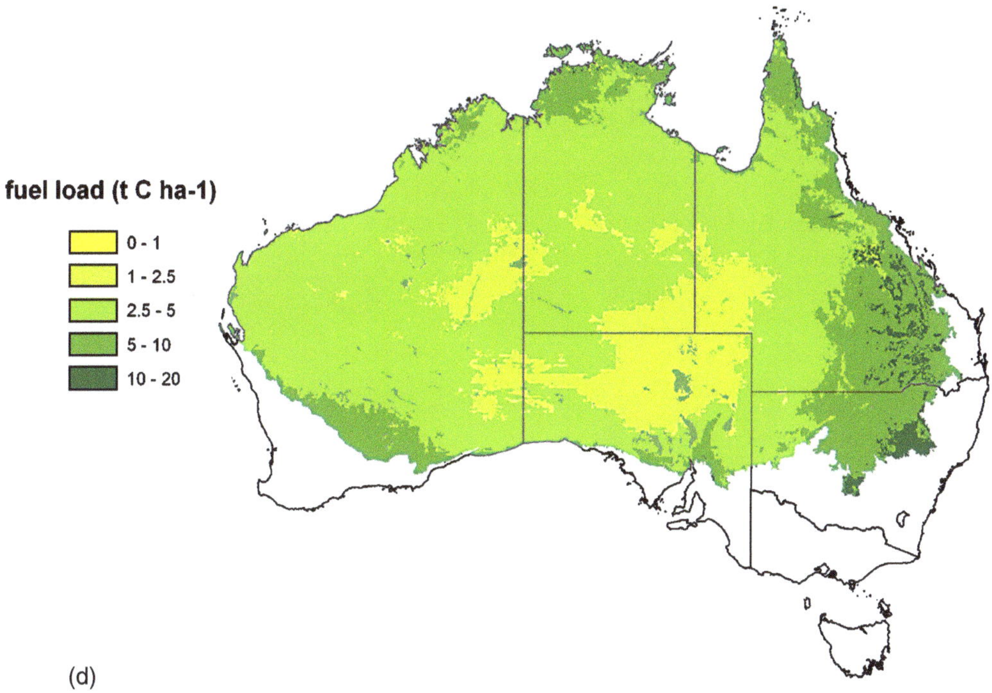

(d)

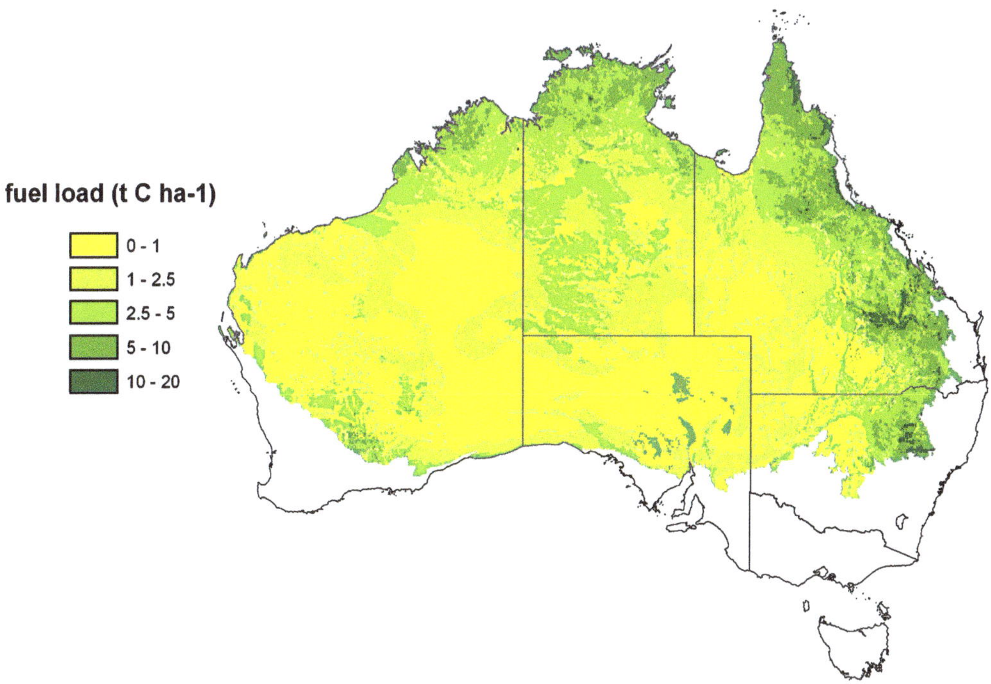

Figure 11.4. (Continued).

Table 11.1. Comparison between (a) mean fuel loads in the savanna woodland measured by Russell-Smith *et al.* (2009), with values predicted by VAST 1.2, and by BIOS2 for June 2006, and (b) burned fuel estimated by Meyer *et al.* (2008a) and the national methodologies (DoE 2012)

	Fuel load (t ha^{-1})			Burned fuel (t–ha^{-1})	
Vegetation class	**VAST 1.2**	**BIOS2**	**CFI/National**	**VAST 1.2**	**CFI/National**
Eucalypt open forest	10.6	12.7	11.9	5.5	5.3
Eucalypt woodland	7.2	8.7	7.5	4.2	3.9
Sandstone woodland	7.1	8.3	10.6	4.1	4.9
Melaleuca woodland	8.4	8.2	7.5	4.4	4.0
Shrub	10.3	8.3	10.7	4.9	5.7
Sandstone heath	8.6	12.4	10.7	5.0	5.2
Grass	5.6	5.7	3	3.6	2.2
Wetlands	9.0	8.3	3	4.8	2.2
Other vegetation types	13.2	8.6	3	5.7	2.2

and BIOS2 are highly consistent with the observed fuel loads; for most vegetation/fuel classes the differences are less than 20%, and overall bias between model and observations is small, particularly with BIOS2. The exceptions to this are the grass, wetlands and non-native vegetation, where the national and CFI inventory fuel loads are assumed to comprise only grass fuels. The validity of this assumption has not been tested; however, because fires in these classes are infrequent, the contribution of these classes to total emissions is small. Patchiness and BEF values used to estimate burned fuel carbon are sourced from Russell-Smith *et al.* (2009); for VAST, an annual average is assumed, following Meyer *et al.* (2008a). Burned fuel loads estimated by the national inventory and VAST for the woodlands and shrublands differ by less than 20% (Table 11.1). Rangeland fuel loads estimated by VAST, BIOS2 and the national methodology (which are based largely on expert judgment; Meyer 2004) are similar in magnitude; however, the relative differences are considerably greater in the rangelands than in the woodlands.

Reconciling regional and national accounts

In the previous section, comparison of the available fire area and fuel load estimates confirmed that the regional fire areas amended by hotspots give reliable estimates of fire area and distribution, and the regional-scale fuel load estimates are reliable. The global fire products do not provide fuel load estimates; however, estimates of total carbon emissions are provided in GFED3.1. These are aggregated to the national reporting regions and compared to the national inventory estimates of annual total carbon emissions for 1998–2010 (Fig. 11.5). The correlation between GFED and national inventory estimates for both rangelands and woodlands is high ($r = 0.92$ and 0.66 respectively), although, compared to the national inventory, GFED overestimates the emissions by 8% in the rangelands and 35% in the woodlands. In contrast, relative to Landgate, GFED underestimates the fire area in the savanna woodland by 11% and overestimates fires in the rangelands by 13%. This implies that, while GFED estimates of burned fuel loads are similar to national estimates in the rangelands, the estimates are 45% larger in the woodlands. Therefore, given that national methodology fuel loads are measured and are also consistent with regional averages from VAST and BIOS2, which account for the spatial variation in vegetation type, cover and dynamics through assimilation of observation of vegetation distribution by remote sensing, it appears that, regionally, GFED may be inaccurate.

However, both GFED and the national inventory may be underestimating actual carbon emissions in the savanna woodlands. We assume that amending Landgate fire area with hotpots corrects most errors of omission, and hence GFED underestimates actual fire area by ~35% in the savanna woodlands and ~40% in the rangelands, while the national inventory underestimates the burned area in the same regions by 24% and 68% respectively. The estimated carbon emissions from the spatial grids defined in Meyer *et al.* (2008a) therefore are slightly larger than the GFED and national inventory estimates in the rangelands and intermediate in the woodlands.

This comparison of regional and global estimates of fire area and carbon emissions highlights the need for detailed and verified regional observations. While global datasets are critical for interpreting the global systems, their algorithms are invariably a compromise between regional and global accuracy. To date, none of the global datasets accurately represents the Australian vegetation or fire regimes and therefore regional analysis of fire activity and emission parameters remains essential for accurate regional accounting.

Top-down accounting for reconciling bottom-up GHG methodologies

The three primary greenhouse gases emitted by biomass combustion, carbon dioxide (CO_2), methane (CH_4) and nitrous oxide (N_2O), have sources and sinks other than biomass combustion. Most of these are poorly quantified, and the gases themselves are generally difficult to measure at the precision required for emission quantification. However, biomass combustion also emits many other chemical species, some of which can be measured with precision in surface air at ground stations or in the atmospheric column from satellite sensors. Some of these chemical species are emitted at rates highly correlated with the primary greenhouse gases, do not have many other potential sources

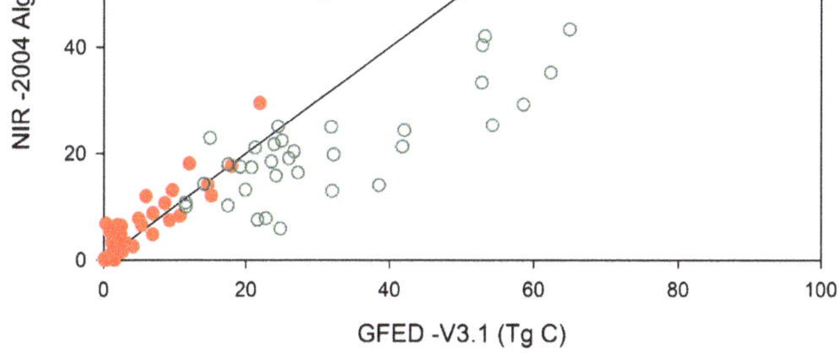

Figure 11.5. Comparison between national estimates of annual total carbon emission in the rangelands and savanna woodlands compared to GFED3.1 estimates for the same periods and regions.

or sinks, and can be measured with precision due to their optical properties. These include particulate matter (aerosol), carbon monoxide (CO), the odd nitrogen oxides (nitric oxide, NO and nitrogen dioxide, NO_2), and some of the volatile organic compounds such as formaldehyde (HCHO). Others are chemically reactive, combining to produce secondary reaction products in the atmosphere that can also be measured with precision by both ground-based instrumentation or by satellite optical sensors. Ozone, which is produced only by photochemistry, is the key indicator species for secondary reactions.

Emission of particulate matter (PM) is a promising tracer, given its relative ease of detection. PM concentration in the surface layer of the atmosphere is a primary measure of urban air quality, and in Australia all capital cities have air quality monitoring stations that continuously record PM10, PM2.5 (PM with an aerodynamic diameter less than 10 μm and 2.5 μm respectively), or both. Most PM produced by open combustion of biomass falls in the PM2.5 size range (Luhar *et al.* 2008). Darwin, the capital city of Australia's Northern Territory, is relatively free from industrial and vehicular pollution sources, but during the dry season it is significantly impacted by smoke from bushfires in the surrounding region. Therefore it is a good case for testing whether emissions estimated using standard bottom-up emissions models lead to atmospheric concentrations of PM2.5 that are consistent with observations. It appears that they are.

Using the atmospheric dispersion model TAPM, and the spatial PM2.5 emission model (Meyer *et al.* 2008a) at a horizontal resolution of 10 km, Luhar *et al.* (2008) found that observed and modelled daily mean surface PM2.5 at Darwin during the 2004 dry season were consistent in magnitude and significantly correlated ($r = 0.39$). Meyer *et al.* (2008b) and Luhar *et al.* (2009) extended this modelling study to the 5 years 2003–2007. As an example, Fig. 11.6a shows the time series of the daily averaged observed concentration of PM2.5 near the ground at Darwin for the year 2007, and that modelled using TAPM incorporating the bottom-up emissions calculated using the methodology of Meyer *et al.* (2008a). The daily ground-level concentrations are important for air quality applications (e.g. for compliance with Air National Environment Protection Measure) and also provide a more stringent test of the validity of estimated emissions than column integrated values, which have an increased process uncertainty due to an additional degree of freedom associated with vertical plume movement. The modelled time series is in good agreement with the data, with the model performing reasonably well at predicting the observed peaks. This can be seen better in the scatter plot in Fig. 11.6b, where the correlation coefficient is 0.69 and the slope of the best fit line is 0.92.

Restricting the comparison to surface concentrations limits the footprint over which the validation applies to a relatively small region. This is because surface PM2.5 concentrations are mostly determined by the behaviour of smoke plumes confined within the atmospheric mixing layer; smoke that has dispersed into the lower troposphere is less likely to be mixed back to the surface. Evidence of this is seen in Fig. 11.6a, which shows seasonal peak of surface PM2.5 concentration occurring in the early dry season (May and June). In the more populous region of the Northern Territory within ~200 km radius of Darwin, fires occur mainly in the early dry season, in contrast to the rest of the northern savanna, which is dominated by late dry season fires.

Meyer *et al.* (2013) present an inverse analysis of smoke dispersion that illustrates this point. Luhar *et al.* (2008) extended their analysis to compare predicted and observed aerosol in the atmospheric column. The primary measurement is via the AOD, the optical extinction along a vertical path through the entire atmosphere. This is measured directly

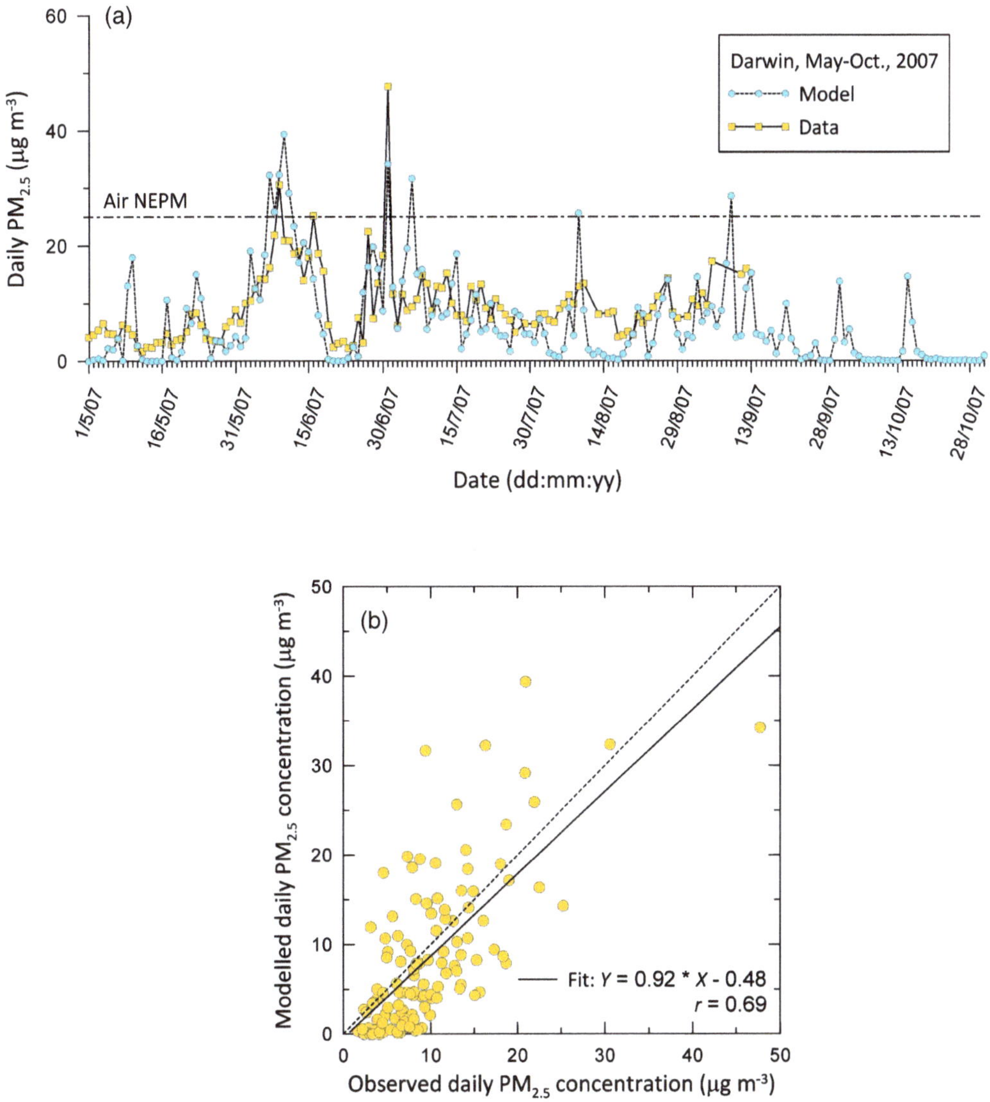

Figure 11.6. (a) Time series of the observed and modelled daily averaged PM2.5 concentrations at Darwin for the period 1 May–31 October 2007; (b) scatter plot of the observed and modelled daily averaged PM2.5 concentrations.

from the surface using a sun photometer; modelling AOD from atmospheric concentrations of PM requires knowledge of the particle size distribution and, particularly, how this varies with humidity and pressure through the atmospheric column. Using an empirical parameterisation based on observations of aerosol microphysics at Darwin, Luhar *et al.* (2008) were able to match 75% and 80% of the daily AOD observations at Darwin and Jabiru within a factor of two, and to achieve correlations of 0.33 and 0.66 respectively. The relationship between AOD and gaseous emissions in Darwin has also been studied by Paton-Walsh *et al.* (2010), who found a generally tight correlation between the two with occasional perturbations due to advection of air from Indonesia affected by biomass burning.

AOD can also be inferred from satellite sensors, e.g. MODIS, of Along Track Scanning Radar (ATSR), or lidar backscattering, e.g. the CALIPSO satellite (Winker *et al.* 2010). These products extend the possible range of comparison between observation and predicted atmospheric concentrations from a small number of points to long transects or large regions. However, although simple empirical relationships between PM2.5 and AOD can be reliable at the locations where aerosol microphysical data are available, the relationships become more problematic across large spatial domains. For this, chemical transport models that describe the aerosol chemistry and microphysics are needed.

Chemical transport modelling framework

Fig. 11.7 shows the modelling framework used to simulate particle processes in northern Australia. The framework consists of a prognostic meteorological model, CSIRO's Cubic Conformal Atmospheric Model (CCAM, McGregor and Dix 2001), a chemical transport model (CTM, Cope *et al.* 2009b), and modules for modelling the emissions of particles and particle precursors from sources including smoke, dust, sea salt and vegetation.

CCAM has been used widely in Australia, for regional downscaling studies of climate change (Corney *et al.* 2013), and coupled to the CTM for regional and local atmospheric chemical constituent modelling studies, such as an evaluation of the sources and sinks and transport of mercury in Australia (Cope *et al.* 2009b). The CTM has been extensively used in Australia for modelling urban and regional scale air quality, both historical and under climate change (Cope *et al.* 2009c). For the purposes of the current study, the CTM has been configured to model both primary particle emissions and the production of secondary particles from the gas phase reactions of inorganic and organic chemical precursors. The following particle source groups have been considered in the current study.

- Sea salt emissions are modelled by assuming that bubble-bursting is the dominant mechanism (Gong 2003). The parameterised emissions vary with near-surface wind speed (which is provided by CCAM). The emissions of PM2.5 and PM2.5–10 were modelled in the current study.
- The emissions of dust are modelled as a function of vegetation cover, soil type, soil moisture and the surface shear stress (which is provided on an hourly basis by CCAM). A two-stage approach is used (Shao 2001), in which horizontal sand saltation is simulated first, followed by vertical dust liftoff. Dust size categories of 0–2.5 μm, 2.5–10 μm, 10–20 μm, 20–40 μm and 40–60 μm were modelled.
- The emission of particles and gas phase particle precursors from fires were modelled according to the approach described earlier in this report. Note that the CTM models both the primary particle emissions and secondary particles generated from the reaction of semi-volatile organic compounds (sVOCs), sulfate, nitric acid and ammonia in the smoke plumes.
- Particles are generated from gas phase reactions of sVOCs (such as isoprene) emitted by vegetation. Biogenic VOC emissions are parameterised as a function of vegetation type, leaf density, leaf temperature and incident radiation (Guenther *et al.* 1993).
- Emissions of primary particles and secondary particle precursors from anthropogenic sources have not been considered in this study. Thus we are assuming that the

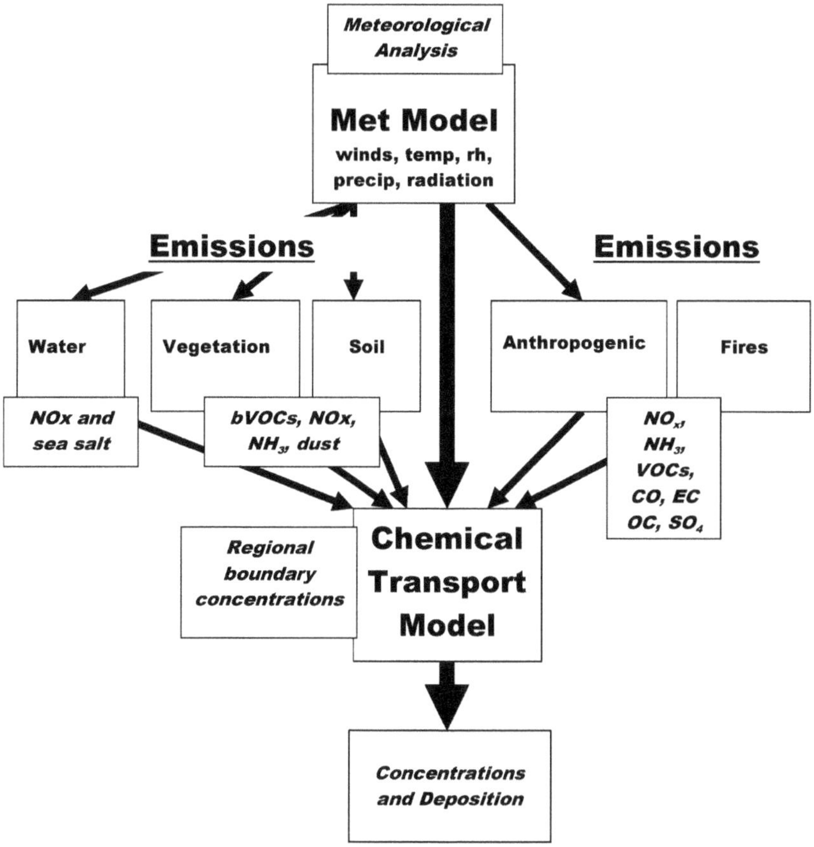

Figure 11.7. Schematic diagram of the modelling system, showing the major components used for modelling the meteorology, particle and particle precursor emission and chemical transport processes.

production of particles from anthropogenic emissions is small compared to other sources such as fires and dust.

- In this study we have not considered the advection of smoke from land masses to the north of Australia.

In order to model all of the particle production processes above, the CTM tracks the following aerosol chemical components:

- elemental carbon (EC)
- organic carbon (OC), both emitted and formed by secondary reactions
- sulfate
- nitrate
- ammonium
- metallic ions (sodium, calcium, magnesium, potassium).

Gas phase chemistry is modelled using the Carbon Bond 2005 mechanism (Yarwood *et al.* 2005), which consists of 51 species and 156 chemical reactions. Additionally, the generation of secondary inorganic aerosols (including ammonium nitrate and ammonium sulfate) is modelled using the ISORROPIA-II model (Fountoukis and Nenes 2007). The production of secondary organic aerosol is modelled using the volatility basis set approach as described in Tsimpidi *et al.* (2010). The CTM was configured to model PM2.5 and larger particle size categories in the current study, with particular focus on the characterisation of secondary particle production. This is supported by a study (Qin and Mitchell 2009) which suggested that aged smoke formed a major component of the fine particle mass in the Top End.

The modelling system was configured to simulate particle production over northern Australia at 50 km resolution, and then at 10 km resolution for a nested sub-domain covering part of the Top End (Fig. 11.8). The model domain extends from the surface to over 10 km above ground. The model integration period covered the dry seasons (May–November inclusive) of 2004–09 inclusive and represents a compromise between the need to model a long-term period in order to examine inter-annual changes in particle loading, and computer resource limitations.

Output from the CTM includes hourly concentrations of 31 particle species which were then available for comparison with speciated mass-based observations. Additionally, the particle species are used to estimate the AOD according to the method used in the GOCART aerosol model (Colarco *et al.* 2010). The GOCART approach parameterises AOD as a function of particle chemistry (including sulfate, sea salt, dust, elemental carbon, aged and fresh organic carbon) and relative humidity. Fig. 11.9 shows the mass extinction coefficients and demonstrates how the coefficients (and hence AOD) can be influenced by relative humidity for hydrophilic particles. Fig. 11.9 also provides an example of the modelled AOD for the two model domains. Note how the higher resolution Top End

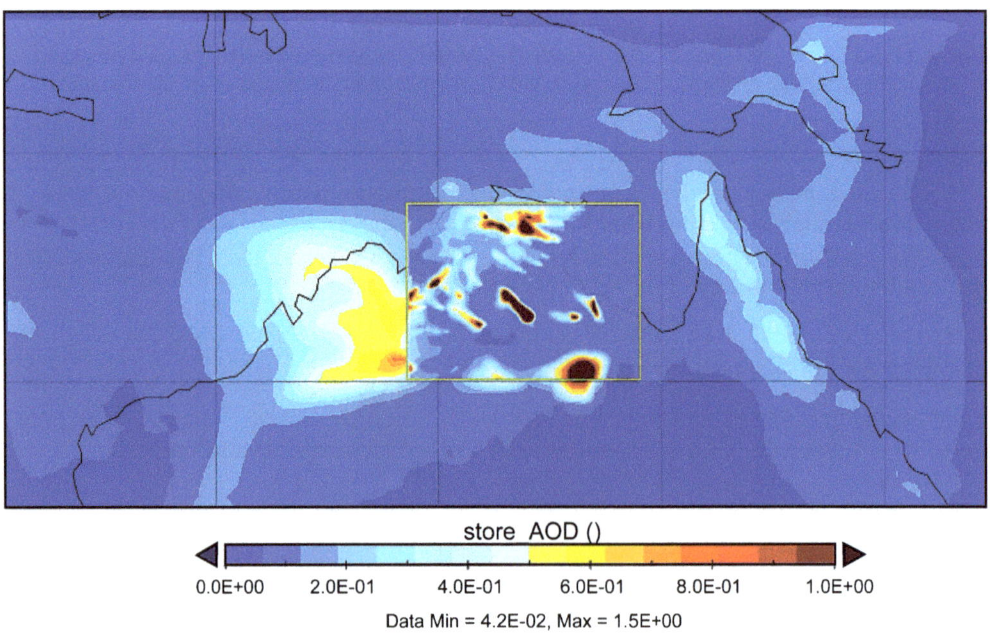

Figure 11.8. Showing the 0.5° and (inset) 0.1° computational domains used to model AOD in the Top End. The shaded contours provide an example of the modelled hourly AOD in the two domains.

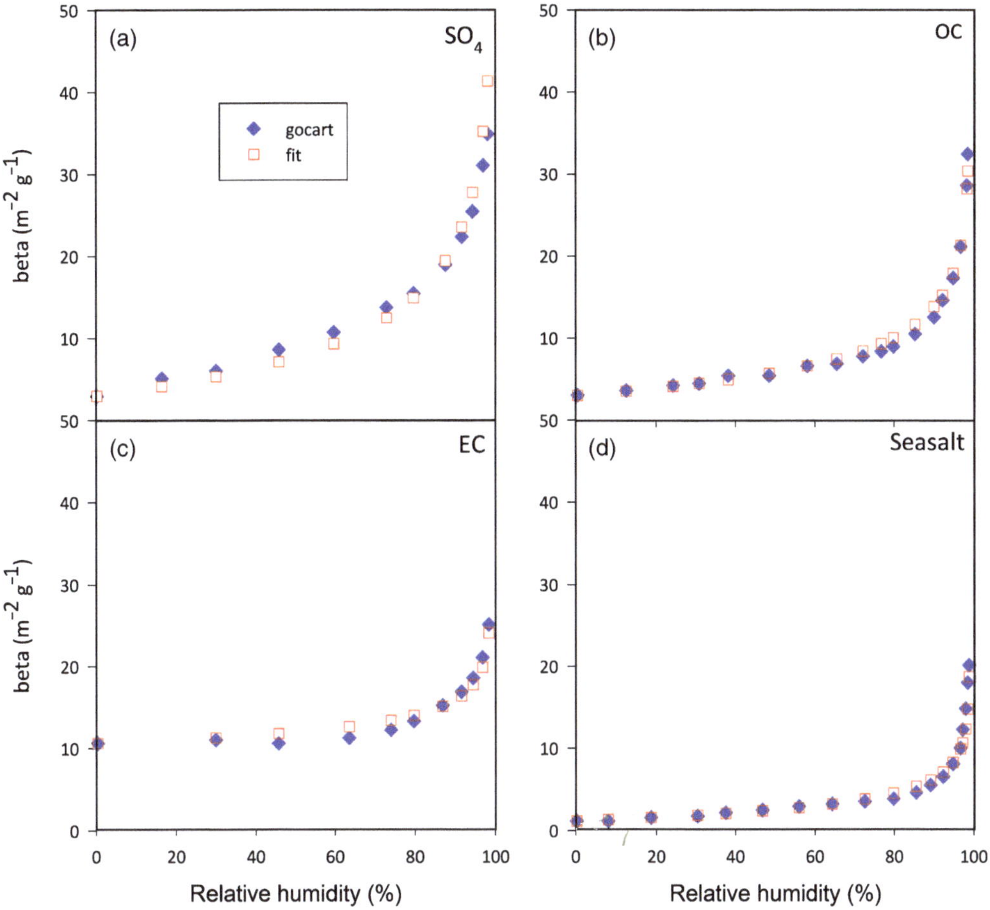

Figure 11.9. Mass extinction coefficients for (a) sulfate; (b) hydrophilic organic carbon; (c) elemental carbon: (d) seasalt. Shown are the original GOCART curves and equivalent (fitted) CTM computational functions.

domain (~10 km × 10 km grid spacing) is better able to resolve the variation of AOD within individual smoke plumes.

Assessment of the accuracy of regional predictions

In this section we explore the accuracy of the bottom-up accounts by comparing the observed atmospheric concentrations with the concentrations estimated from the chemical transport model using the spatial field of emission rates (Option 1).

An important component of the modelling task is the assessment of model performance. Such an assessment provides insight into whether the methodology has the capability to identify biases in the emission fluxes of gases and particles from fires. However, one of the challenges in assessing model performance in northern Australia is the relative scarcity of datasets suitable for undertaking a diagnostic investigation of model performance. Because of this, we have focused on the use of surface- and some satellite-based observations of AOD to make this initial assessment. AOD is useful in this respect because it is by definition a column-integrated measurement of particle loading, and thus has the advantage of being representative of a larger spatial footprint than near-surface point observations.

Surface-based observations of AOD are available from sites located at Lake Argyle, Jabiru and Darwin in the Top End. These sites are maintained as part of the CSIRO AeroSpan (Aerosol characterisation via Sun Photometry: Australian Network), and are affiliated with the Aerosol Robotic Network (AERONET) operated by NASA. AeroSpan sites have been used extensively for analysing the aerosol characteristics in the Australian region (Mitchell and Forgan 2003; Qin and Mitchell 2009; Rotstayn *et al.* 2009). Of particular relevance to this study, the climatology of Australian tropical aerosol was recently published using a combination of CSIRO AeroSpan data and aerosol data from Bureau of Meteorology instruments (Mitchell *et al.* 2013).

Fig. 11.10 shows the observed and modelled daily AOD for the three Top End AeroSpan sites for 2006 and 2009. The observation time series show a similar trend with the AOD being low (generally <0.2) early in the dry season and increasing midway through the season with daily peaks of up to 0.6. The observed AOD generally peaks in October and begins to decrease in November (after day 300), corresponding to the onset of the wet season. The model has generally done well in reproducing the observed trend, although it exhibits a larger day-to-day variability (this is particularly evident at Darwin), and it has a bias to over-predict the AOD early in the dry season. With respect to the variability, an analysis of the modelled AOD indicates that several the peaks are caused by dust storms in the model and that the frequency of dust storms may be over-predicted. In other work we have found that the modelled incidence of dust storms is very sensitive to the parameterised soil moisture in the top centimetre of the model soil layer. In the current work, the ratio of the soil surface moisture to the average moisture in the first soil layer in CCAM is 0.1. Experience now suggests that this ratio may be too low (with an optimal value being 0.15–0.2), and may lead to an over-estimate of the dust storm frequency and thus may be the cause of some of the 'noise' in the modelled AOD time series.

Fig. 11.11 shows the scatter plots of the observed and modelled daily AOD and indicates that the performance is particularly good at Lake Argyle and at Jabiru in 2009. However, the model does show a tendency to over-predict lower AOD values at Jabiru in 2006. Overall the fraction of daily AOD predictions which lie within a factor of two of the observed values varies between 51% (Jabiru 2006) and 82% (Darwin 2009). The correlation coefficients vary between a minimum of 0.13 (Darwin 2006) and 0.56 (Lake Argyle 2009).

The potential reasons given for the tendency of the model to under-predict the lower concentrations included the omission of: contribution from non-biomass burning sources such as background maritime haze, mineral dust and other pollution sources (e.g. motor vehicles); possible generation of secondary organic aerosols; possible recirculation of aerosol once it leaves the model domain; and contributions from fire sources outside the model domain.

Fig. 11.12 shows the observed (Mitchell *et al.* 2013) and modelled AOD climatology for the three AeroSpan stations in the Top End. Here the observed AOD climatology has been generated from 10–14 years of data while the model climatology has been generated from 6 years of data. Fig. 11.12 demonstrates that the model has been able to successfully capture the marked rise in AOD which occurs in September. Moreover the model has also been able to capture the magnitude of the rise, with AOD typically reaching 0.25 to 0.3 after this month. Prior to September the model does well at predicting the mean AOD at Jabiru and Lake Argyle, while over-predicting the AOD at Darwin.

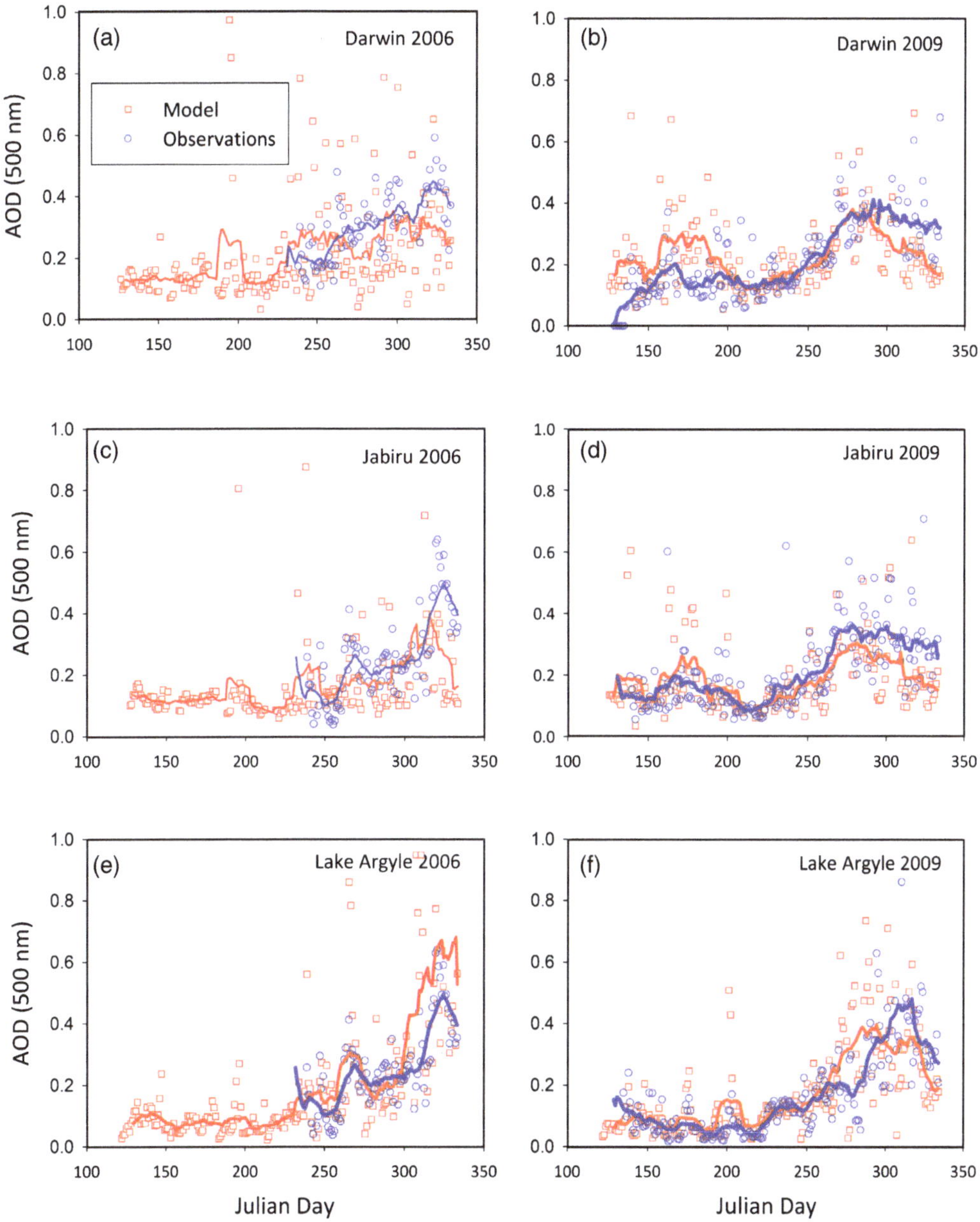

Figure 11.10. Observed and modelled aerosol optical depth for the Top End dry season (May–November inclusive) for 2006 (left panels) and 2009 (right panels). Model results are shown in red, and observed results in blue. The points depict daily averages, while the lines depict 15-day running means. Successive panels show results for Darwin (upper panels), Jabiru (middle panels) and Lake Argyle (lower panels).

Fig. 11.13 gives the predicted breakdown of the modelled AOD into sea salt, dust, fresh smoke particles and secondary particles (which is principally due to aged smoke in the model). It suggests that the only source which doesn't contribute significantly to the AOD is sea salt, with a contribution of only a few per cent.

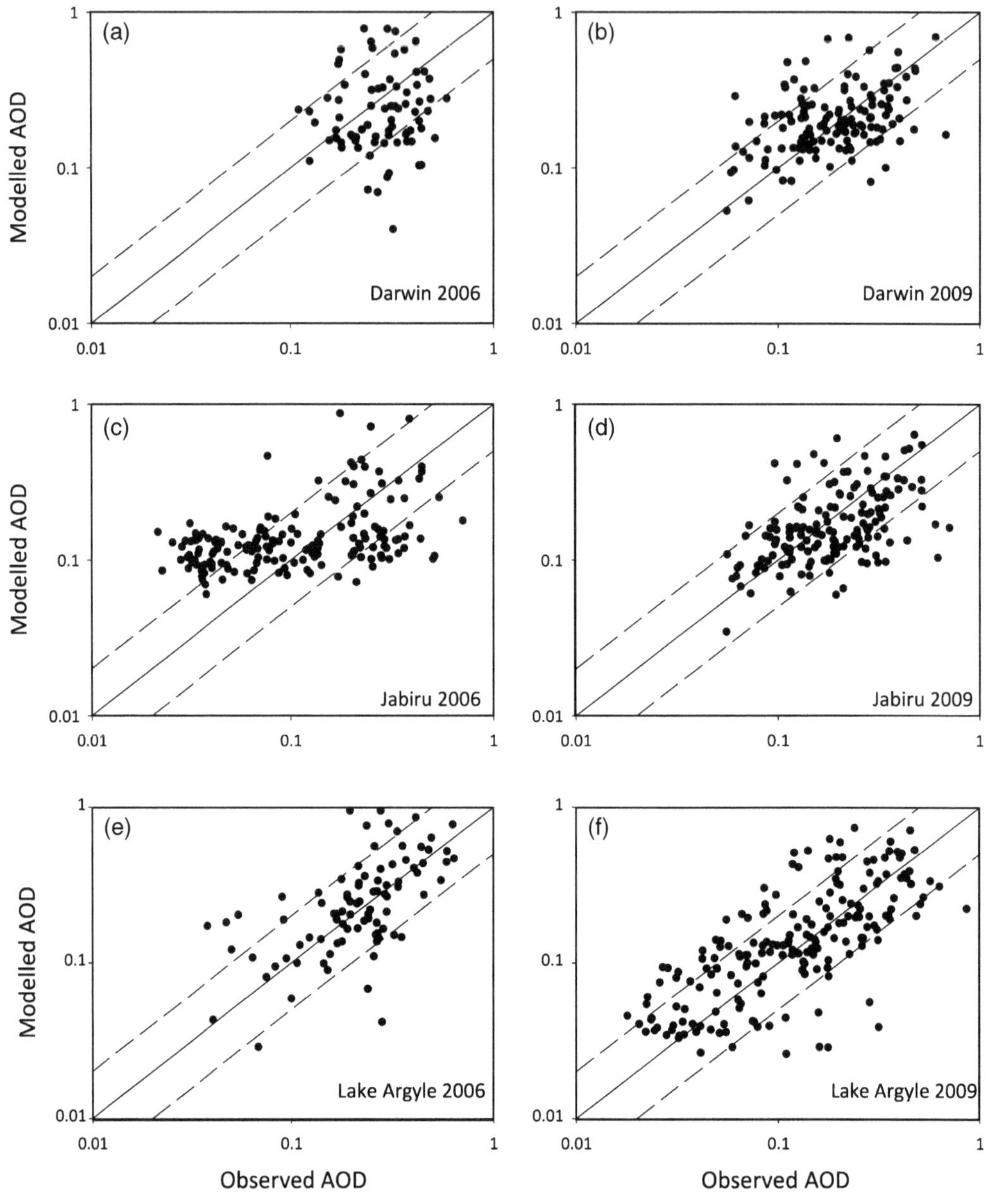

Figure 11.11. Scatter plots of observed and modelled daily average aerosol optical depth. The solid diagonal is the 1:1 line. The dashed lines show ± factor of two. (a), (c), (e) 2006; (b), (d), (f) 2009; (a), (b) Darwin; (c), (d) Jabiru; (e), (f) Lake Argyle.

Windblown dust is modelled to be a significant contributor to AOD at all of the sites (typically 25%–30%), with an increasing contribution later in the dry season in response to a drying of the soil and an increased likelihood of dust mobilisation under windy conditions. However, earlier discussion suggested that the incidence of dust storms may be over-predicted in these simulations.

It is interesting to note that secondary particles generated from biogenic VOCs are modelled to contribute up to 20% of the total AOD. This result is consistent with other studies which have found that most of the secondary organic aerosol (SOA) mass may have a biogenic origin (Kanakidou *et al.* 2005).

Primary and secondary particles generated from smoke emissions are modelled to contribute ~50% of the AOD at Darwin and Jabiru and ~40–50% of the AOD at Lake Argyle. Within the smoke component, most of the AOD contribution is predicted to come from secondary aerosol production, with secondary inorganic and secondary organic aerosol having about an equal contribution.

Qin and Mitchell (2009) have characterised aerosol types over the Australian continent by undertaking a cluster analysis of aerosol optical properties obtained from an inversion of clear-sky radiance distributions. The cluster analysis, coupled with satellite imagery and back trajectories, identified four major classes of aerosol: aged smoke, fresh smoke, coarse dust, and a super-absorptive aerosol. Fig. 11.14 shows the seasonal distribution of each class for Darwin, Jabiru, Lake Argyle, and Tinga Tingana (a station in the arid zone). The plots for the tropical stations show the dominance of fresh and aged smoke classes at these sites. Note that only a low frequency of dust cases was found at these sites, which may provide additional evidence that our model is over-predicting the frequency of dust events.

Further analysis of the data underlying the aerosol typology of Qin and Mitchell (2009) allows estimation of the monthly average AOD for days when the dominant particle class is aged or fresh smoke. Figs 11.15 and 11.16 show these data together with a model-based estimate of the same parameters. In the case of the modelled AOD, the dominant particle class on any given day has been assigned on the basis of the source type that makes the largest contribution to the modelled AOD; i.e. fresh smoke, sea salt, biogenic, or aged (secondary organic and secondary inorganic) smoke. Note that neither the observed nor modelled AOD has been plotted for the months where low tallies precluded a robust

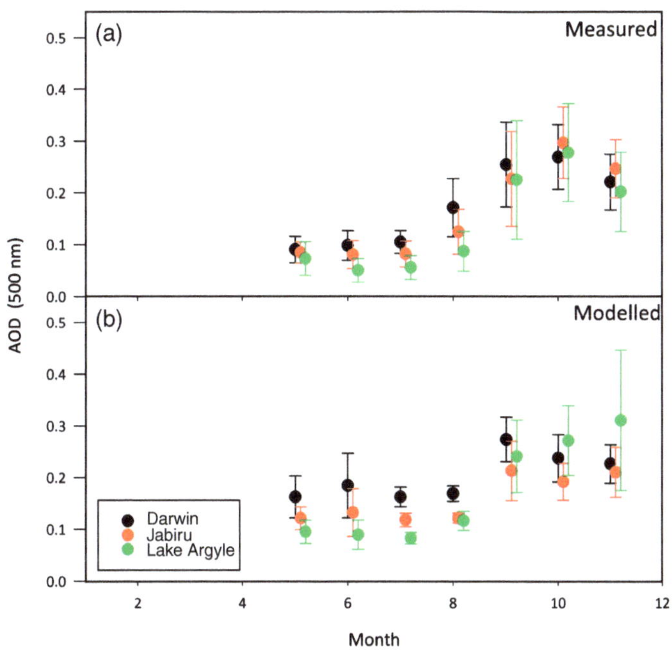

Figure 11.12. (a) Observed AOD climatology (monthly means and standard deviation) for (generally) the period 1998–2012 (Mitchell *et al.* 2013); (b) modelled AOD for the period 2004–09.

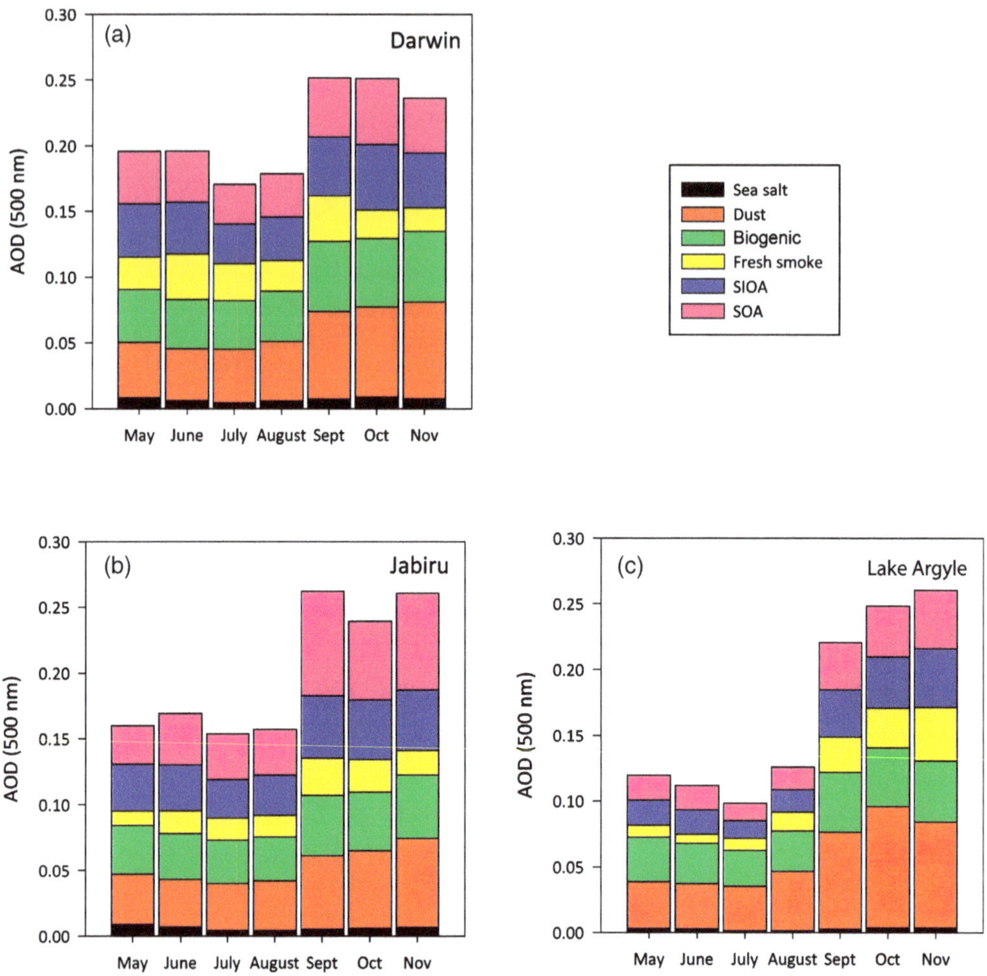

Figure 11.13. Monthly average (for the period 2004–09) modelled breakdown of AOD into source/chemical components. SIOA = secondary inorganic aerosol. SOA = secondary organic aerosol. SIOA + SOA ≅ aged smoke component.

estimate of a monthly average. Both Figures indicate that the model has generally matched the observed AOD contribution from fresh and aged smoke. This is a promising outcome, particularly given the very different approaches used to define the aerosol classes. An important outcome of the comparison is the large contribution of aged smoke (predicted to be caused by secondary particle formation in the model) to the aerosol optical depth.

The CALIPSO satellite flies in a formation of four satellites as part of the A-Train constellation (Stephens *et al.* 2002). CALIPSO's primary instrument is a dual-wavelength (1064 nm and 532 nm), dual-polarisation, backscatter lidar (CALIOP – Cloud Aerosol Lidar with Orthogonal Polarization) (Winker *et al.* 2007), which has been providing a global dataset of range-resolved profiles of the vertical distribution and optical properties (Young and Vaughan 2009) of cloud and aerosol layers since mid-June 2006. In addition to the heights of cloud and aerosol layers, information on the shape of the particles in a layer can be obtained from the ratio of the signals in the two polarisation channels, allowing

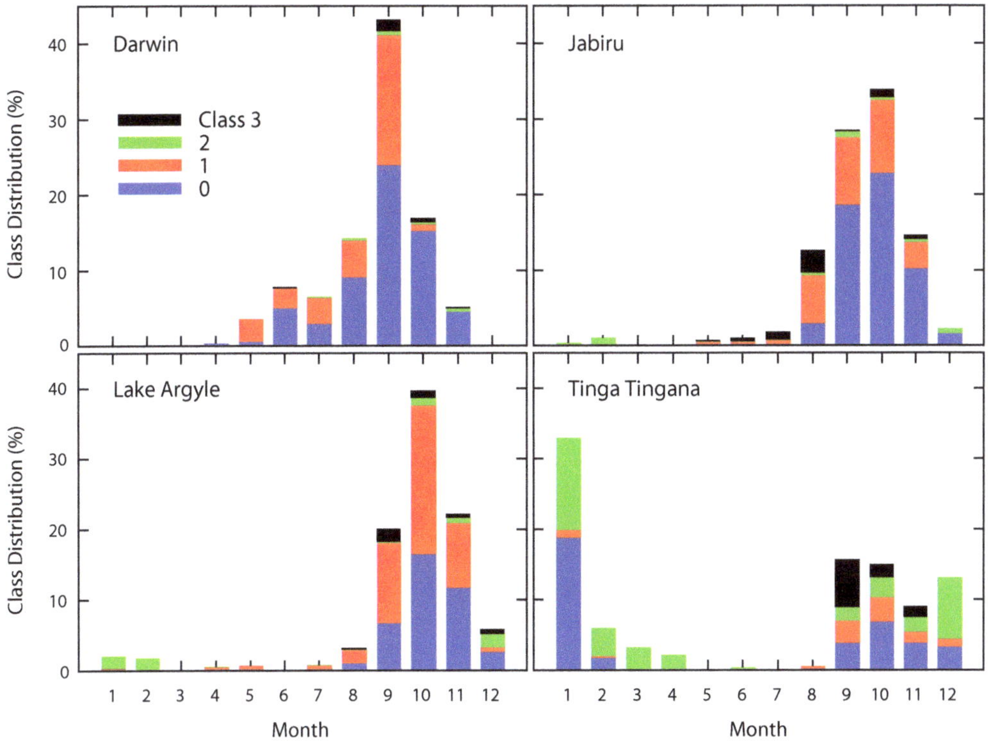

Figure 11.14. Seasonal distribution of four significant aerosol classes (Qin and Mitchell 2009). These classes have been associated with aged smoke (0), fresh smoke (1), coarse dust (2), and super-absorbing (3).

spherical particles, as found in hydrated (e.g. marine) aerosol layers, to be distinguished from irregular dust particles. The depolarisation ratio, together with the integrated attenuated backscatter, latitude and surface type, allows a degree of classification of aerosol features into sub-types, in particular the identification of dust and smoke layers (Omar *et al.* 2009).

The nature of the CALIPSO's orbit and the details of CALIOP's sampling regime dictate the extent to which the satellite data can be used to support models or surface-based measurements. The polar, circular, sun-synchronous orbit has an inclination of 98.2° which, when combined with CALIPSO's altitude of 705 km, gives a 16-day repeat cycle. The equatorial crossing time for the ascending node is around 1330 local time and the descending node crossing is 12 h later. CALIOP has a footprint at the surface of ~90 m and profiles are measured every 330 m. Consecutive profiles of various measured or retrieved parameters, when plotted together on height vs. along-track distance axes, are referred to as 'curtain files'. Note that, because of constraints imposed by available power, data transmission bandwidth and limitations to system weight and complexity, CALIOP only makes near-nadir measurements and does not scan across track to produce a swath. So, although CALIOP data can provide a wealth of information on the location, type, and optical properties of aerosol layers detected along its path, these paths do not always intersect aerosol plumes of interest on a given day. The best use of CALIOP data, therefore, is in the

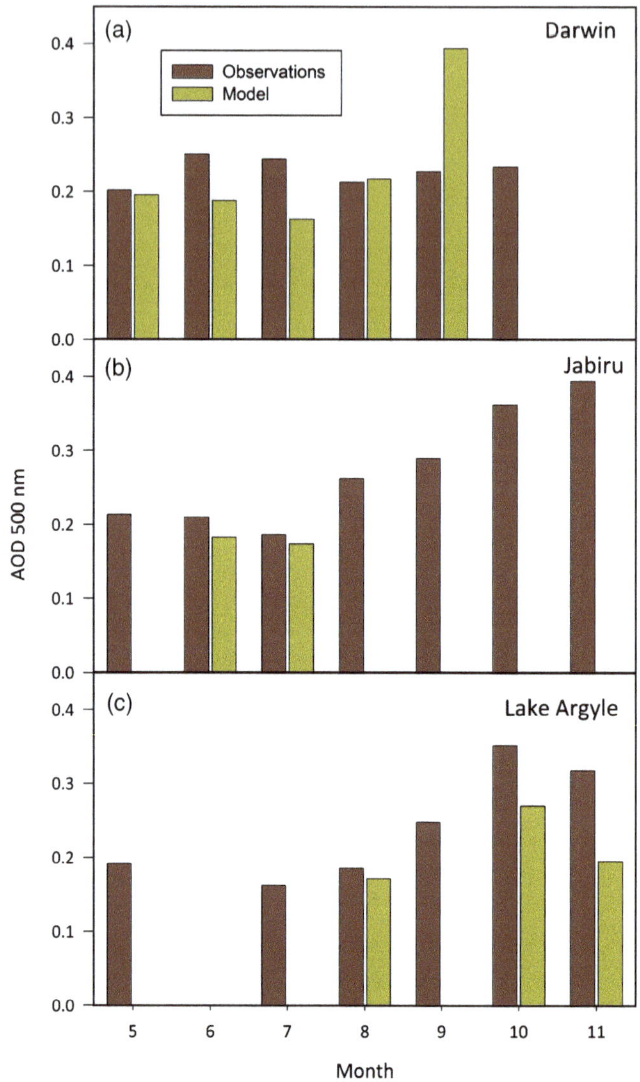

Figure 11.15. Observed and modelled monthly average AOD for air masses classified as dominated by fresh smoke. Results have not been plotted when insufficient daily events exist for the calculation.

comparison with data from other sources or with model predictions, for events of interest at the times and locations of the satellite overpasses.

The data products from CALIPSO/CALIOP that are of most interest to this study are the Vertical Feature Mask (e.g. files of the form CAL_LID_L2_VFM-ValStage1-V3-01.2009-10-21T16-58-40ZN.hdf) and the Aerosol Layer Products (e.g. CAL_LID_L2_05kmALay-Prov-V3-01.2009-10-21T16-58-40ZN.hdf). For this comparison we are using version 3.01 data. The former give the location in terms of height and latitude–longitude coordinates of all the layers detected, along with identification of the type and subtype. (See the CALIPSO data quality statement for more details: http://eosweb.larc.

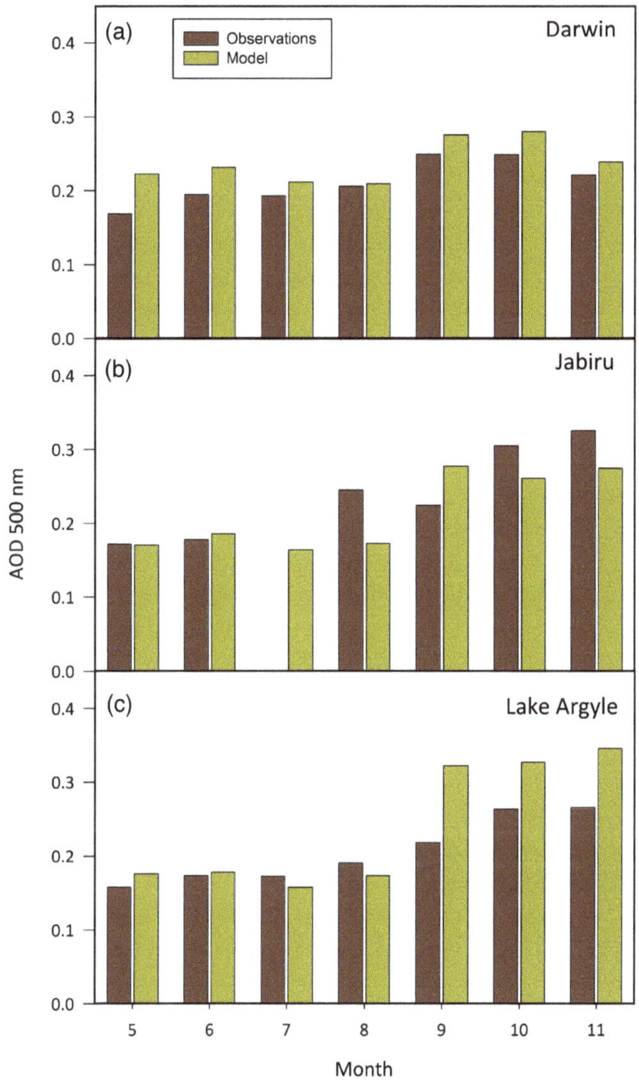

Figure 11.16. Observed and modelled monthly average AOD for air masses classified as dominated by aged smoke.

nasa.gov/PRODOCS/calipso/Quality_Summaries/CALIOP_L2VFMProducts_3.01.html). The latter provide location information along with the optical depths of individual layers and of the total atmospheric column at that location reduced to a resolution of 60 m vertically and 5 km along track: <http://eosweb.larc.nasa.gov/PRODOCS/calipso/Quality_Summaries/CALIOP_L2LayerProducts_3.01.html>).

Important factors to be considered in interpreting the satellite data are those of sensitivity and the effects of noise on the signal, especially during the day. CALIPSO senses the atmosphere by sending out pulses of laser light and measuring the minute fraction that is scattered back to the satellite from any given volume of air, which is roughly 705 km away. During the day, sunlight scattered from the surface into the field of view of CALIOP's receiver can reduce the signal-to-noise ratio (SNR) significantly, especially over relatively bright regions where there is little vegetation cover (e.g. parts of northern Australia). As a consequence, the automatic algorithms that are used to detect aerosol and cloud layers in CALIOP's backscatter profiles (Vaughan *et al.* 2009) do not always detect a layer, or if they do, sometimes do not detect the whole extent of the layer, even if the layer is visible in the level 1 attenuated backscatter data. This is because the automated algorithms are set up to give the best overall performance over the whole globe, and are not especially optimised to specific regions in Australia. (However, we have developed techniques that allow us to extract information from the 'raw' level 1 attenuated backscatter profiles.) The determination of the aerosol subtypes is also affected by SNR and is more reliable in the night-time data than in the daytime data.

In Fig. 11.17 we plot the results of measurements by CALIOP on 21 October 2009 at around 1658 UTC. The path is over Western Australia. The 532 nm attenuated scattering ratio is plotted in Fig. 11.17a and is the ratio of the measured attenuated backscatter coefficient to that expected from a purely molecular atmosphere. In Fig. 11.17b we plot the Cloud-Aerosol Vertical Feature Mask. Here dark blue indicates clear air, mid-blue indicates clouds, cyan aerosol layers, and the colours orange to magenta indicate surface-related signals. Figure 11.17c is the corresponding aerosol subtype Vertical Feature Mask. Here dark blue indicates regions where the CALIOP algorithm finds aerosols matching its Clean Marine sub-type, while mid-blue indicates Dust regions. Polluted Continental (a mixture of Smoke and Clean Continental) aerosols are found in the cyan regions, Clean Continental in the yellow regions, Polluted Dust (a mixture of Smoke and Dust subtypes) in the orange regions and Smoke in the red regions. We note that there is a complex distribution of aerosol subtypes. To the north, over the Timor Sea, the aerosol is predominantly of Clean Marine type. Moving towards the Australian coast, a considerable contribution from smoke can be seen and this mixes with dust to form the Polluted Dust type seen over the continent. One feature that stands out is the considerable mixing of the types and the lack of well-defined boundaries between them.

By comparison, in Figs 11.18a, 11.18b and 11.18c, we plot the modelled PM concentrations, along the CALIPSO ground track, of smoke, dust, and sea salt aerosols. Fig. 11.18 shows that the model also predicts mixing between the various aerosol types, which lends support to the model results in Fig. 11.13. It is interesting to see too that there is a step change in modelled aerosol concentrations at ~4 km above ground level, which is consistent with the observed. Likely the location of this step change represents the vertical extent of the convective boundary layer. Note however that the model does not predict the gradual increase (from 4 km at 10°S to 5 km at 20°S) in the boundary layer. This may be a result of the relatively coarse resolution (1 km) of the model layers above 4 km.

The modelled smoke concentration distribution shown in Fig. 11.18a suggests the presence of an elevated peak stretching from 14 to 18°S, which is consistent with the observed distribution shown in Fig. 11.17c. Similarly, the modelled dust and polluted dust are seen to extend along the entire length of the sample (Fig. 11.18b), which is consistent with the observed. The sea salt, on the other hand, may be constrained closer to the surface in the model than is suggested by the observations.

The attenuated scattering ratio data in Fig. 11.17a are processed further by the CALIPSO algorithms to retrieve the particulate volume extinction coefficient (Young and Vaughan 2009) in regions where aerosols are found. The integral through the depth of a layer of the extinction coefficient is the optical depth. The sum of the optical depths of all layers at a particular location is the column optical depth, which we plot in Fig. 11.19 for the data

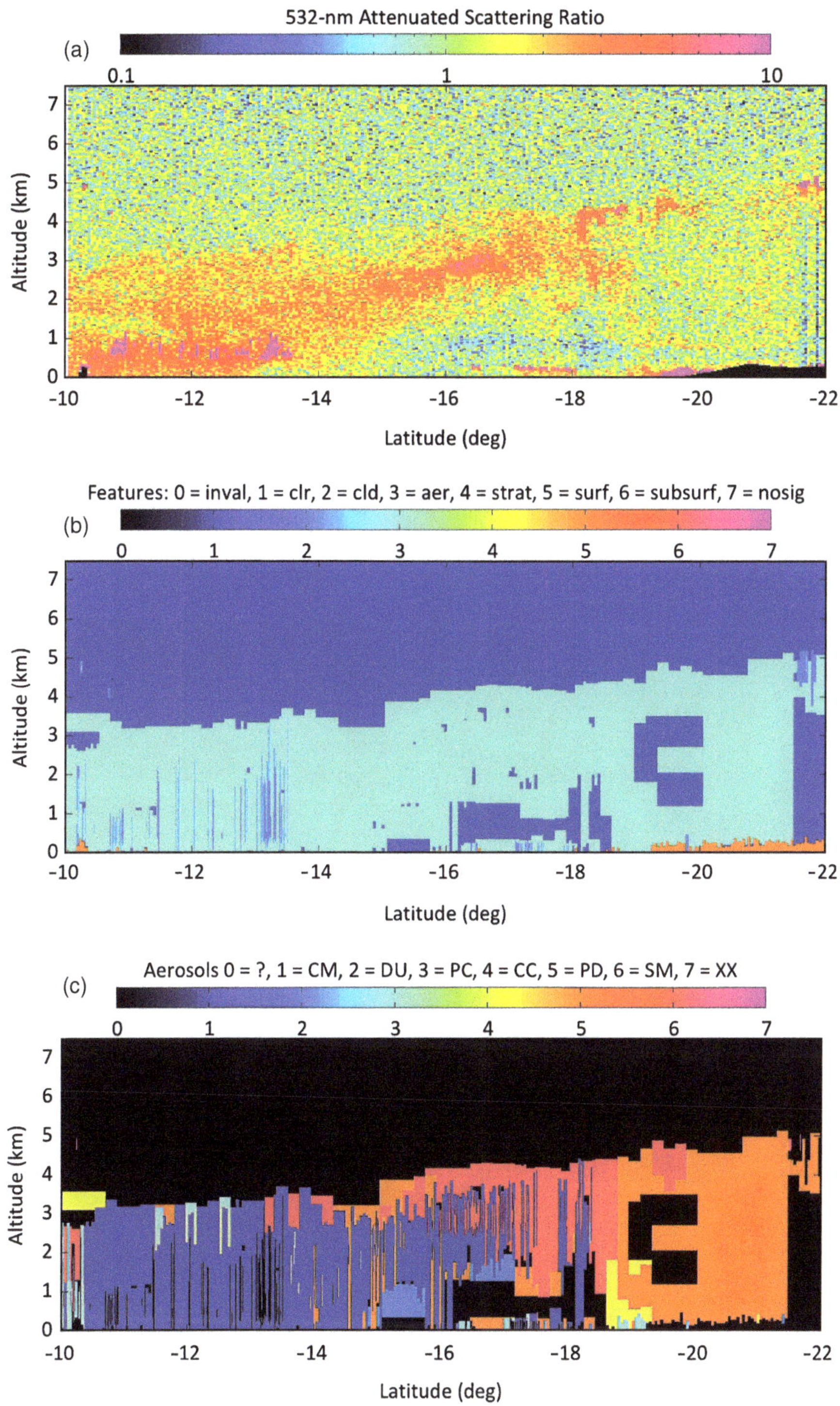

Figure 11.17. Results of CALIOP measurements on 21 October 2009 (a) The CALIOP 532 nm attenuated scattering ratio; (b) CALIOP's cloud-aerosol Vertical Feature Mask; (c) the corresponding aerosol sub-type mask. (See text for details.)

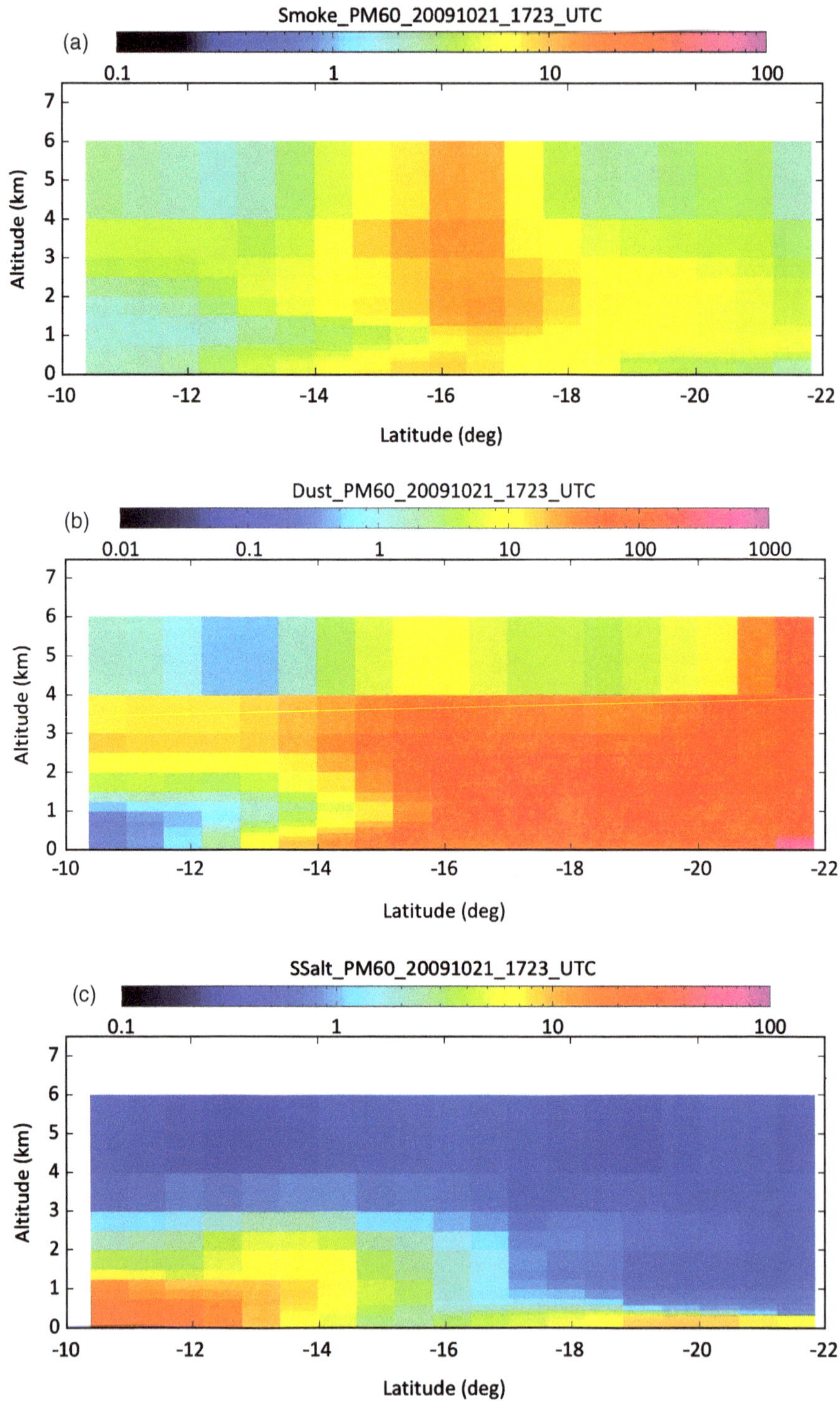

Figure 11.18. Chemical transport model forecasts of PM concentrations (μg m^{-3}) corresponding to the same time and locations as in the 21 October 2009 CALIOP measurements in Fig. 11.17: (a) fire smoke, (b) dust, (c) sea salt.

shown in Figs 11.17a, 11.17b and 11.17c. The sharp changes in optical depth, especially between latitudes 10°S and 17°S, result from the changes in the proportions of the detected aerosol type from Clean Marine to Smoke in this region, as seen by the mixing of dark blue and red regions in Fig. 11.17c.

The modelled column optical depth is also shown in Fig. 11.19 for the Australian domain and matches the observed AOD well for latitudes 10°S to 14°S. The model predicts that the aerosol is dominated by smoke for the latitudes 14°S to 17°S and does not show the same oscillating behaviour as the observations because of the relatively coarse model domain (50 km grid spacing) – thus smoothing the aerosol spatial gradients.

Options for operational methods for estimating regional emissions: constraining and verifying CFI projects

In this chapter we have addressed Option 1 of our three comparative approaches (to compare bottom-up accounts with modelled concentrations using spatial field of emission). As a first-order comparison the results are encouraging. The good agreement between both observed surface PM2.5 concentrations and column AOD with model predictions gives confidence that the daily PM2.5 emission estimates are reasonably accurate, at least to within a factor of ±2, and that there are no large regional biases in the estimates. We find similar accuracy from two independent transport models: TAPM, in which PM2.5 is modelled as a stable tracer; and CCAM/CTM, where full aerosol chemistry is considered. It is unlikely that the agreement between predictions and observations arises from compensating errors. The next step is to assess sensitivity of top-down accounting to uncertainties in the spatial emission fields; in particular to the footprints to which each class of atmospheric tracer observation applies and the degree of temporal smoothing that arises from long-distance transport.

The precision of top-down accounting is determined not only by the precision and resolution of the inputs, but also by the accuracy of model parameterisations. Currently, they are not perfect. For example, used in isolation, total AOD does not provide a strong

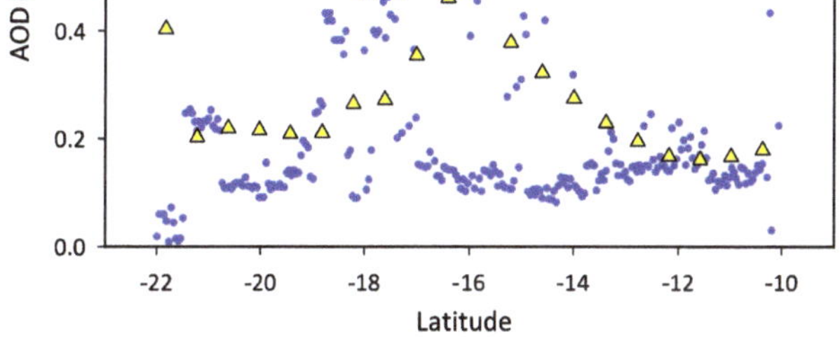

Figure 11.19. CALIOP 532 nm column optical depths obtained from the CALIOP Version 3.01 aerosol layer products file. Also shown is the modelled aerosol optical depth for the 50 × 50 km^2 Australian modelling domain (MODEL).

constraint for emissions verification because estimation of column AOD requires knowledge of the relative contributions of smoke, dust, and marine sources to total aerosol composition. The chemical transport model requires further refinement to achieve this with accuracy. Inversion techniques to separate AOD into particle classes (Qin and Mitchell 2009) show promise, particularly when combined with independent observations of particle classes (i.e. CALIPSO) or trace gases with a strong biomass combustion source such as carbon monoxide (Paton-Walsh *et al.* 2010). However, the models are improving through better parameterisations and through model–data fusion, and progress towards effective inversion techniques is rapid. Constraining and verifying project level and regional emissions estimates (Option 2) requires this next generation of transport models.

Option 3 has not been explored in this work. This option requires a spatially and temporally complete set of atmospheric observations across the region (data that are not currently available) and has become more problematic with the malfunctioning of the Schiamachy satellite.

References

Barrett DJ (2011) Timescales and dynamics of carbon in Australia's savannas. In *Ecosystem Function in Savannas: Measurement and Modeling at Landscape to Global Scales*. (Eds MJ Hill, NP Hanan) pp. 347–366. CRC Press: Boca Raton, FL.

Colarco P, da Silva A, Chin M, Diehl T (2010) Online simulations of global aerosol distributions in the NASA GEOS-4 model and comparisons to satellite and ground-based aerosol optical depth. *Journal of Geophysical Research* **115**, D14207. doi:10.1029/2009JD012820

Commonwealth of Australia (2013) Carbon Credits (Carbon Farming Initiative) Reduction of Greenhouse Gas Emissions through Early Dry Season Savanna Burning—1.1) Methodology Determination 2013. www.comlaw.gov.au/Series/F2013L01165

Cope M, Hibberd M, Lee S, Malfroy H, McGregor J, Meyer CP, Morrison A, Nelson P (2009a) 'The transportation and fate of mercury in Australia: atmospheric transport modelling and dispersion.' Final report to Australian Government Department of Environment, Water, Heritage & the Arts, RFT 100/0607.

Cope M, Lee S, Noonan J, Lilley B, Hess D, Azzi M (2009b) 'Chemical Transport Model technical description.' Centre for Australian Weather and Climate Research: Melbourne.

Cope M, Lee S, Physick W, Abbs D, McGregor J, Nguyen KA (2009c) 'Methodology for determining the impact of climate change on ozone levels in an urban area.' Clean Air Research Project, Final report to Australian Government Department of Environment, Water, Heritage and the Arts.

Corney S, Grose M, Bennett JC, White C, Katzfey J, McGregor J, Holz G, Bindoff NL (2013) Performance of downscaled regional climate simulations using a variable-resolution regional climate model: Tasmania as a test case. *Journal of Geophysical Research, D, Atmospheres* **118**(21). doi:10.1002/2013JD020087

DoE (Australian Government Department of the Environment) (2012) 'National Inventory Report 2012.' Volume 1. The Australian Government Submission to the United Nations Framework Convention on Climate Change Australian National Greenhouse Accounts, Commonwealth of Australia.

Fountoukis C, Nenes A (2007) ISORROPIA II: a computationally efficient thermodynamic equilibrium model for $K^+ - Ca^{2+} - Mg^{2+} - NH_4^+ - Na^+ - SO_4^{2-} - NO_3^- - Cl^- - H_2O$ aerosols. *Atmospheric Chemistry and Physics* 7, 4639–4659. doi:10.5194/acp-7-4639-2007

Giglio L, Loboda T, Roy DP, Quayle B, Justice CO (2009) An active-fire based burned area mapping algorithm for the MODIS sensor. *Remote Sensing of Environment* **113**, 408–420. doi:10.1016/j.rse.2008.10.006

Gong SL (2003) A parameterisation of sea-salt aerosol source function for sub- and super-micron particles. *Global Biogeochemical Cycles* **17**(4), 1097. doi:10.1029/2003GB002079

Gras JL, Meyer CP, Weeks IA, Gillett RW, Galbally IE, Todd J, Carnovale F, Joynt RC, Hinwood A, Berko H, Brown S (2002) 'Emissions from domestic solid fuel burning appliances [wood-heaters, open fireplaces].' Technical report no. 5. Environment Australia: Canberra.

Guenther A, Zimmerman P, Harley P, Monson R, Fall R (1993) Isoprene and monoterpene emission rate variability: model evaluation and sensitivity analysis. *Journal of Geophysical Research* **98D**(12), 609–612, 617.

Haverd V, Raupach MR, Briggs PR, Canadell JG, Isaac P, Pickett-Heaps C, Roxburgh SH, van Gorsel E, Rossel RAV, Wang Z (2013) Multiple observation types reduce uncertainty in Australia's terrestrial carbon and water cycles. *Biogeosciences* **10**(3), 2011–2040. doi:10.5194/bg-10-2011-2013

IPCC (2006) Agriculture, forestry and other land use. In *Guidelines for National Greenhouse Gas Inventories.* (Eds S Eggleston, L Buendia, K Miwa, T Ngara, K Tanabe) Vol. 4. Institute for Global Environmental Strategies (IGES) for the Intergovernmental Panel on Climate Change: Hayama, Japan.

Kanakidou M, Seinfeld JH, Pandis SN, Barnes I, Dentener FJ, Facchini MC, Dingenen RV, Ervens B, Nenes A, Nielsen CJ, Swietlicki E, Putaud JP, Balkanski Y, Fuzzi S, Horth J, Moortgat GK, Winterhalter R, Myhre CEL, Tsigaridis K, Vignati E, Stephanou EG, Wilson J (2005) Organic aerosol and global climate modelling: a review. *Atmospheric Chemistry and Physics* **5**, 1053–1123. doi:10.5194/acp-5-1053-2005

Luhar AK, Mitchell RM, Meyer CP, Qin Y, Campbell S, Gras JL, Parry D (2008) Biomass burning emissions over northern Australia constrained by aerosol measurements: II – Model validation, and impacts on air quality and radiative forcing. *Atmospheric Environment* **42**(7), 1647–1664. doi:10.1016/j.atmosenv.2007.12.040

Luhar A, Meyer M, Lee S, Cope M, Parry D (2009) Modelling fine particulate matter over Darwin (Australia) due to extensive savannah burning during 2003–2007. In *Proceedings of the 9th International Conference on Southern Hemisphere Meteorology and Oceanography* 9–13 February 2009. American Meteorological Society, Melbourne, Australia.

McGregor JL, Dix MR (2001) The CSIRO conformal-cubic atmospheric GCM. *UTAM Symposium on Advances in Mathematical Modelling of Atmosphere and Ocean Dynamics,* 2000. Limerick, Ireland, Kluwer Academic.

Meyer CP (2004) 'Establishing a consistent time-series of greenhouse gas emission estimates from savanna burning in Australia.' Rep. C/0958. CSIRO Atmospheric Research: Aspendale. www.cmar.csiro.au/e-print/open/meyercp_2004a.pdf

Meyer CP (2011) 'Review of the national greenhouse gas emissions methodology for savanna burning.' Final report to the Australian Government Department of Climate Change and Energy Efficiency. CSIRO Marine and Atmospheric Research: Aspendale.

Meyer CP, Lee S, Cope M (2013) 'Smoke impacts from prescribed burning in Victoria; developing a risk climatology.' 20th International Congress on Modelling and Simulation, Adelaide, Australia, 1–6 December 2013. www.mssanz.org.au/modsim2013/A3/meyer2.pdf

Meyer CP, Luhar AK, Mitchell RM (2008a) Biomass burning emissions over northern Australia constrained by aerosol measurements: I – Modelling the distribution of hourly emissions. *Atmospheric Environment* **42**(7), 1629–1646. doi:10.1016/j.atmosenv.2007.10.089

Meyer CP, Reisen F, Luhar AK, Powell JC, Lee S, Cope ME, Keywood MD, Galbally IE, Linfoot S, Parry D, McCaw DL, Tolhurst K (2008b) 'Particles, ozone and air toxic levels in rural communities during prescribed burning seasons.' Final report to Australian Government Department of the Environment, Water, Heritage and the Arts. CSIRO Marine and Atmospheric Research: Aspendale.

Mitchell RM, Forgan BW (2003) Aerosol measurements in the Australian Outback: intercomparison of sun photometers. *Journal of Atmospheric and Oceanic Technology* **20**, 54–66. doi:10.1175/1520-0426(2003)020<0054:AMITAO>2.0.CO;2

Mitchell RM, Forgan BW, Campbell SK, Qin Y (2013) The climatology of Australian tropical aerosol: evidence for regional correlation. *Geophysical Research Letters* **40**(10), 2384–2389. doi:10.1002/grl.50403

Omar AH, Winker D, Kittaka C, Vaughan M, Liu Z, Hu Y, Trepte C, Rogers R, Ferrare R, Kuehn R, Hostetler C (2009) The CALIPSO automated aerosol classification and lidar ratio selection algorithm *Journal of Atmospheric and Oceanic Technology* **26**, 1994–2014. doi:10.1175/2009JTECHA1231.1

Paton-Walsh C, Deutscher NM, Griffith DWT, Forgan BW, Wilson SR, Jones NB, Edwards DP (2010) Trace gas emissions from savanna fires in northern Australia. *Journal of Geophysical Research, D, Atmospheres* **115** doi:10.1029/2009JD013309

Qin Y, Mitchell RM (2009) Characterisation of episodic aerosol types over the Australian continent. *Atmospheric Chemistry and Physics* **9**(6), 1943–1956. doi:10.5194/acp-9-1943-2009

Rotstayn LD, Keywood MD, Forgan BW, Gabric AJ, Galbally IE, Gras JL, Luhar AK, McTainsh GH, Mitchell RM, Young SA (2009) Possible impacts of anthropogenic and natural aerosols on Australian climate: a review. *International Journal of Climatology* **29**(4), 461–479. doi:10.1002/joc.1729

Roy DP, Boschetti L, Justice CO, Ju J (2008) The Collection 5 MODIS burned area product – global evaluation by comparison with the MODIS active fire product. *Remote Sensing of Environment* **112**, 3690–3707. doi:10.1016/j.rse.2008.05.013

Russell-Smith J, Murphy BP, Meyer CP, Cook GD, Maier S, Edwards AC, Schatz J, Brocklehurst P (2009) Improving estimates of savanna burning emissions for greenhouse accounting in northern Australia: limitations, challenges, applications. *International Journal of Wildland Fire* **18**(1), 1–18. doi:10.1071/WF08009

Seiler W, Crutzen PJ (1980) Estimates of gross and net fluxes of carbon between the biosphere and the atmosphere from biomass burning. *Climatic Change* **2**, 207–247. doi:10.1007/BF00137988

Shao Y (2001) A model for mineral dust emission. *Journal of Geophysical Research* **106**, 20239–20254. doi:10.1029/2001JD900171

Stephens GL, Vane DG, Boain RJ, Mace GG, Sassen K, Wang Z, Illingworth AJ, O'Connor EJ, Rossow WB, Durden SL, Miller SD, Austin RT, Benedetti A, Mitrescu Cthe CloudSat Science Team (2002) The CloudSat mission and the A-Train: A new dimension of space-based observations of clouds and precipitation. *Bulletin of the American Meteorological Society* **83**, 1771–1790. doi:10.1175/BAMS-83-12-1771

Tsimpidi AP, Karydis VA, Zavala M, Lei W, Molina L, Ulbrich IM, Jimenez JL, Pandis SN (2010) Evaluation of the volatility basis-set approach for the simulation of organic aerosol formation in the Mexico City metropolitan area. *Atmospheric Chemistry and Physics* **10**, 525–546. doi:10.5194/acp-10-525-2010

van der Werf GR, Randerson JT, Giglio L, Collatz GJ, Mu M, Kasibhatla PS, Morton DC, DeFries RS, Jin Y, van Leeuwen TT (2010) Global fire emissions and the contribution of deforestation, savanna, forest, agricultural, and peat fires (1997–2009). *Atmospheric Chemistry and Physics* **10**(23), 11707–11735. doi:10.5194/acp-10-11707-2010

Vaughan MA, Powell KA, Winker DM, Hostetler CA, Kuehn RE, Hunt WH, Getzewich BJ, Young SA, Liu Z, McGill MJ (2009) Fully automated detection of cloud and aerosol layers in the CALIPSO Lidar measurements. *Journal of Atmospheric and Oceanic Technology* **26**(10), 2034–2050. doi:10.1175/2009JTECHA1228.1

Winker DM, Hunt WH, McGill MJ (2007) Initial performance assessment of CALIOP. *Geophysical Research Letters* **34**(19), L19803. doi:10.1029/2007GL030135

Winker DM, Pelon J, Coakley JA, Ackerman SA, Charlson RJ, Colarco PR, Flamant P, Fu Q, Hoff RM, Kittaka C, Kubar TL, Le Treut H, McCormick MP, Mégie G, Poole L, Powell K, Trepte C, Vaughan MA, Wielicki BA (2010) The CALIPSO Mission: a global 3D view of aerosols and clouds. *Bulletin of the American Meteorological Society* **91**(9), 1211–1229. doi:10.1175/2010BAMS3009.1

Yarwood G, Rao S, Yocke M, Whitten GZ (2005) 'Updates to the Carbon Bond Mechanism: CB05.' US EPA Final Report. www.camx.com/publ/pdfs/CB05_Final_Report_120805.pdf

Young SA, Vaughan MA (2009) The retrieval of profiles of particulate extinction from Cloud-Aerosol Lidar Infrared Pathfinder Satellite Observations (CALIPSO) data: algorithm description. *Journal of Atmospheric and Oceanic Technology* **26**(6), 1105–1119. doi:10.1175/2008JTECHA1221.1

12

Fire or water: which limits tree biomass in Australian savannas?*

Brett P Murphy, Adam C Liedloff and Garry D Cook

Summary

The processes that allow the coexistence of trees and grasses in tropical savannas have long intrigued ecologists. Early theories had a strong focus on climatic controls, but over the last two decades a conceptual model has emerged suggesting that savanna trees are subject to a fire-mediated recruitment bottleneck. Under this model, frequent savanna fires prevent recruitment of saplings into the tree layer, and this bottleneck maintains tree biomass below a climate-determined carrying capacity. We propose that the relevance of this conceptual model, which is based largely on African savanna research, has been substantially overemphasised in the northern Australian context. In this region, the overall abundance of trees in savannas is very strongly controlled by water availability, not fire. There is considerable evidence that the dominant trees of Australian savannas – eucalypts – are one of the most fire-tolerant tree lineages on Earth, and have a remarkable capacity to grow rapidly through the 'fire trap' and reach fire-resistant sizes. The extreme fire tolerance of savanna eucalypts makes them relatively unresponsive to reductions in fire frequency and intensity. In contrast, other groups of trees in Australian savannas typically have a much lower capacity to escape the fire trap, and therefore respond positively in abundance to reductions in fire frequency and intensity. From an applied perspective, there have been suggestions that managers could sequester substantial quantities of carbon (thereby offsetting anthropogenic greenhouse gas emissions) by reducing the frequency and intensity of savanna fires and, hence, increasing tree biomass. We contend that if tree biomass in Australian savannas is strongly limited by water availability, then the potential for fire management to increase tree biomass is relatively limited, at least in relation to the dominant eucalypt component. There may be potential to increase the abundance of the more fire-sensitive non-eucalypt tree component, but the biomass carrying capacity of non-eucalypts in this eucalypt-dominated system remains poorly understood, including whether there is a trade-off between eucalypt and non-eucalypt biomass.

* This chapter is an edited version of an article that was first published in 2015 in *International Journal of Wildland Fire*, Vol. 24(1), pp. 1–13. doi:10.1071/WF14092

Introduction

Savannas dominate the seasonally dry tropical landscapes of Africa, Australia and South America, and to a lesser extent South-East Asia, covering around 20% of the Earth's land surface. They are the most flammable biome on Earth, with around 20% burning in a typical year (Dwyer *et al.* 2000), and global fire activity is strongly concentrated in the tropical savannas (Fig. 12.1) (van der Werf *et al.* 2008). The biomass dynamics of tropical savannas have received enormous attention in recent years, with strong focus on the factors that determine the abundance of woody biomass, and the relative importance of climate and disturbance (Scholes and Archer 1997; Sankaran *et al.* 2004; Sankaran *et al.* 2005; Bucini and Hanan 2007; Sankaran *et al.* 2008; Lehmann *et al.* 2014). Much of the interest

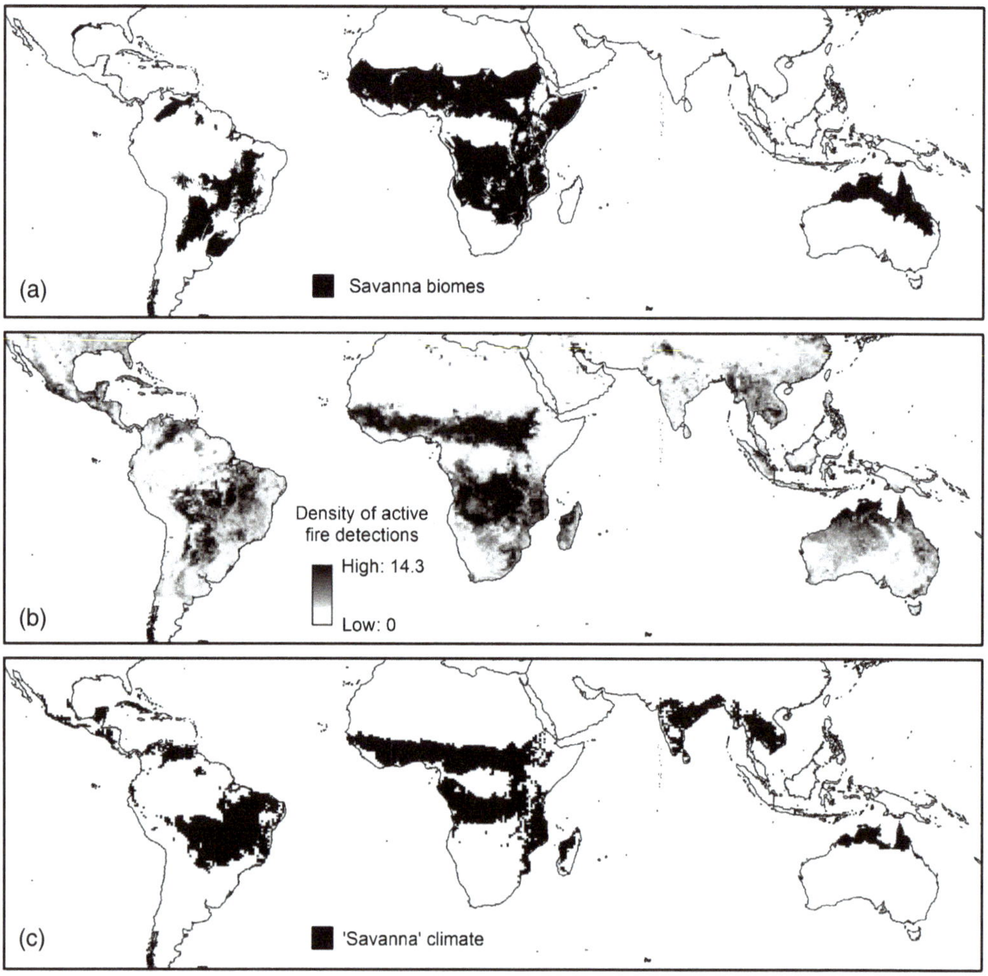

Figure 12.1. The global distribution of (a) savannas, (b) fire activity and (c) the classic savanna climate, warm with intensely seasonal rainfall. Savanna distribution is inferred from a global ecoregions map (Olson *et al.* 2001). Fire activity is inferred from the density of active fire detections from the MODIS satellite sensors (in units: detections km^{-2} year^{-1}) (Bowman *et al.* 2014). Distribution of the 'savanna' climate is inferred from the 'equatorial savannah' classes (As and Aw) of the Köppen-Geiger climate classification (Kottek *et al.* 2006).

in savanna dynamics has been driven by the prominent role that this biome plays in the global carbon cycle, estimated to contribute around 30% of global terrestrial net primary production (NPP), roughly equivalent to the NPP of tropical forests (Grace *et al.* 2006). However, because of the very high frequency of fires in savannas, much of the carbon captured is quickly released back to the atmosphere, such that savannas sequester and store much less carbon than other highly productive biomes such as tropical forests (Grace *et al.* 2006). Given that savanna fires result in the loss of carbon from vegetation to the atmosphere, it has been suggested that reducing fire frequencies could lead to substantial increases in the amount of carbon stored in these systems (Grace *et al.* 2006; Beringer *et al.* 2007; Murphy *et al.* 2010; Murphy *et al.* 2014) – a key management objective in the 21st century.

The dynamic nature of woody biomass in savannas is exemplified by a global trend of woody thickening (i.e. increasing woody plant abundance) that has been observed across many savanna ecosystems (e.g. Roques *et al.* 2001; Burrows *et al.* 2002; Hansen *et al.* 2002). Indeed, Murphy *et al.* (2014) have recently demonstrated that a weak thickening trend can be observed across northern Australian savannas over the last half-century. With the intensification of land use in savannas, changed patterns of fire and grazing and increasing atmospheric carbon dioxide concentration ([CO_2]), there is an urgent need for a mechanistic understanding of how fire, climate and [CO_2] determine woody biomass abundance in savannas. However, the roles of environmental controls in structuring the vegetation of tropical savannas remain far from well understood. Thus, understanding both the trends in savanna woody biomass and the drivers of changes to carbon storage in these systems is highly relevant.

The last decade has seen the emergence of extensive fire management projects that seek to generate carbon credits in northern Australian savanna landscapes (Russell-Smith *et al.* 2009; Russell-Smith *et al.* 2013). Current projects generate carbon credits by abating non-CO_2 greenhouse gases methane and nitrous oxide, but some authors have suggested that an important co-benefit of reductions in the frequency and intensity of fires may be an increase in the amount of carbon stored in both soils (Richards *et al.* 2011) and tree biomass (Murphy *et al.* 2014). The dynamics of soil carbon stocks in relation to fire remain are relatively poorly understood and difficult to demonstrate (Beyer *et al.* 2011; Richards *et al.* 2011), so the use of fire to manage soil carbon in northern Australian savannas seems a distant prospect.

In this chapter we focus solely on the carbon stored in live tree biomass, the dominant biomass pool in northern Australian savannas (Murphy *et al.* 2014). Any attempt to predict the long-term effects of fire management on biomass requires consideration of the theoretical controls of tree abundance in savannas, so we review both the global and Australian literature to evaluate the extent to which these savannas are limited by climate or fire, and the extent to which reductions in fire frequency and intensity are likely to lead to long-term increases in tree biomass. We identify areas where the current state of knowledge is poor, and where this poor state of knowledge is likely to impact our ability to predict the effects of fire management on biomass carbon stocks.

Climatic controls

The extent to which climate controls the distribution of vegetation is a longstanding question in ecology. Early vegetation ecologists observed that at large spatial scales, water availability exerts primary control over the distribution of rainforest (wet), savanna (seasonally wet–dry) and grasslands (dry) (e.g. Schimper 1903; Holdridge 1947). Indeed, under the Köppen-Geiger climate classification system, the Earth's tropical climates with high rainfall seasonality are described as 'savanna' climates (Kottek *et al.* 2006), and these correspond closely to the Earth's major savanna regions (Fig. 12.1). The advent of global-scale vegetation and climate datasets has spawned several recent analyses of the distribution of savannas (Lehmann *et al.* 2011; Murphy and Bowman 2012), confirming that the transition from savanna to rainforest is closely related to increasing mean annual rainfall and decreasing rainfall seasonality. In the last decade, considerable attention has also turned to the climatic controls of tree biomass *within* savannas, and the question of how trees and grasses are able to coexist without one excluding the other (e.g. Sankaran *et al.* 2005; Bucini and Hanan 2007; Bond 2008).

Walter's (1971) 'root-niche separation' hypothesis proposed that only trees are able to access water deep in the soil profile, with grasses only able to access water in the top-most layers of the soil profile. Hence, as soil moisture at depth increases (i.e. in high rainfall areas) so too does tree abundance. A key prediction of the Walter hypothesis is that tree abundance should increase steadily with increasing rainfall – a pattern that was confirmed for a set of northern Australian sites by Williams *et al.* (1996) and Cook *et al.* (2002). Across a 300–1600 mm rainfall gradient, tree basal area was very strongly correlated with mean annual rainfall ($r^2 = 0.90$).

The northern Australian results of Cook *et al.* (2002) differ slightly from those of Sankaran *et al.* (2005), whose seminal paper showed that mean annual rainfall imposes a very strong upper bound to tree cover in African savannas, but with enormous variation below the upper bound. Lehmann *et al.* (2014) have recently confirmed Australian (and South American) savannas clearly conform to the African pattern, with similarly enormous variation in tree basal area below a rainfall-determined upper bound (Fig. 12.2). While these papers clearly demonstrated that rainfall plays a key role in limiting tree biomass in savannas, broadly in line with the Walter hypothesis, the very high level of variability beneath the rainfall-determined upper bound suggested that other controls were also important. These authors have generally interpreted the high level of variability as evidence that disturbance, primarily fire, reduces tree biomass below the climate-determined carrying capacity. However, it is unclear why this variation is more likely to be caused by disturbance rather than other resource-based factors, such as soil depth (which affects water storage capacity) or nutrient availability. Indeed, by focusing on a single soil type (sands), Cook *et al.* (2002) found a remarkable correlation between mean annual rainfall and tree basal area in northern Australia, in very clear contrast to the findings of Lehmann *et al.* (2014). Williams *et al.* (1996) developed a relationship accounting for the interactions between soil type and rainfall in tree basal area.

Disturbance-based controls

The advent of remote-sensing technologies and global fire histories has revealed the vast extent of fire activity in the tropical savannas (van der Werf *et al.* 2008; Bowman *et al.* 2014). A longstanding and unresolved question is the extent to which flammability is an emergent property of grassy savanna vegetation, or savannas are an emergent state of seasonal climates that promote flammability – a classic 'chicken and egg' problem. However, there is an emerging body of work suggesting that the distributions of savannas and the other dominant biome in the tropics, rainforests, are controlled by both climate and disturbance, primarily fire (Hirota *et al.* 2011; Lehmann *et al.* 2011; Staver *et al.* 2011;

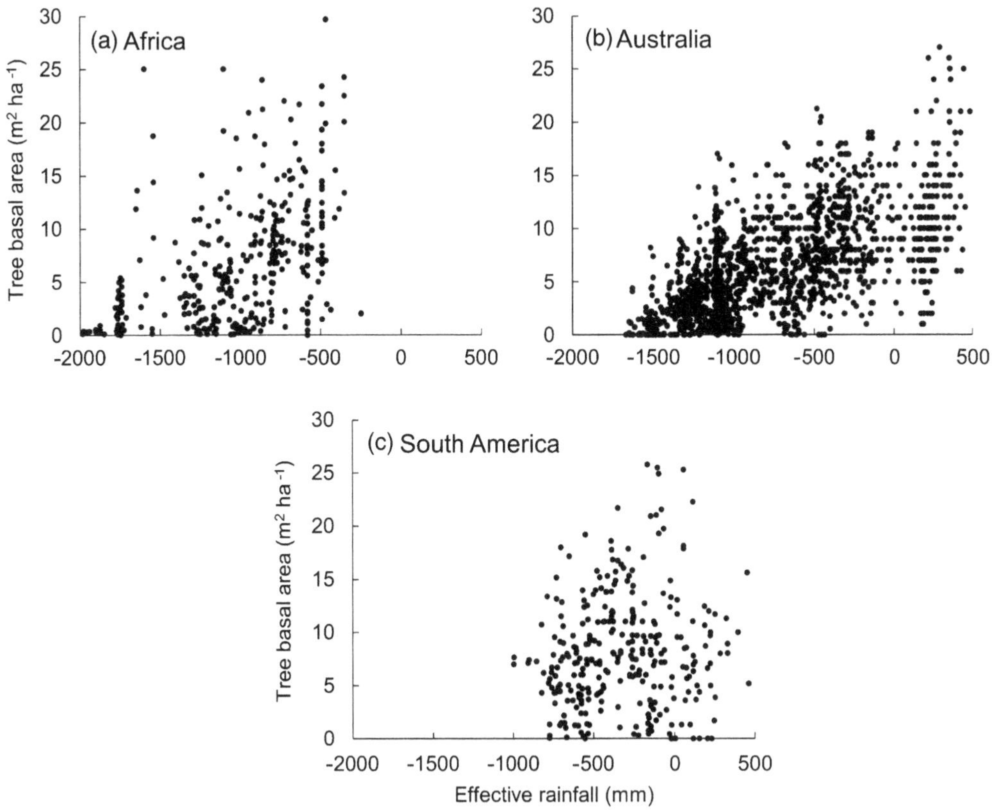

Figure 12.2. The observed patterns of tree basal area in the savannas of three continents: (a) Africa, (b) Australia and (c) South America. Effective rainfall is mean annual rainfall minus mean annual potential evapotranspiration. Adapted from Lehmann *et al.* (2014), with permission from AAAS.

Murphy and Bowman 2012). This work has led to the development of a broad conceptual model combining two key concepts: (1) savanna and rainforest are alternative stable states, maintained by a strong negative feedback between the forest canopy and fire activity; (2) it is the interaction between tree growth rates and fire frequency that determines the likelihood of a forest canopy forming to displace flammable savanna (Lehmann *et al.* 2011; Murphy and Bowman 2012). According to the model, any factor that increases tree growth rates, such as elevated water availability or [CO_2], will increase the likelihood of seedlings of forest trees maturing between subsequent savanna fires and forming a closed forest canopy, thereby excluding grasses and reducing the flammability of the vegetation. Additionally, any factor that decreases fire frequency is likely to increase the chance of forest formation, e.g. rockiness, topographic fire protection, insularity.

Perhaps the most compelling evidence that fire controls the distribution of savanna and rainforest is the results of fire-exclusion experiments that have shown rainforest species invading fire-protected savanna, especially in high rainfall areas (San Jose and Farinas 1983; Moreira 2000), with a complete biome shift from grassy to closed canopy sometimes occurring within a few decades (Trapnell 1959; Swaine *et al.* 1992; Louppe *et al.* 1995; Peterson and Reich 2001). There are, however, exceptions to this generalisation, with some fire-exclusion experiments failing to produce shifts from savanna to rainforest after several decades, especially under low rainfall conditions, pointing to additional limiting factors (Bond *et al.* 2003; Higgins *et al.* 2007).

Not only is fire widely believed to control the distribution of savannas, but there is substantial evidence that fire limits tree biomass *within* savannas. Higgins *et al.* (2000) proposed a model whereby tree densities in savannas are limited by the effects of fire on critical life stages – a so-called fire-mediated demographic bottleneck. In mesic savannas, the critical life-stages are thought to be saplings (Sankaran *et al.* 2004), and repeated fires can maintain juvenile trees (saplings and smaller) in a suppressed state, potentially for decades. Only by growing through the sapling stage, and attaining sufficient height (Bond and Midgley 2000) or bark thickness (Hoffmann *et al.* 2009; Lawes *et al.* 2011a) are savanna trees able to escape the 'fire trap' and reach maturity. Under this conceptual model, reductions in fire frequency and/or intensity would diminish the demographic bottleneck, allow more trees to mature, and tree biomass to progress closer to its climate-determined upper bound.

Cook and colleagues (Cook and Heerdegen 2001; Cook *et al.* 2002; Liedloff and Cook 2007; Liedloff and Cook 2011) have argued that tree carrying capacity in Australian savannas is set by water stress, which is determined by dry season length. In various papers, Fensham and colleagues (Fensham and Fairfax 2007; Fensham *et al.* 2009) show that tree stands in many Queensland savannas are ultimately limited by water supply, with periodic droughts substantially reducing live tree populations. Liedloff and Cook (2007) argue that in the monsoonal north, water limits tree populations through more frequent dieback events that affect much smaller proportions of the adult tree populations. Prior and Eamus (2000) observed that dieback of eucalypt saplings due to water stress is common during the 5 months of the dry season. For 1–3 m high *Eucalyptus tetrodonta* saplings, Prior *et al.* (1997) found that only 16% of tagged stems were still alive at the end of each dry season due to water stress. Thus, water limitation clearly creates a demographic bottleneck restricting recruitment of trees into the canopy.

Evidence from long-term fire experiments

Dramatic increases in woody biomass have been observed following long-term fire exclusion in a range of savannas globally (San José *et al.* 1998; Tilman *et al.* 2000; Higgins *et al.* 2007). Much of the work on this topic comes from several savanna fire experiments in Africa (summarised in detail by Furley *et al.* 2008). For example, Brookman-Amissah *et al.* (1980) described massive negative effects of annual fire regimes, especially of high intensity fires late in the dry season, on rates of recovery of woody biomass following tree clearing in a Ghanaian savanna. Similar findings have been obtained in mesic savannas in other parts of west and southern Africa (Trapnell 1959; Rose Innes 1972). In a 40-year study in the semi-arid savannas of Kruger National Park, South Africa, complete fire exclusion resulted in very large increases in woody biomass, triennial and biennial fires resulted in smaller increases, and annual fires resulted in declines in woody biomass (Higgins *et al.* 2007).

Two similar, although shorter term, fire experiments have been undertaken in northern Australian savannas. The Munmarlary experiment, in Kakadu National Park, ran for 23 years (1973–1996), and consisted of four fire treatments: fire exclusion; annual in the early dry season (EDS: July or earlier); biennial in the EDS; and annual in the late dry season (LDS: August or later) (Russell-Smith *et al.* 2003). In general, eucalypts – the dominant group of trees in Australia savannas – were not released from the understorey as a result of 23 years of fire suppression, although the more fire-sensitive tree species were significantly released into the midstorey. Russell-Smith *et al.* (2003) noted that structural changes in response to any of the fire treatments were relatively minor. Eucalypts did not appear to be limited by fire but by strong overstorey root competition, and recruitment of eucalypts into the overstorey typically occurs following the death of a large eucalypt.

The Kapalga fire experiment, also in Kakadu National Park, ran for 5 years (1990–94), and consisted of three fire treatments: fire exclusion; annual EDS; and annual LDS (Andersen *et al.* 2003). In contrast to Munmarlary, there were significant effects of treatment on mature eucalypts, with tree basal area increasing in the fire exclusion (+3.5%) and annual EDS (+0.5%) treatments, and declining very substantially (–27%) in the annual LDS treatment (Williams *et al.* 1999). As noted by Russell-Smith and Edwards (2006) and Cook *et al.* (Chapter 13), the very strong discrepancy between the muted effects of fire at Munmarlary and the large effects at Kapalga appears due the higher intensity of the Kapalga fires, even those in the EDS. This occurred because the Kapalga plots were burnt by 'fronting' fires, with maximum rate of spread, and hence higher intensity, than 'flanking' and 'backing' fires. The fire intensities experienced at Munmarlary were more typical of ambient fire regimes in Kakadu National Park (Russell-Smith and Edwards 2006).

A third 'natural experiment' has taken place in the mesic savannas of the Northern Territory, at Solar Village near Darwin (Woinarski *et al.* 2004). An area of savanna was subject to fire exclusion for 23 years and an adjacent area burnt frequently during that time (approximately annually). At Solar Village, there were no statistically significant changes in eucalypt or total tree basal area as a result of fire exclusion; however, there was a massive increase (+280%) in overall tree density. Importantly, eucalypt density declined significantly (–35%) as a result of fire exclusion.

Evidence from three fire exclusion experiments in northern Australia suggests that reductions in fire frequency or intensity have a minor impact on the population structure of eucalypts, the dominant group of trees in northern Australian savannas. Recruitment of eucalypts into the overstorey appears not to be limited by fire but by root competition with large overstorey trees. However, fire suppression can clearly lead to recruitment of relatively fire-sensitive non-eucalypt species into the midstorey. The extent to which increases in the non-eucalypt component can increase tree biomass is unclear, given that most non-eucalypt species are smaller in stature.

Demonstrated short-term impacts of fire on northern Australian savanna trees

Important insights into the short-term effects of fire on demographic processes of Australian savanna trees – such as recruitment, mortality and growth – have been derived from the buffalo exclusion (e.g. Werner 2005) and fire manipulation (e.g. Williams *et al.* 1999) experiments at Kapalga, and the 'natural experiment' provided by analysis of the Three Parks vegetation and fire monitoring dataset from Kakadu, Nitmiluk and Litchfield national parks (e.g. Prior *et al.* 2009).

The prevailing view of savanna tree dynamics focuses on the importance of a fire-mediated recruitment bottleneck, with very limited recruitment of saplings into the adult tree size classes (Higgins *et al.* 2000; Sankaran *et al.* 2004; Bond 2008). Fires are assumed to maintain juvenile trees in a suppressed state – referred to as 'Gullivers' by Bond and Van

Wilgen (1996) – potentially for many decades. These suppressed individuals are repeatedly 'topkilled' by fire and can only recruit into the tree layer by reaching fire-resistant height ('escape height') or obtaining sufficiently thick bark (Lawes *et al.* 2011a). Indeed, there is much evidence from northern Australian savannas that fire, especially high-intensity fire, strongly limits recruitment of adult trees (Prior *et al.* 2009; Prior *et al.* 2010; Werner and Prior 2013). Counterintuitively, Bond *et al.* (2012) found that the recruitment of eucalypts into large size classes (≥4 m height) was actually promoted by high-intensity fire – presumably by killing mature individuals and creating gaps, though the recruitment of non-eucalypts was reduced by fire.

As well as limiting recruitment of at least some groups of trees, intense fires can also kill both tree stems ('topkill') and whole trees. For example, Lonsdale and Braithwaite (1991) observed that a single very high-intensity fire in Kakadu National Park reduced live tree basal area by 14%. Stems that are most susceptible to topkill are generally the smallest, but in northern Australian savannas large trees are also susceptible, primarily because they tend to be hollow due to termite damage, and fires can burn the tree from the inside out (Williams *et al.* 1999; Cook *et al.* 2005). Resistance to fire seems to peak in stems of around 15–30 cm in diameter at breast height (DBH) (Williams *et al.* 1999; Cook *et al.* 2005; Prior *et al.* 2009). It seems likely that susceptibility to topkill is only indirectly related to DBH; it is most directly related to either stem height (Bond *et al.* 2012) or bark thickness (Lawes *et al.* 2011a). The relative importance of stem height and bark thickness in determining 'escape' from fire is a critical unresolved question in northern Australian savanna ecology.

The negative effects of fire on subsequent growth of trees have been reported from savannas in northern Australian and elsewhere globally (Hoffmann 2002; Werner 2005; Prior *et al.* 2006; Chidumayo 2007; Murphy *et al.* 2010; Chapter 13). Werner (2005) and Prior *et al.* (2006) described the effects of ambient and experimental fires at Kapalga, and found strong negative effects of high-intensity fires on height growth of saplings and radial growth of adults (expressed as annual increment of DBH). Murphy *et al.* (2010) analysed the Three Parks dataset and found similar effects for adult trees. In that case, two high-severity fires in a 5-year period reduced DBH increment to less than 1 mm year^{-1} for that period, from ~2 mm year^{-1} in the absence of fire. Murphy *et al.* concluded that this growth reduction reflects a period of lost productivity and substantial carbon and nutrient costs associated with rebuilding the canopy following defoliation by high-severity fire (i.e. leaf-scorch to full canopy height: Russell-Smith and Edwards 2006), an interpretation consistent with published ecophysiological studies (Beringer *et al.* 2003; Cernusak *et al.* 2006; Beringer *et al.* 2007).

It is important to note that of the three studies that found a substantial negative effect of high-intensity fires on DBH increment in northern Australian savannas (Werner 2005; Prior *et al.* 2006; Murphy *et al.* 2010), none considered the possibility that the apparent reduction in growth might be simply due to a reduction in bark thickness (i.e. due to bark being burnt away by fires). Burrows *et al.* (2010) have pointed out the problems with assessing the impact of even low-intensity fires on tree growth rates if bark thickness is not taken into account. For example, in the southern Australian rough-barked eucalypt species *Eucalyptus obliqua*, *E. marginata*, and *Corymbia calophylla*, individual fires with intensities greater than 1000 kW m^{-1} result in a reduction in bark thickness of 7–9 mm (Gill *et al.* 1986; Burrows *et al.* 2001; Chatto *et al.* 2003). Given that the reduction in DBH increment observed by Murphy *et al.* (2010) to accompany four severe fires per decade was only ~1 mm year^{-1} (10 mm decade^{-1}) relative to unburnt areas, reduction in bark thickness might easily account for the apparent reduction in DBH increment. This is not to say that severe fires do not or cannot cause a decrease in growth rates, but that without data on bark thickness, the large negative effects of fire on DBH increment reported in northern Australia must be viewed cautiously.

How important is the fire-mediated recruitment bottleneck in Australian savannas?

Several authors have emphasised the importance of a fire-mediated recruitment bottleneck exists in limiting tree populations in northern Australian savannas (Werner 2005; Prior *et al.* 2010; Werner 2012; Werner and Prior 2013). One of the most thorough treatments of this question was that by Prior *et al.* (2010), who analysed data from the Kapalga fire experiment to conclude that the annual fire treatments (especially the early dry season treatment) severely reduced the abundance of eucalypt saplings and rates of recruitment into adult size classes. However, these findings apply to a regime of annual fronting fires of moderate to high intensity – a regime not typical of the broader landscape where frequency and typical intensity are much lower. Despite arguing that savanna eucalypts are subject to a fire-mediated recruitment bottleneck, Prior *et al.* (2010) also point out that recruitment of eucalypts into the overstorey is strongly limited by competition with existing large trees, so the existence of a fire-mediated recruitment bottleneck may be a moot point – a greater abundance of saplings is unlikely to lead to a denser overstorey.

Although there is substantial evidence that, in at least some northern Australian situations, fire reduces the abundance of large juvenile stems (e.g. saplings), and limits their accession to larger size classes, it does not necessarily follow that recruitment into adult size classes is strongly limited by fire. Given that adult trees are long-lived with low annual mortality rates, even low recruitment rates are likely to be sufficient to maintain the adult tree population (Fensham and Bowman 1992). We consider that the importance of the fire-mediated recruitment bottleneck – which may well be highly relevant to other savanna regions, e.g. Africa where the theory originated – has been overemphasised in northern Australia's eucalypt-dominated savannas (e.g. Prior *et al.* 2010), a conclusion also reached by Fensham (2012). There is much evidence that the dominant group in these savannas – eucalypts – are relatively weak competitors and recruitment into adult size classes is strongly suppressed by competition with existing adults. For example, Wilson and Bowman (1987) describe dense regrowth of saplings in mesic eucalypt savanna near Darwin following extensive canopy damage by a severe tropical cyclone, despite a regime of frequent fires. Similarly, following widespread destruction of the savanna overstorey by Severe Tropical Cyclone Monica in 2005 (Cook and Goyens 2008), there was abundant regrowth of large saplings within 3–4 years, despite very frequent fires (Fig. 12.3). Fensham and Bowman (1992) demonstrated large increases in recruitment of saplings in mesic eucalypt savanna on Melville Island following removal of the overstorey. They recorded one *Eucalyptus miniata* sapling growing from 1 to 8 m in 3 years in the absence of an overstorey.

Furthermore, eucalypts appear to have a remarkable capacity to grow quickly through the savanna fire trap. For example, Bond *et al.* (2012) showed that under a range of fire treatments applied as part of the Kapalga fire experiment, eucalypt recruitment into the adult layer was greatest under a regime of annual high-intensity savanna fires, and lowest in unburnt savannas. They attributed this to high-intensity fires killing mature trees and creating gaps in the overstorey. Eucalypts are able to grow through the fire trap apparently

Figure 12.3. Widespread destruction of the eucalypt overstorey caused by Severe Tropical Cyclone Monica, which crossed the Arnhem Land coast near Maningrida on 24 April 2006. Close to where Monica crossed the coast, around 80% of mature trees were windthrown (Cook and Goyens 2008), but within a few years there was prolific regeneration of eucalypt saplings and poles, indicating that they had been released from competition with the overstorey. (Photo: Garry Cook)

because of their phenomenal ability to resprout prolifically from epicormic buds. In eucalypts and related taxa, these buds are deeply buried at the level of the vascular cambium, where they are protected by the maximum bark thickness (Burrows 2002). They are well documented as having lower rates of topkill than other groups in northern Australian

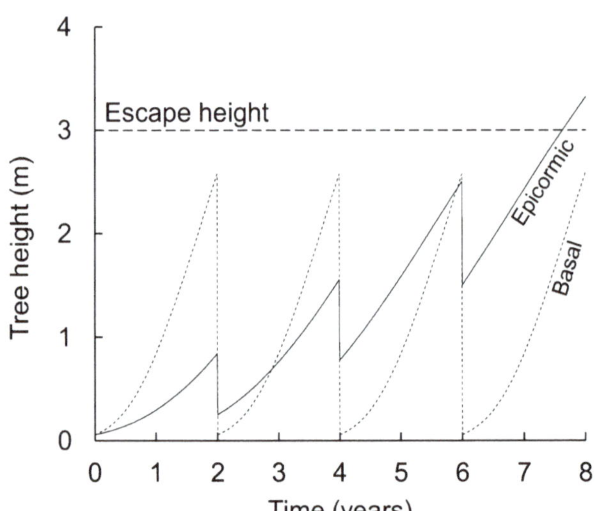

Figure 12.4. The probability of a sapling reaching 'escape height' is strongly affected by its ability to resprout following fire. Strong epicormic resprouters (such as eucalypts) are able to resprout from well above ground level, losing only a proportion of their pre-fire height, and are more likely to reach escape height despite frequent fires. An obligate basal resprouter, which is topkilled by fire, loses all of its aboveground biomass and must grow from ground level after every fire, and consequently is much less likely to reach escape height under a regime of frequent fires, even with a rapid growth rate between fires.

savannas (Williams *et al.* 1999; Lawes *et al.* 2011a). By avoiding topkill, strong epicormic resprouters such as eucalypts are able to surpass their pre-fire height rapidly (Fig. 12.4).

Although there appears to be little evidence that the abundance of mature eucalypts is limited by fire in northern Australian savannas, there is compelling evidence that fire strongly limits the non-eucalypt tree component. Bond and co-workers (Bond 2008; Bond *et al.* 2012) suggest that non-eucalypts in Australian savannas conform more closely to the recruitment bottleneck conceptual model, with recruitment into the adult population strongly limited by fire. Indeed, there are numerous examples from northern Australia where complete fire exclusion leads to substantial increases in the recruitment of non-eucalypts into the mid- and overstorey, but with the dominant eucalypt component showing little response (Russell-Smith *et al.* 2003; Woinarski *et al.* 2004; Bond *et al.* 2012).

Strong tree–tree competition is largely neglected by the prevailing hypotheses in the global savanna literature (Hanan and Lehmann 2011). Scholes and Archer (1997) describe a 'balanced competition' hypothesis, whereby competition between the superior competitor becomes self-limiting at a biomass insufficient to eliminate the inferior competitor. Such a conceptual model seems highly consistent with our understanding of tree dynamics in northern Australia's mesic savannas. In this case, the dominant eucalypts would be superior competitors to grasses, such that both can persist even in the absence of fire. However, a range of other tree species, often with rainforest and pantropical affinities, are able to recruit beneath the eucalypt overstorey in the absence of fire, and it is these species that would eventually exclude grasses, as demonstrated in various northern Australian field studies (Russell-Smith *et al.* 2003; Woinarski *et al.* 2004), paving the way for a biome switch to closed forest. It seems unlikely that savanna eucalypts alone would ever completely exclude grasses to the point that the understorey becomes non-flammable. It is clear that in the northern Australian context, tree–tree competition has been severely overlooked as a key driver of tree biomass dynamics, especially in relation to the dominant eucalypt component.

Are northern Australian savannas special?

A key emerging question is whether there are fundamental differences between the major tropical regions in terms of the distribution of savannas and tree biomass within savannas, in relation to climate and other environmental factors (Bond 2008; Lehmann *et al.* 2009; Lehmann *et al.* 2011; Murphy and Bowman 2012). Some authors have suggested that biogeographic factors – such as phylogenetic history – may contribute strongly to the distribution of vegetation at continental and global scales (Bond 2008; Murphy and Bowman 2012).

Indeed, there has been some suggestion that northern Australian savannas have higher and more homogeneous tree biomass than other climatically similar savanna regions. Bond and co-workers (Bond 2008; Bond *et al.* 2012) have suggested that this is due to the dominance of Australian savannas by eucalypts, a group of trees widely regarded as being the most fire-tolerant in the world. They contend that the extreme fire-resistance of eucalypts, primarily due to their ability to resprout prolifically from epicormic buds, makes them effectively oblivious to the fire trap. Eucalypts could therefore be expected to be more strongly controlled by resource availability than by fire activity (Bond 2008). This hypothesis is supported by the recent intercontinental comparison of savanna woody tree basal area by Lehmann *et al.* (2014). They assembled a large dataset of tree basal area observations from Africa, Australia and South America, and developed structural equation models to explain variation in both tree basal area and fire frequency. They found that the negative effect of fire on tree basal area in Australian savannas was weaker than in Africa, and attributed this to differences in woody plant traits, fuel loads, and flammability of savanna grasses. Resolving the true upper bound to savanna tree biomass in relation to critical resource gradients (e.g. rainfall, soil depth, nutrient availability) is critical to predicting how tree biomass responds to fire management in northern Australian savannas. For example, if savanna tree populations are currently very close to carrying capacity, as suggested by Bond (2008), we could expect tree biomass to be relatively insensitive to changes in fire management.

Other disturbance-based controls of savanna tree biomass

Wind disturbance

It is likely that the importance of wind disturbance in structuring tree populations in northern Australian savannas is underappreciated. Wind damage can have a range of short- and long-term impacts on trees, varying with wind strength, soil properties, the exposure of the vegetation and tree species. Branches can be lost, trees defoliated, trunks snapped or whole trees uprooted. Winds of sufficient strength to cause significant structural damage can arise from localised storm activity and affect relatively small patches at the scale of several hectares (Williams and Douglas 1995; Franklin *et al.* 2010), while individual cyclones can affect thousands of square kilometres (Cook and Goyens 2008) (Fig. 12.3). Northern Australia is one of the Earth's few savanna regions strongly affected by tropical cyclones.

Although the impacts of wind damage from cyclones on tropical rainforests has long been recognised (Webb 1958), the effects on savannas have had little research. Nevertheless, historical records of tree damage through the Australian savanna zone date back to the 1820s (Murphy 1984; Cook and Goyens 2008). The impacts on savannas differ markedly from those on rainforests (Turton 2008) due to the frequency of fire as a disturbance and the stress of seasonal drought. Within the Australian savanna biome, fire-sensitive monsoon rainforest patches exist as an archipelago within the frequently burnt savanna matrix. Wind damage followed by frequent fires can reduce the size of or eliminate monsoon forest patches (Panton 1993; Bowman and Panton 1994). Woinarski and Westaway (2008) found that damage by cyclones had a major negative impact on the abundance of large trees in the mesic Northern Territory, and suggested that recovery would take many decades. Cook and Goyens (2008) estimated that in the Darwin region, significant windthrow events (>20% of trees felled) have typical intervals of around 10 years, at least as far as 100 km inland. Given that large, old trees – which strongly dominate a site in terms of biomass storage – tend to be most susceptible to windthrow (Cook and Goyens 2008), the impact of wind disturbance of biomass dynamics in coastal savannas may be of a similar order of magnitude to that of fire disturbance.

Herbivory

The role of vertebrate herbivores in controlling tree biomass in savannas remains poorly understood, although there is evidence that grazing and browsing exert a strong control on some aspects of vegetation structure. Grazing (consumption of grass) can favour woody vegetation by reducing grass biomass, thereby reducing both tree–grass competition and fire frequency and intensity, which have been shown to increase tree recruitment and growth (Werner 2005). Such a 'grazing effect' explains 'bush encroachment' following heavy overgrazing by domestic stock (van Langevelde *et al.* 2003). Conversely, browsing (consumption of woody plants) can limit woody plants, maintaining open, grassy systems. Indeed, browsing has been shown experimentally to reduce woody plant abundance in savannas (Augustine and McNaughton 2004), largely due to its strong negative effect on height growth, especially in conjunction with fire (Staver *et al.* 2009). There is some evidence that the swamp buffalo (*Bubalus bubalis*) contributed to woody thickening in the savannas of Kakadu National Park. Petty *et al.* (2007) suggested that substantial woody thickening occurred in areas around floodplains with high buffalo densities in the 1960s and 1970s, before their near-complete eradication in Kakadu by the late 1980s. Indeed, work in Kakadu has demonstrated that buffalo promote the growth and survival of juvenile savanna trees by reducing the abundance of grasses, which compete with juvenile trees and provide fuel for fires (Werner 2005; Petty *et al.* 2007). However, it seems likely that the impacts of vertebrate herbivores are relatively insignificant in Australian savannas, given the complete absence of natives herbivores >50 kg, and the patchy distribution of large exotic herbivores such as cattle, buffalo and donkeys.

Although the effects of vertebrate herbivores on tree populations are poorly understood in northern Australian savannas, the effects of termites are very conspicuous. Most trees in northern Australian savannas are hollow, with termites consuming the equivalent of ~25% of wood production each year (Cook *et al.* 2005), and termite activity increases with increasing total basal area of trees (Jamali *et al.* 2013). The increased susceptibility of large trees to being topkilled by fire is most likely due to the effect of this hollowing (Williams *et al.* 1999; Cook *et al.* 2005). Hence, there is a strong interactive effect of termites and fire activity in these savannas.

Woody thickening in savannas

Several studies have demonstrated high rates of woody thickening in savannas around the world (see Murphy *et al.* 2014 for references). Consistent with this global trend, Murphy *et al.* (2014) have recently used a meta-analysis to demonstrate that there is a weak, though statistically significant, trend of woody thickening across of range of savannas landscapes in northern Australia. A frequently cited cause of thickening in savannas is overgrazing by domestic livestock (e.g. Archer *et al.* 1995; Sharp and Whittaker 2003). By reducing grass biomass, overgrazing can reduce fire frequency and intensity and increase resource availability – changes which could be expected to favour the establishment of woody plants. Given that there are reports of intensification of grazing by domestic livestock in recent decades in various parts of northern Australia (Garnett and Crowley 1995; Sharp and Whittaker 2003), this seems a plausible explanation for the observed pattern of thickening. However, the general consistency of the pattern across a range of sites with different land-use histories (i.e. conservation versus pastoralism), albeit without replication, suggests that localised changes in land use are not solely responsible, and that processes operating at regional to continental scales are also important. Plausible explanations include increased [CO_2] (Bond and Midgley 2012; Higgins and Scheiter 2012); increased rainfall, with northern Australia subject to a trend of increasing rainfall over the latter half of the 20th century (Smith 2004); and increased soil moisture due to decreased evaporation (Roderick and Farquhar 2004).

If savannas are to be managed to increase carbon storage in woody biomass, with the objective of generating carbon credits, underlying global trends in carbon stock (e.g. increasing woody biomass due to nutrient deposition or rising [CO_2]) must be characterised, in order to demonstrate 'additionality'. This is essential to accurately attributing observed increases in carbon stocks to management actions. While there is potential for the use of satellite imagery and historical aerial photographs to characterise rates of vegetation change over the last 50 years (e.g. Donohue *et al.* 2009; Murphy *et al.* 2014), such efforts may be complicated by accelerating rates of vegetation thickening (e.g. Scheiter and Higgins 2009).

Managing fire to increase tree biomass in Australian savannas

Work on the Arnhem Plateau in the Northern Territory has demonstrated that land managers can use prescribed burning to substantially reduce the frequency of high-intensity LDS fires and, to a lesser extent, overall fire frequencies (Price *et al.* 2012; Russell-Smith *et al.* 2013). We will now describe the three main processes by which management-imposed reductions in fire frequency and intensity could potentially affect tree biomass in Australian savannas, and the likely magnitude of these effects.

Relaxing the demographic bottleneck

The fire-mediated demographic bottleneck hypothesis (Higgins *et al.* 2000) implies that most stands of savanna trees are below their long-term carrying capacity. As we have described earlier, there is evidence that biomass of the dominant trees in Australian savannas – eucalypts – is not strongly constrained by a fire-mediated demographic bottleneck. In many situations, eucalypt populations may be close to a carrying capacity largely determined by water availability, so it seems unlikely that fire management would result in a substantial increase in eucalypt biomass. However, there is convincing evidence that the abundance of more fire-sensitive trees of the midstorey – typically broad-leaved deciduous species – is strongly constrained by a fire-mediated demographic bottleneck (Lawes *et al.* 2011b; Bond *et al.* 2012), so it is plausible that biomass of these species would increase as a result of fire management. The extent to which this could occur is essentially unknown, but there is evidence of a tradeoff between eucalypt and non-eucalypt abundance in these savannas (Lawes *et al.* 2011b), so long-term increases in non-eucalypt biomass may simply be offset by decreases in eucalypt biomass. In sum, relaxation of the demographic bottleneck is most likely to lead to shifts in species composition rather than substantial increases in tree biomass, consistent with a range of fire exclusion studies (Russell-Smith *et al.* 2003; Woinarski *et al.* 2004).

Increasing growth rates

The negative effect of intense fires on tree growth rates (Werner 2005; Prior *et al.* 2006; Murphy *et al.* 2010) implies that under a lower intensity, managed fire regime, the biomass of a stand will recover faster following the death of a mature tree. Hence, the overall biomass carrying capacity is not changed by the management-imposed reduction in fire intensity, but the stand would be closer to carrying capacity for a greater proportion of time, and biomass (averaged over time) would tend to be higher.

Long-term shifts in stand structure

Because fire-driven mortality is greatest for small trees (Cook *et al.* 2005), decreasing fire frequency and intensity is likely to lead to changes in stand structure in the long term. A plausible change is an increase in the density of saplings and small trees, which will compete for resources with large trees, slowing their growth and increasing mortality rates. A similar effect is well documented in southern Australian woodlands and forests (e.g. Close *et al.* 2009), but has received little attention in the savannas. Hence, it seems likely that tree biomass would be reduced by a shift in dominance from large to small trees. Cook *et al.* (Chapter 13) describe an allometric effect whereby, for a given level of water availability at a site, large trees store biomass more efficiently than small trees. This effect arises because tree biomass is proportional to basal area raised to the power of 1.3 (O'Grady *et al.* 2000), while water use is thought to scale to basal area raised to the power of 1 (Cook *et al.* 2002). Hence, reducing fire frequency and intensity may act to slightly decrease tree biomass in the long term by increasing the relative abundance small trees.

Conclusions

Over the last 30 years the mesic savannas of the Northern Territory have seen a sustained research effort with the broad aim of understanding the drivers of the dynamics of tree populations, much of it building on the datasets and knowledge provided by the Kapalga and Munmarlary fire experiments, and large field studies such as the North Australian Tropical Transect, and Three Parks vegetation and fire monitoring program. Much of the focus has been on fire effects, and there is now a widespread acceptance that the flammable eucalypt woodlands of Australia's monsoon tropics have stronger functional affinities with the savannas of Africa and South America than with the eucalypt woodlands and forests of Australia's temperate south. However, there is increasing evidence that Australia's tropical savannas are unique, due to domination by an extremely fire-resistant group of tree – eucalypts. Bond *et al.* (2012) suggest that, given their epicormic sprouting capacity, eucalypts are 'super trees' and the first group of trees to crack the fire problem. As a result, the abundance of mature eucalypts within many savanna sites appears to be relatively insensitive to fire activity – eucalypt biomass is more closely controlled by water availability. It is only the fire-sensitive non-eucalypts of the midstorey – mostly broadleaved, deciduous species – that respond very positively to fire exclusion.

It is likely that management-driven reductions in fire frequency and intensity will have a range of short- to long-term effects on tree populations in northern Australian savannas. We would certainly expect an increase in the biomass of fire-sensitive tree species, but whether this is offset by a reduction in the abundance of fire-resistant eucalypts, as suggested by the findings of Lawes *et al.* (2011b), remains highly uncertain. Shifts in stand structures are likely, with a reduction in fire frequency and intensity likely to result in an increase in the density of small trees (of fire-sensitive size classes) at the expense of the larger trees in fire-resistant size classes, and this is likely to result in a reduction in tree biomass in the long term. Given that there are likely to be a range of both short- and long-term effects of fire management, the most appropriate tools for exploring likely changes in savanna tree populations are process-based demographic models, such as Liedloff and Cook's (2007) 'FLAMES' model, parameterised largely using data from Kapalga, or more generic savanna models such as Scheiter and Higgins' (2009) 'adaptive dynamic global vegetation model', developed for African savannas. Future work on developing such models for northern Australia should focus on incorporating a broader range of tree functional types.

Acknowledgements

Funding from the Australian Government Department of the Environment and the Australian Research Council (DE130100434) supported this work. We thank Rod Fensham, Jeremy Russell-Smith and Karen King for providing helpful comments on an early version of the manuscript.

References

Andersen AN, Cook GD, Williams RJ (2003) *Fire in Tropical Savannas: The Kapalga Fire Experiment.* Springer: New York.

Archer S, Schimel DS, Holland EA (1995) Mechanisms of shrubland expansion: land use, climate or CO_2? *Climatic Change* **29**, 91–99. doi:10.1007/BF01091640

Augustine DJ, McNaughton SJ (2004) Regulation of shrub dynamics by native browsing ungulates on East African rangeland. *Journal of Applied Ecology* **41**, 45–58. doi:10.1111/j.1365-2664.2004.00864.x

Beringer J, Hutley LB, Tapper NJ, Coutts A, Kerley A, Grady AP (2003) Fire impacts on surface heat, moisture and carbon fluxes from a tropical savanna in northern Australia. *International Journal of Wildland Fire* **12**, 333–340. doi:10.1071/WF03023

Beringer J, Hutley LB, Tapper NJ, Cernusak LA (2007) Savanna fires and their impact on net ecosystem productivity in North Australia. *Global Change Biology* **13**, 990–1004. doi:10.1111/j.1365-2486.2007.01334.x

Beyer S, Kinnear A, Hutley LB, McGuinness K, Gibb K (2011) Assessing the relationship between fire and grazing on soil characteristics and mite communities in a semi-arid savanna of northern Australia. *Pedobiologia* **54**, 195–200. doi:10.1016/j.pedobi.2011.03.002

Bond WJ (2008) What limits trees in C_4 grasslands and savannas? *Annual Review of Ecology Evolution and Systematics* **39**, 641–659. doi:10.1146/annurev.ecolsys.39.110707.173411

Bond WJ, Midgley GF (2000) A proposed CO_2-controlled mechanism of woody plant invasion in grasslands and savannas. *Global Change Biology* **6**, 865–869. doi:10.1046/j.1365-2486.2000.00365.x

Bond WJ, Midgley GF (2012) Carbon dioxide and the uneasy interactions of trees and savannah grasses. *Philosophical Transactions of the Royal Society of London. Series B, Biological Sciences* **367**, 601–612. doi:10.1098/rstb.2011.0182

Bond WJ, Van Wilgen BW (1996) *Fire and Plants.* Chapman and Hall, London.

Bond WJ, Midgley GF, Woodward FI (2003) What controls South African vegetation – climate or fire? *South African Journal of Botany* **69**, 79–91.

Bond WJ, Cook GD, Williams RJ (2012) Which trees dominate in savannas? The escape hypothesis and eucalypts in northern Australia. *Austral Ecology* **37**, 678–685. doi:10.1111/j.1442-9993.2011.02343.x

Bowman DMJS, Panton WJ (1994) Fire and cyclone damage to woody vegetation on the north coast of Northern Territory, Australia. *The Australian Geographer* **25**, 32–35. doi:10.1080/00049189408703096

Bowman DMJS, Murphy BP, Williamson GJ, Cochrane MA (2014) Pyrogeographic models, feedbacks and the future of global fire regimes. *Global Ecology and Biogeography* **23**, 821–824. doi:10.1111/geb.12180

Brookman-Amissah J, Hall JB, Swaine MD, Attakorah JY (1980) A re-assessment of a fire protection experiment in north-eastern Ghana savanna. *Journal of Applied Ecology* **17**, 85–99. doi:10.2307/2402965

Bucini G, Hanan NP (2007) A continental-scale analysis of tree cover in African savannas. *Global Ecology and Biogeography* **16**, 593–605. doi:10.1111/j.1466-8238.2007.00325.x

Burrows GE (2002) Epicormic strand structure in *Angophora*, *Eucalyptus* and *Lophostemon* (Myrtaceae) – implications for fire resistance and recovery. *New Phytologist* **153**, 111–131. doi:10.1046/j.0028-646X.2001.00299.x

Burrows ND, Ward BG, Robinson AD (2001) Bark as fuel in a moderate intensity jarrah fire. *CaLMScience* **3**, 405–409.

Burrows WH, Henry BK, Back PV, Hoffmann MB, Tait LJ, Anderson ER, Menke N, Danaher T, Carter JO, McKeon GM (2002) Growth and carbon stock change in eucalypt woodlands in northeast Australia: ecological and greenhouse sink implications. *Global Change Biology* **8**, 769–784. doi:10.1046/j.1365-2486.2002.00515.x

Burrows N, Ward B, Robinson A (2010) Fire regimes and tree growth in low rainfall jarrah forest of south-west Australia. *Environmental Management* **45**, 1332–1343. doi:10.1007/s00267-010-9490-6

Cernusak LA, Hutley LB, Beringer J, Tapper NJ (2006) Stem and leaf gas exchange and their responses to fire in a north Australian tropical savanna. *Plant, Cell & Environment* **29**, 632–646. doi:10.1111/j.1365-3040.2005.01442.x

Chatto K, Bell T, Kellas J (2003) 'Effects of Repeated Low-Intensity Fire on Tree Growth and Bark in a Mixed Eucalypt Foothill Forest in South-Eastern Australia.' Victorian Government Department of Sustainability and Environment: Melbourne.

Chidumayo E (2007) Growth responses of an African savanna tree, *Bauhinia thonningii* Schumacher, to defoliation, fire and climate. *Trees* **21**, 231–238. doi:10.1007/s00468-006-0115-x

Close DC, Davidson NJ, Johnson DW, Abrams MD, Hart SC, Lunt ID, Archibald RD, Horton B, Adams MA (2009) Premature decline of *Eucalyptus* and altered ecosystem processes in the absence of fire in some Australian forests. *Botanical Review* **75**, 191–202. doi:10.1007/s12229-009-9027-y

Cook GD, Goyens CMAC (2008) The impact of wind on trees in Australian tropical savannas: lessons from Cyclone Monica. *Austral Ecology* **33**, 462–470. doi:10.1111/j.1442-9993.2008.01901.x

Cook GD, Heerdegen RG (2001) Spatial variation in the duration of the rainy season in monsoonal Australia. *International Journal of Climatology* **21**, 1723–1732. doi:10.1002/joc.704

Cook GD, Williams RJ, Hutley LB, O'Grady AP, Liedloff AC (2002) Variation in vegetative water use in the savannas of the North Australian Tropical Transect. *Journal of Vegetation Science* **13**, 413–418. doi:10.1111/j.1654-1103.2002.tb02065.x

Cook GD, Liedloff AC, Eager RW, Chen X, Williams RJ, O'Grady AP, Hutley LB (2005) The estimation of carbon budgets of frequently burnt tree stands in savannas of northern Australia, using allometric analysis and isotopic discrimination. *Australian Journal of Botany* **53**, 621–630. doi:10.1071/BT04150

Donohue RJ, McVicar TR, Roderick ML (2009) Climate-related trends in Australian vegetation cover as inferred from satellite observations, 1981–2006. *Global Change Biology* **15**, 1025–1039. doi:10.1111/j.1365-2486.2008.01746.x

Dwyer E, Pinnock S, Grégoire J-M, Pereira JMC (2000) Global spatial and temporal distribution of vegetation fire as determined from satellite observations. *International Journal of Remote Sensing* **21**, 1289–1302. doi:10.1080/014311600210182

Fensham RJ (2012) Fire regimes in Australian tropical savanna: perspectives, paradigms and paradoxes. In *Flammable Australia: Fire Regimes, Biodiversity and Ecosystems in a Changing World.* (Eds RA Bradstock, AM Gill, RJ Williams) pp. 173–193. CSIRO Publishing, Melbourne.

Fensham RJ, Bowman DMJS (1992) Stand structure and the influence of overwood on regeneration in tropical eucalypt forest on Melville Island. *Australian Journal of Botany* **40**, 335–352. doi:10.1071/BT9920335

Fensham RJ, Fairfax RJ (2007) Drought-related tree death of savanna eucalypts: species susceptibility, soil conditions and root architecture. *Journal of Vegetation Science* **18**, 71–80. doi:10.1111/j.1654-1103.2007.tb02517.x

Fensham RJ, Fairfax RJ, Ward DP (2009) Drought-induced tree death in savanna. *Global Change Biology* **15**, 380–387. doi:10.1111/j.1365-2486.2008.01718.x

Franklin DC, Gunton RM, Schatz J, Lawes MJ (2010) Resprouting responses of trees in a fire-prone tropical savanna following severe tornado damage. *Austral Ecology* **35**, 685–694. doi:10.1111/j.1442-9993.2009.02075.x

Furley PA, Rees RM, Ryan CM, Saiz G (2008) Savanna burning and the assessment of long-term fire experiments with particular reference to Zimbabwe. *Progress in Physical Geography* **32**, 611–634. doi:10.1177/0309133308101383

Garnett ST, Crowley GM (1995) *Ecology and Conservation of the Golden-Shouldered Parrot.* Queensland Government Department of Environment and Heritage: Brisbane.

Gill AM, Cheney NP, Walker J, Tunstall BR (1986) Bark losses from two eucalypt species following fires of different intensities. *Australian Forest Research* **16**, 1–8.

Grace J, San Jose J, Meir P, Miranda HS, Montes RA (2006) Productivity and carbon fluxes of tropical savannas. *Journal of Biogeography* **33**, 387–400. doi:10.1111/j.1365-2699.2005.01448.x

Hanan NP, Lehmann CER (2011) Tree–grass interactions in savannas: paradigms, contradictions, and conceptual models. In *Ecosystem Function in Savannas: Measurement and Modeling at Landscape to Global Scales.* (Eds MJ Hill, NP Hanan) pp. 39–53. CRC Press: Boca Raton, FL.

Hansen MC, DeFries RS, Townshend JRG, Carroll M, Dimiceli C, Sohlberg R (2002) Development of a MODIS tree cover validation data set for Western Province, Zambia. *Remote Sensing of Environment* **83**, 320–335. doi:10.1016/S0034-4257(02)00080-9

Higgins SI, Scheiter S (2012) Atmospheric CO_2 forces abrupt vegetation shifts locally, but not globally. *Nature* **488**, 209–212. doi:10.1038/nature11238

Higgins SI, Bond WJ, Trollope WSW (2000) Fire, resprouting and variability: a recipe for grass–tree coexistence in savanna. *Journal of Ecology* **88**, 213–229. doi:10.1046/j.1365-2745.2000.00435.x

Higgins SI, Bond WJ, *et al.* (2007) Effects of four decades of fire manipulation on woody vegetation structure in savanna. *Ecology* **88**, 1119–1125. doi:10.1890/06-1664

Hirota M, Holmgren M, Van Nes EH, Scheffer M (2011) Global resilience of tropical forest and savanna to critical transitions. *Science* **334**, 232–235. doi:10.1126/science.1210657

Hoffmann WA (2002) Direct and indirect effects of fire on radial growth of cerrado savanna trees. *Journal of Tropical Ecology* **18**, 137–142. doi:10.1017/S0266467402002080

Hoffmann WA, Adasme R, Haridasan M, de Carvalho MT, Geiger EL, Pereira MAB, Gotsch SG, Franco AC (2009) Tree topkill, not mortality, governs the dynamics of savanna–forest boundaries under frequent fire in central Brazil. *Ecology* **90**, 1326–1337. doi:10.1890/08-0741.1

Holdridge LR (1947) Determination of world plant formations from simple climatic data. *Science* **105**, 367–368. doi:10.1126/science.105.2727.367

Jamali H, Livesley SJ, Hutley LB, Fest B, Arndt SK (2013) The relationships between termite mound CH_4/CO_2 emissions and internal concentration ratios are species specific. *Biogeosciences* **10**, 2229–2240. doi:10.5194/bg-10-2229-2013

Kottek M, Grieser J, Beck C, Rudolf B, Rubel F (2006) World map of the Köppen-Geiger classification updated. *Meteorologische Zeitschrift* **15**, 259–263. doi:10.1127/0941-2948/2006/0130

Lawes MJ, Adie HR, Russell-Smith J, Murphy BP, Midgley JJ (2011a) How do small savanna trees avoid stem mortality by fire? The roles of stem diameter, height and bark thickness. *Ecosphere* **2**, art42. doi:10.1890/ES10-00204.1

Lawes MJ, Murphy BP, Midgley JJ, Russell-Smith J (2011b) Are the eucalypt and non-eucalypt components of Australian tropical savannas independent? *Oecologia* **166**, 229–239. doi:10.1007/s00442-010-1829-4

Lehmann CER, Ratnam J, Hutley LB (2009) Which of these continents is not like the other? Comparisons of tropical savanna systems: key questions and challenges. *New Phytologist* **181**, 508–511. doi:10.1111/j.1469-8137.2009.02734.x

Lehmann CER, Archibald SA, Hoffmann WA, Bond WJ (2011) Deciphering the distribution of the savanna biome. *New Phytologist* **191**, 197–209. doi:10.1111/j.1469-8137.2011.03689.x

Lehmann CER, Anderson TM, Sankaran M, Higgins SI, Archibald S, *et al.* (2014) Savanna vegetation–fire–climate relationships differ among continents. *Science* **343**, 548–552. doi:10.1126/science.1247355

Liedloff AC, Cook GD (2007) Modelling the effects of rainfall variability and fire on tree populations in an Australian tropical savanna with the FLAMES simulation model. *Ecological Modelling* **201**, 269–282. doi:10.1016/j.ecolmodel.2006.09.013

Liedloff AC, Cook GD (2011) The interaction of fire and rainfall variability on tree structure and carbon fluxes in savannas: application of the FLAMES model. In *Ecosystem Function in Savannas: Measurement and Modeling at Landscape to Global Scales.* (Eds MJ Hill and NP Hanan) pp. 293–308. CRC Press: Boca Raton, FL.

Lonsdale WM, Braithwaite RW (1991) Assessing the effects of fire on vegetation in tropical savannas. *Australian Journal of Ecology* **16**, 363–374. doi:10.1111/j.1442-9993.1991.tb01064.x

Louppe D, Ouattara N, Coulibaly A (1995) The effect of brush fires on vegetation: the Aubreville fire plots after 60 years. *Commonwealth Forestry Review* **74**, 288–292.

Moreira AG (2000) Effects of fire protection on savanna structure in Central Brazil. *Journal of Biogeography* **27**, 1021–1029. doi:10.1046/j.1365-2699.2000.00422.x

Murphy K (1984) *Big Blow up North: A History of Tropical Cyclones in Australia's Northern Territory.* University Planning Authority: Darwin.

Murphy BP, Bowman DMJS (2012) What controls the distribution of tropical forest and savanna? *Ecology Letters* **15**, 748–758. doi:10.1111/j.1461-0248.2012.01771.x

Murphy BP, Russell-Smith J, Prior LD (2010) Frequent fires reduce tree growth in northern Australian savannas: implications for tree demography and carbon sequestration. *Global Change Biology* **16**, 331–343. doi:10.1111/j.1365-2486.2009.01933.x

Murphy BP, Lehmann CER, Russell-Smith J, Lawes MJ (2014) Fire regimes and woody biomass dynamics in Australian savannas. *Journal of Biogeography* **41**, 133–144. doi:10.1111/jbi.12204

O'Grady AP, Chen X, Eamus D, Hutley LB (2000) Composition, leaf area index and standing biomass of eucalypt open forests near Darwin in the Northern Territory, Australia. *Australian Journal of Botany* **48**, 629–638. doi:10.1071/BT99022

Olson DM, Dinerstein E, Wikramanayake ED, Burgess ND, Powell GVN, *et al.* (2001) Terrestrial ecoregions of the world: a new map of life on Earth. *Bioscience* **51**, 933–938. doi:10.1641/0006-3568(2001)051[0933:TEOTWA]2.0.CO;2

Panton WJ (1993) Changes in post World War II distribution and status of monsoon rainforests in the Darwin area. *The Australian Geographer* **24**, 50–59. doi:10.1080/00049189308703088

Peterson DW, Reich PB (2001) Prescribed fire in oak savanna: fire frequency effects on stand structure and dynamics. *Ecological Applications* **11**, 914–927. doi:10.1890/1051-0761(2001)011[0914:PFIOSF]2.0.CO;2

Petty AM, Werner PA, Lehmann CER, Riley JE, Banfai DS, Elliott LP (2007) Savanna responses to feral buffalo in Kakadu National Park, Australia. *Ecological Monographs* **77**, 441–463. doi:10.1890/06-1599.1

Price OF, Russell-Smith J, Watt FA (2012) The influence of prescribed fire on the extent of wildfire in savanna landscapes of western Arnhem Land, Australia. *International Journal of Wildland Fire* **21**, 297–305. doi:10.1071/WF10079

Prior LD, Eamus D (2000) Seasonal changes in hydraulic conductance, xylem embolism and leaf area in *Eucalyptus tetrodonta* and *Eucalyptus miniata* saplings in a north Australian savanna. *Plant, Cell & Environment* **23**, 955–965. doi:10.1046/j.1365-3040.2000.00612.x

Prior LD, Eamus D, Duff GA (1997) Seasonal and diurnal patterns of carbon assimilation, stomatal conductance and leaf water potential in *Eucalyptus tetrodonta* saplings in a wet–dry savanna in Northern Australia. *Australian Journal of Botany* **45**, 241–258. doi:10.1071/BT96017

Prior LD, Brook BW, Williams RJ, Werner PA, Bradshaw CJA, Bowman DMJS (2006) Environmental and allometric drivers of tree growth rates in a north Australian savanna. *Forest Ecology and Management* **234**, 164–180. doi:10.1016/j.foreco.2006.06.034

Prior LD, Murphy BP, Russell-Smith J (2009) Environmental and demographic correlates of tree recruitment and mortality in north Australian savannas. *Forest Ecology and Management* **257**, 66–74. doi:10.1016/j.foreco.2008.08.015

Prior LD, Williams RJ, Bowman D (2010) Experimental evidence that fire causes a tree recruitment bottleneck in an Australian tropical savanna. *Journal of Tropical Ecology* **26**, 595–603. doi:10.1017/S0266467410000362

Richards AE, Cook GD, Lynch BT (2011) Optimal fire regimes for soil carbon storage in tropical savannas of northern Australia. *Ecosystems* **14**, 503–518. doi:10.1007/s10021-011-9428-8

Roderick ML, Farquhar GD (2004) Changes in Australian pan evaporation from 1970 to 2002. *International Journal of Climatology* **24**, 1077–1090. doi:10.1002/joc.1061

Roques KG, O'Connor TG, Watkinson AR (2001) Dynamics of shrub encroachment in an African savanna: relative influences of fire, herbivory, rainfall and density dependence. *Journal of Applied Ecology* **38**, 268–280. doi:10.1046/j.1365-2664.2001.00567.x

Rose Innes R (1972) Fire in West African vegetation. *Proceedings of Tall Timbers Fire Ecology Conference* **11**, 175–199.

Russell-Smith J, Edwards AC (2006) Seasonality and fire severity in savanna landscapes of monsoonal northern Australia. *International Journal of Wildland Fire* **15**, 541–550. doi:10.1071/WF05111

Russell-Smith J, Whitehead PJ, Cook GD, Hoare JL (2003) Response of *Eucalyptus*-dominated savanna to frequent fires: lessons from Munmarlary, 1973–1996. *Ecological Monographs* **73**, 349–375. doi:10.1890/01-4021

Russell-Smith J, Whitehead PJ, Cooke P (2009) *Culture, Ecology and Economy of Fire Management in North Australian Savannas: Rekindling the* Wurrk *Tradition*. CSIRO Publishing: Melbourne.

Russell-Smith J, Cook GD, Cooke PM, Edwards AC, Lendrum M, Meyer CP, Whitehead PJ (2013) Managing fire regimes in north Australian savannas: applying Aboriginal approaches to contemporary global problems. *Frontiers in Ecology and the Environment* **11**, e55–e63. doi:10.1890/120251

San Jose JJ, Farinas MR (1983) Changes in tree density and species composition in a protected *Trachypogon* savanna, Venezuela. *Ecology* **64**, 447–453. doi:10.2307/1939963

San José JJ, Montes RA, Farinas MR (1998) Carbon stocks and fluxes in a temporal scaling from a savanna to a semi-deciduous forest. *Forest Ecology and Management* **105**, 251–262. doi:10.1016/S0378-1127(97)00288-0

Sankaran M, Ratnam J, Hanan NP (2004) Tree–grass coexistence in savannas revisited – insights from an examination of assumptions and mechanisms invoked in existing models. *Ecology Letters* **7**, 480–490. doi:10.1111/j.1461-0248.2004.00596.x

Sankaran M, Hanan NP, Scholes RJ, Ratnam J, Augustine DJ, *et al.* (2005) Determinants of woody cover in African savannas. *Nature* **438**, 846–849. doi:10.1038/nature04070

Sankaran M, Ratnam J, Hanan N (2008) Woody cover in African savannas: the role of resources, fire and herbivory. *Global Ecology and Biogeography* **17**, 236–245. doi:10.1111/j.1466-8238.2007.00360.x

Scheiter S, Higgins SI (2009) Impacts of climate change on the vegetation of Africa: an adaptive dynamic vegetation modelling approach. *Global Change Biology* **15**, 2224–2246. doi:10.1111/j.1365-2486.2008.01838.x

Schimper AFW (1903) *Plant-Geography upon a Physiological Basis.* Clarendon Press: Oxford.

Scholes RJ, Archer SR (1997) Tree–grass interactions in savannas. *Annual Review of Ecology and Systematics* **28**, 517–544. doi:10.1146/annurev.ecolsys.28.1.517

Sharp BR, Whittaker RJ (2003) The irreversible cattle-driven transformation of a seasonally flooded Australia savanna. *Journal of Biogeography* **30**, 783–802. doi:10.1046/j.1365-2699.2003.00840.x

Smith I (2004) An assessment of recent trends in Australian rainfall. *Australian Meteorological Magazine* **53**, 163–173.

Staver AC, Bond WJ, Stock WD, van Rensburg SJ, Waldram MS (2009) Browsing and fire interact to suppress tree density in an African savanna. *Ecological Applications* **19**, 1909–1919. doi:10.1890/08-1907.1

Staver AC, Archibald S, Levin SA (2011) The global extent and determinants of savanna and forest as alternative stable states. *Science* **334**, 230–232. doi:10.1126/science.1210465

Swaine MS, Hawthorne WD, Orgie TK (1992) The effects of fire exclusion on savanna vegetation at Kpong, Ghana. *Biotropica* **24**, 166–172. doi:10.2307/2388670

Tilman D, Reich P, Phillips H, Menton M, Patel A, Vos E, Peterson D, Knops J (2000) Fire suppression and ecosystem carbon storage. *Ecology* **81**, 2680–2685. doi:10.1890/0012-9658(2000)081[2680:FSAECS]2.0.CO;2

Trapnell CG (1959) Ecological results of woodland burning experiments in Northern Rhodesia. *Journal of Ecology* **47**, 129–168. doi:10.2307/2257252

Turton SM (2008) Landscape-scale impacts of Cyclone Larry on the forests of northeast Australia, including comparisons with previous cyclones impacting the region between 1858 and 2006. *Austral Ecology* **33**, 409–416. doi:10.1111/j.1442-9993.2008.01896.x

van der Werf GR, Randerson JT, Giglio L, Gobron N, Dolman AJ (2008) Climate controls on the variability of fires in the tropics and subtropics. *Global Biogeochemical Cycles* **22**(3), GB3028.

van Langevelde F, van de Vijver C, Kumar L, van de Koppel J, de Ridder N, *et al.* (2003) Effects of fire and herbivory on the stability of savanna ecosystems. *Ecology* **84**, 337–350. doi:10.1890/0012-9658(2003)084[0337:EOFAHO]2.0.CO;2

Walter H (1971) *Ecology of Tropical and Subtropical Vegetation.* Oliver and Boyd: Edinburgh.

Webb LJ (1958) Cyclones as an ecological factor in tropical lowland rainforest, north Queensland. *Australian Journal of Botany* **6**, 220–230. doi:10.1071/BT9580220

Werner PA (2005) Impact of feral water buffalo and fire on growth and survival of mature savanna trees: an experimental field study in Kakadu National Park, northern Australia. *Austral Ecology* **30**, 625–647. doi:10.1111/j.1442-9993.2005.01491.x

Werner PA (2012) Growth of juvenile and sapling trees differs with both fire season and understorey type: trade-offs and transitions out of the fire trap in an Australian savanna. *Austral Ecology* **37**, 644–657. doi:10.1111/j.1442-9993.2011.02333.x

Werner PA, Prior LD (2013) Demography and growth of subadult savanna trees: interactions of life history, size, fire season, and grassy understory. *Ecological Monographs* **83**, 67–93. doi:10.1890/12-1153.1

Williams RJ, Douglas M (1995) Windthrow in a tropical savanna in Kakadu National Park, northern Australia. *Journal of Tropical Ecology* **11**, 547–558. doi:10.1017/S026646740000910X

Williams RJ, Duff GA, Bowman DMJS, Cook GD (1996) Variation in the composition and structure of tropical savannas as a function of rainfall and soil texture along a large-scale climatic gradient in the Northern Territory, Australia. *Journal of Biogeography* **23**, 747–756. doi:10.1111/j.1365-2699.1996.tb00036.x

Williams RJ, Cook GD, Gill AM, Moore PH (1999) Fire regime, fire intensity and tree survival in a tropical savanna in northern Australia. *Australian Journal of Ecology* **24**, 50–59. doi:10.1046/j.1442-9993.1999.00946.x

Wilson BA, Bowman DMJS (1987) Fire, storm, flood and drought: The vegetation ecology of Howards Peninsula, Northern Territory, Australia. *Australian Journal of Ecology* **12**, 165–174. doi:10.1111/j.1442-9993.1987.tb00937.x

Woinarski JCZ, Westaway J (2008) 'Hollow Formation in the *Eucalyptus miniata–E. tetrodonta* Open Forests and Savanna Woodlands of Tropical Northern Australia.' Final report to Land and Water Australia, Native Vegetation Program, Project TRC-14.

Woinarski JCZ, Risler J, Kean L (2004) Response of vegetation and vertebrate fauna to 23 years of fire exclusion in a tropical *Eucalyptus* open forest, Northern Territory, Australia. *Austral Ecology* **29**, 156–176. doi:10.1111/j.1442-9993.2004.01333.x

13

Predicting the effects of fire management on carbon stock dynamics using statistical and process-based modelling

Garry D Cook, Adam C Liedloff and Brett P Murphy

Summary

This chapter presents the results of two approaches to predicting the long-term effects of fire regimes on the carbon stocks in Australian tropical savannas. The first approach applies a statistical model to the Three Parks data from Kakadu, Litchfield and Nitmiluk national parks to project observations of carbon stock changes in live trees over 5-year periods to a century. The second applies the FLAMES model to predict the effect of a changed fire regime on equilibrium carbon stocks in live trees, fine fuel and coarse woody debris. We conclude that a process-based modelling approach, such as FLAMES, provides a more robust and conservative prediction of the positive effect of fire management on carbon stocks than does simple extrapolation of a statistical model of short-term changes in carbon stocks, which cannot account for homeostatic and long-term demographic effects.

Introduction

Here we describe the use of two modelling approaches to predict the impact of changing fire regimes on above-ground carbon stocks in northern Australia's mesic savannas (>1000 mm mean annual rainfall). The first approach was to use a statistical model which extrapolates the relationships observed in the Three Parks dataset between carbon stock change over short periods (2–7 years) and the frequency of fires of varying severity. This approach follows the analysis presented by Murphy *et al.* (2009), Murphy *et al.* (2010) and Russell-Smith *et al.* (2014). The second approach was to apply the FLAMES model (Liedloff and Cook 2007), which is a demographic model following the fate of individual trees in a stand subject to set fire and climatic regimes. Discrepancies between the statistical and mechanistic models are discussed and recommendations for a conservative carbon sequestration methodology are presented.

The Three Parks analysis

The Three Parks dataset has been used by Murphy *et al.* (2009) to estimate the likely effects of fire management on live tree carbon stocks in northern Australia's mesic savannas. On the basis of this modelling, Russell-Smith *et al.* (2014) estimated that the carbon sequestration resulting from the change in fire regime achieved in the West Arnhem Land Fire Abatement (WALFA) project area could be worth A$12M annually. Here, we repeat the analysis of Murphy *et al.* (2009) incorporating an additional 9 years of monitoring data (2005–13). We also examine the likely processes that contribute to changes in tree biomass, by separately analysing changes due to the three key demographic processes: recruitment, growth, and mortality.

The Three Parks vegetation and fire monitoring program is described in detail by Russell-Smith *et al.* (2009a). Throughout Kakadu, Litchfield and Nitmiluk national parks, a total of 220 monitoring plots were established in 1994–95 in a range of vegetation types, with detailed vegetation assessments being carried out in 1994–95, 1999–2000 and 2004–05. A limited subset of plots was also reassessed in 2009 and 2013. The monitoring interval ranged between 2 and 7 years, but averaged 5 years. As well as the vegetation assessments, a detailed fire history has been assembled for each of the plots. In this chapter we only examine the plots located in savanna woodland vegetation, in either lowland or sandstone plateau landform types (136 plots in total). This vegetation classification would include both the eucalypt open forest and eucalypt woodland classes of the savanna burning emissions abatement methodology, but does not correspond exactly with that classification. The rainforest, heathland and floodplain plots of the Three Parks study were not included because of their different fire ecology. The Three Parks data did not include any consideration of carbon stocks in either fine or coarse fuel.

Estimation of tree carbon stock and characterisation of site fire histories

During each vegetation assessment, a complete inventory of trees was made at each monitoring plot, and this was used to estimate tree carbon stock. The diameter at breast height (DBH, at 1.3 m) of each live adult tree (trees with DBH ≥5 cm) within 40 × 20 m plots was recorded so that live tree basal area could be calculated. Using the allometric equations provided by Williams *et al.* (2005b), estimates of tree height based on relationships between rainfall, height and DBH (Cook unpublished), and the mean annual rainfall of the plot (derived from interpolated surfaces: Bureau of Meteorology, Canberra), live tree DBH values were converted to tree carbon stock, including only above-ground tree biomass (t C ha^{-1}). The analysis was conducted on the carbon stock at a plot scale. As adult trees were permanently tagged in the monitoring plots, we were able to partition changes in tree carbon stocks into three components: (1) changes in the biomass of trees that survived each ~5-year monitoring period; (2) the biomass attributable to the death of existing trees; and (3) biomass attributable to the recruitment of new trees into the adult size class (DBH ≥5 cm).

The Three Parks dataset also contains detailed fire histories for each monitoring plot. Throughout the study period, plots were usually visited and photographed once or twice each year, so we have a record of whether each plot was burnt in any given year and the severity of the fire. Fires were categorised as low, moderate or high severity, using the severity index developed by Russell-Smith and Edwards (2006a) based largely on the scorch height of leaves. Low severity fires were those with scorch heights of ≤2 m, suggesting Byram fire-line intensities of <<1 MW m^{-1}. Moderate severity fires were those where the canopy was scorched to mid-height, suggesting intensities of 1–2 MW m^{-1} while high severity fires were those where the canopy was scorched to its full height, suggesting intensities of >2 MW m^{-1}. A detailed description of the compilation of the fire histories is provided by Russell-Smith and Edwards (2006a).

Statistical analysis

We examined four response variables: (1) the overall change in carbon stock of live adult trees, as well as change in tree carbon stock that could be attributed to (2) existing trees that survived each ~5-year monitoring period: (3) existing trees that died; and (4) trees that recruited into the adult size class (>5 cm DBH). Change was assessed in each of the four ~5-year monitoring periods, and expressed as an annualised rate:

$$\text{Annual change} = \left(\frac{\text{final carbon stock}}{\text{initial carbon stock}}\right)\left(\frac{1}{\text{time}}\right)$$

where 'time' is the duration of the monitoring period in years.

We examined the effect of six variables on changes in tree carbon stock. Three variables related to fire frequency: frequency of (1) low; (2) moderate; and (3) high severity fires. The remaining three variables were: (4) tree biomass in the plot at the start of the monitoring period (log-transformed); (5) mean annual rainfall at the plot estimated from interpolated surfaces (0.05° resolution; Australian Bureau of Meteorology, Canberra); and (6) landform (i.e. either lowland or sandstone plateau), affecting moisture infiltration and maximum rooting depth.

For each of the five response variables, models representing all combinations of the six explanatory variables were constructed as linear mixed effects models in the program R (version 2.15.1). Mixed effects models allowed us to account for the repeated measurement of each plot (i.e. in each ~5-year monitoring period), and hence the lack of independence of observations. Each response variable was appropriately transformed to ensure normally distributed errors.

Models were evaluated using the corrected form of Akaike's Information Criterion, AIC_c, a model selection index favouring both model fit and simplicity. Lower values of AIC_c indicate greater support for a model, relative to other models in the same candidate set. From AIC_c, Akaike weights (w_i) were calculated for each model, and these are equivalent to the probability of a given model being the best in the candidate set. The importance of each variable was evaluated using w+, the sum of w_i for all models in which that variable occurred. For each variable, w+ is equivalent to the probability of the best model containing that variable. We considered that w+ values of <0.73 were indicative of substantial model selection uncertainty, and that a relationship between the response and the explanatory variable in question was not well supported by the data. A w+ value of 0.73 is equivalent to an AIC_c difference of 2 units between the models containing the variable under examination and those not containing it. An AIC_c difference of 2 units is a common 'rule of thumb' used to assess clear evidence of an effect (Richards 2005).

Observed effects of fire on tree carbon stock

Inspection of changes in tree carbon stock indicates a negative effect of frequent, high severity fires (Fig. 13.1). Generally, plots that were subject to three or four moderate severity fires, or one or more high severity fires, in a 5-year period experienced declines in carbon stock (Fig. 13.1a). For trees, the negative effect of fire was most apparent in changes in carbon stock of surviving individuals (Fig. 13.1b), with frequent high severity fires substantially reducing the rate of accumulation of biomass by individual trees. For changes that were due to the death or recruitment of trees, relationships with fire frequency and severity were less pronounced (Fig. 13.1c, d).

These observations were confirmed by formal model selection, with very strong support for effects of fire frequency on changes in tree carbon stock. There was strong support for the effect of the frequency of severe and, to a lesser extent, moderate fires on changes in tree carbon stock (Fig. 13.2). The effect of the frequency of moderate and high severity fires was clearly apparent in changes in carbon stock of surviving individuals, but there was little support for the effect of fire frequencies on change in carbon stock attributable to either the death or recruitment of trees.

Where there were detectable relationships between change in carbon stock and fire frequency, the relationships were consistently negative, with high severity fires causing substantially greater declines than moderate and low severity fires (Fig. 13.3). A regime of

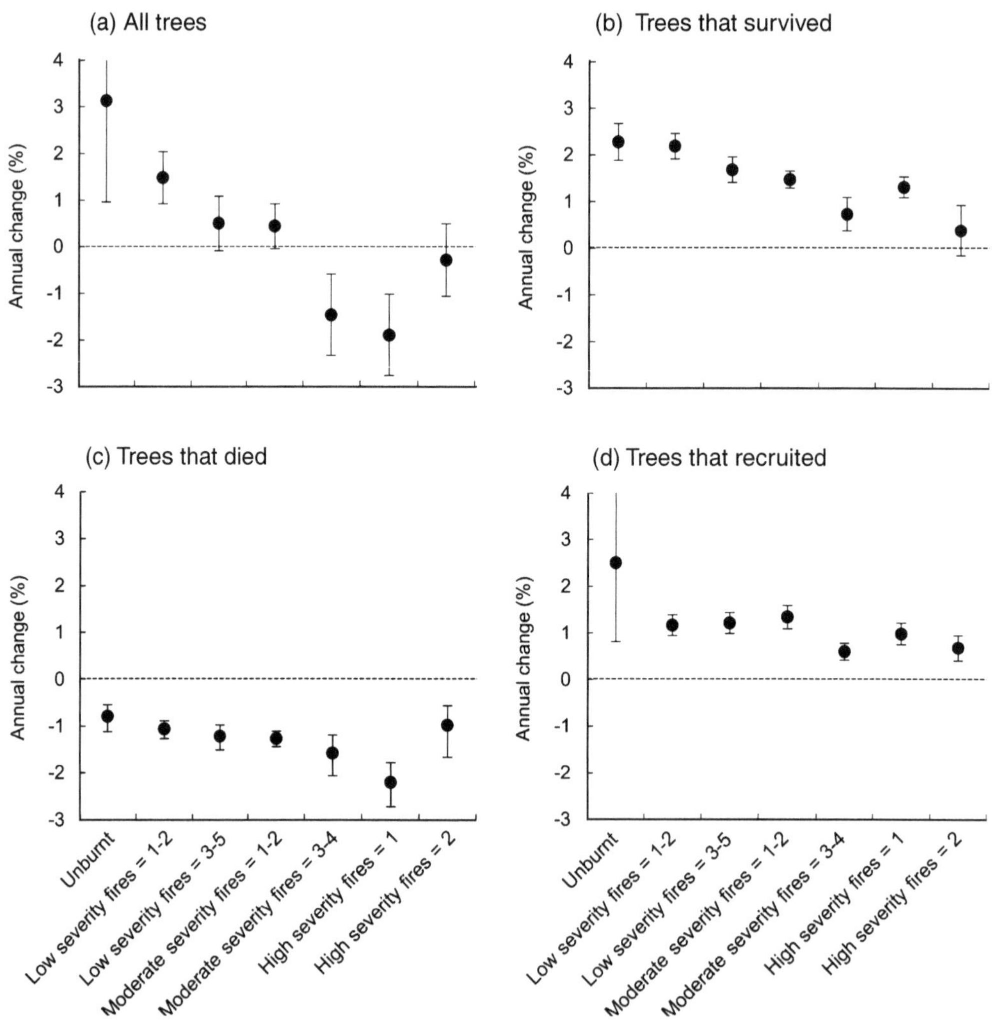

Figure 13.1. Mean annual change (%) in carbon stock (t C ha^{-1}) in (a) live adult tree above-ground biomass, in relation to fire regimes of varying fire frequency and severity. From left to right along the *x*-axis, fires are increasingly frequent and severe. Change in tree carbon stock is separated into that due to: (b) survival of existing trees; (c) death of existing trees; and (d) recruitment of new trees. Standard errors are shown.

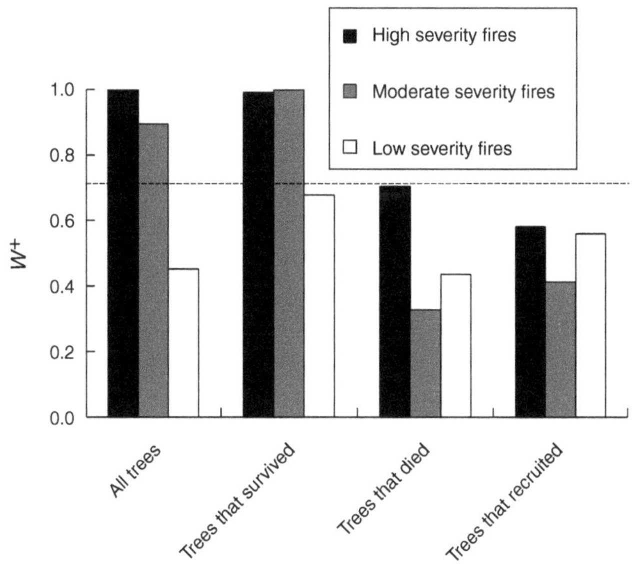

Figure 13.2. Summary of the importance of the three fire frequency variables examined (mild, moderate and severe fire frequency) as determinants of various components of change in tree biomass carbon stock. w+ is equivalent to the probability of a given variable occurring in the best model, and therefore reflects the weight of evidence of a relationship between that variable and change in carbon stock. The dashed line indicates a w+ value of 0.72, considered a useful 'rule of thumb' to assess clear evidence of an effect.

frequent severe fires (≥0.2 fires yr^{-1}) is clearly associated with declines in tree biomass. That frequent high severity fires reduce carbon stocks appears to be due largely to reductions in the rate of biomass accumulation by surviving trees (Fig. 13.3b), rather than tree mortality or recruitment (Fig. 13.3c, d).

Simulating the long-term effects of fire management on tree carbon stocks

Murphy *et al.* (2009) used a simple simulation modelling approach based on modelled short-term changes in tree carbon stock in response to fire frequency and severity to examine the impact of fire regime changes achieved by the WALFA project. The WALFA project has changed annual area burnt in the early and late dry season respectively from 8% and 32% (for the baseline period 1997–2004), to 21% and 11% (for the period 2005–11) (Russell-Smith *et al.* 2013). By comparison, in the Three Parks study, 42% and 13% of plots were burnt early and late respectively in each year on average.

The responses of tree carbon stocks to the various fire regimes were simulated using the updated models of tree carbon stock change derived here and the proportions of low, moderate and high severity fires in early and late dry seasons used by Murphy *et al.* (2009). These simulations show that maintenance of the baseline WALFA regime for a century would reduce carbon stock of the sandstone woodland from 25.1 t C ha^{-1} to 19.2 t C ha^{-1} as the system approached equilibrium with the fire regime. Maintenance of the existing regimes across the Three Parks plots would produce a slight decline in carbon stock to 23.4 t C ha^{-1} (Fig. 13.4). Maintaining the more benign WALFA project regime for a century would lead to an increase in carbon stock to 28.5 t C ha^{-1}. Applying the with-project

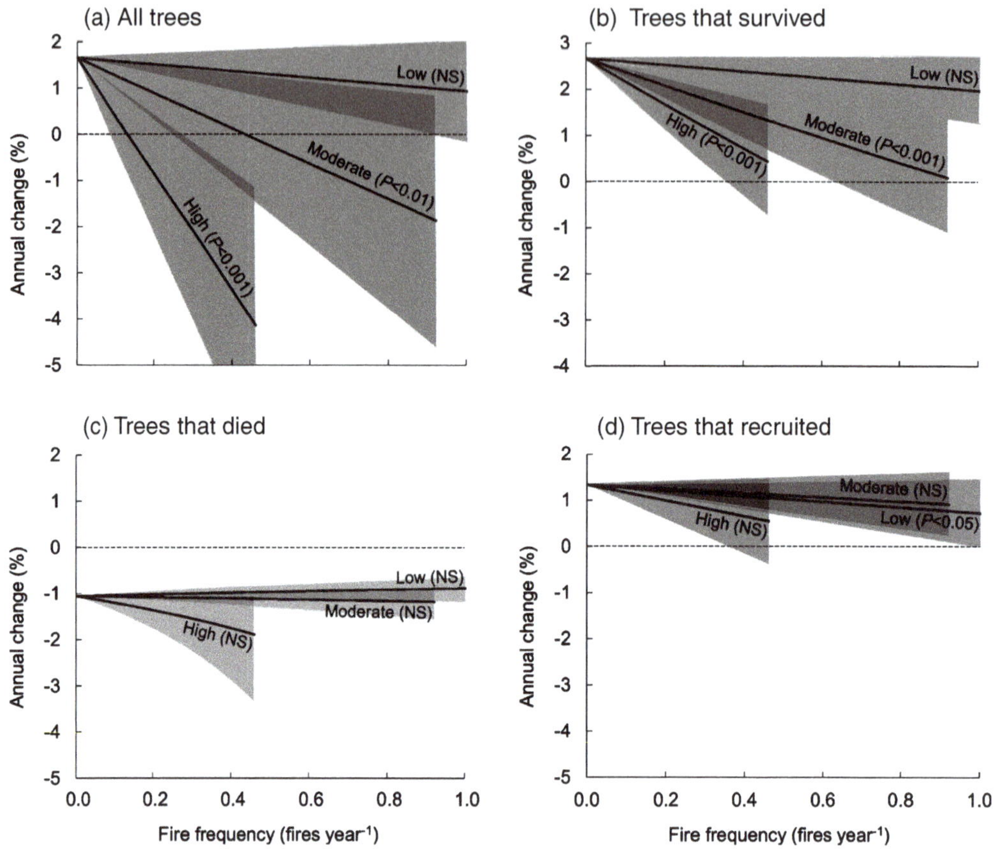

Figure 13.3. The modelled relationship between annual change (%) in carbon stock (t C ha^{-1}) in live adult tree above-ground biomass, in relation to fire regimes of varying fire frequency and severity. The predicted effects are based on multi-model averaging of all candidate models, weighted according to w_i. The shaded areas represent the 95% confidence interval associated with fire frequency.

WALFA fire regime would therefore produce an almost 50% increase in carbon stock over a century compared with the baseline (i.e. 9.3 t C ha^{-1} increase above 19.2 t C ha^{-1}). For lowland woodland sites the carbon stock after 100 years of with-project WALFA fire management was 10.4 t C ha^{-1} greater than for maintaining the baseline scenario. Maintenance of the Three Parks regime produced a slight increase in carbon stock from 25.1 t C ha^{-1} to 26.4 t C ha^{-1}.

The FLAMES analysis

The FLAMES model is an individual-based demographic model of a tree stand with a grassy understorey (Cook *et al.* 2000; Cook and Liedloff 2001; Liedloff and Cook 2007; Liedloff and Cook 2011). In the model, fuel accumulates, decomposes and is consumed by fires, and trees grow, use water, and respond to fires according to their size. The mechanisms simulated in the model are based on field measurements from eucalypt-dominated

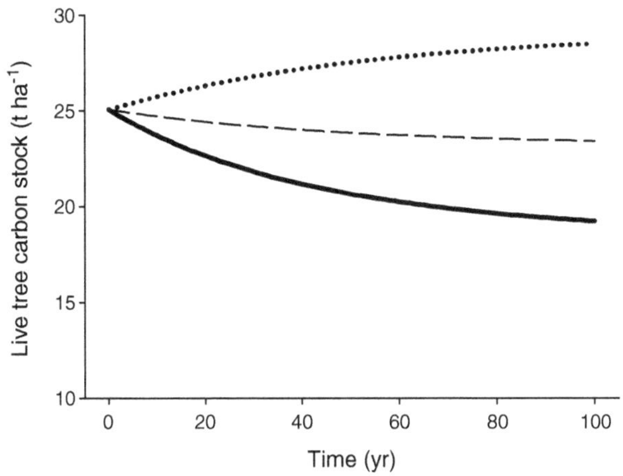

Figure 13.4. The projected change in tree above-ground biomass carbon stocks in sandstone woodlands in the West Arnhem Land Fire Abatement project area, in response to either the baseline WALFA fire management (solid line), baseline Three Parks fire management (dashed line) or with- project WALFA fire management.

savannas of northern Australia (Cook *et al.* 2002; Cook 2003; Cook *et al.* 2005). When fires occur, the Byram fire-line intensity is calculated based on the fuel load and fire weather (Cheney *et al.* 1998). The model also tracks stand density, carbon stocks, water use, gaseous emissions, and a range of other variables.

Parameterisation

We calculated the carbon stock of each tree based on its DBH and its height using the allometry of Williams *et al.* (2005b). Tree height was estimated from DBH and mean annual rainfall following the approach of Cook (unpublished). We chose the allometry of Williams *et al.* (2005b) rather than that of Cook *et al.* (2005) for three reasons: (1) it can take account of varying tree height to DBH relationships along rainfall gradients, which can be a significant source of systematic variation in carbon stocks (Cook unpublished); (2) it is based on a broader range of size classes and sites; and (3) biomass scales with DBH to the power of 2.55, which is closer to the value of 2.67 based on scaling theory than that of the allometry of Cook *et al.* (2005), in which the scaling factor is 2. As a consequence of this choice of scaling factor, carbon stocks will be highly dependent on stand structure, with stands dominated by large trees having a much greater carbon stock than a stand with the same total basal area but with fewer large trees. We believe this to be a much more ecologically realistic description of tree biomass.

There is much empirical evidence that water competition sets the tree carrying capacity in north Australian savannas, with both large trees (Cook *et al.* 2002; Fensham *et al.* 2005; Fensham and Fairfax 2007; Liedloff and Cook 2007; Fensham *et al.* 2009) and juveniles (Fensham and Bowman 1992; Prior *et al.* 1997) being sensitive to water stress caused by competition for water during either the long dry season or major regional droughts. In FLAMES, tree water use scales with DBH to the power of 2, consistent with empirical observations and scaling theory (West *et al.* 1997; Enquist *et al.* 1999; West *et al.* 1999; Cook *et al.* 2002).

The Byram fire-line intensity (Byram 1959) is calculated in FLAMES based on fine fuel loads, heat contents of the fuels and fire rate of spread (Liedloff and Cook 2007). The rate of spread is calculated from fire weather and fuel curing proportion using the equations of Cheney *et al.* (1998). It is assumed in these runs that fires are spreading in an elliptical pattern with the probability of the site being burnt by fronting, flanking or backing fires being determined by the approach of Catchpole *et al.* (1992).

The probability of trees surviving fire was determined by a modification of the algorithms of Cook *et al.* (2005), which were based on field data presented by Williams *et al.* (1999). The algorithm was modified to take account of varying tree height to DBH relationships and is described by Cook (unpublished).

Tree growth rates were expressed as DBH increments and based on data presented by Cook *et al.* (2005). However, because there is some evidence that severe fires reduce growth rates of trees in the subsequent year (Beringer *et al.* 2007; Prior *et al.* 2009; Murphy *et al.* 2010), no DBH increment was added in the year after a tree was scorched by fire to its canopy height or greater using the calculations of Cook (unpublished) and Williams *et al.* (2003a).

Fuel loads were calculated monthly in FLAMES using an approach based on the fuel accumulation model of Olson (1963), which takes account of fuel inputs and decomposition rates. For fine fuels, the input of tree litter and grass was calculated as per Liedloff and Cook (2007), but also accounting for burning efficiency and fire patchiness (Russell-Smith *et al.* 2009c). Coarse woody debris (CWD) in FLAMES is created when trees die, with burning efficiency and patchiness influencing its combustion similar to that of fine fuel. FLAMES does not currently model the creation of CWD due to branch fall from live trees. This is due to the lack of field data on this process. In the absence of any data on the relative proportions of CWD produced by trees in the coarse (≥6–<50 mm) and heavy (≥50 mm) classes defined by Russell-Smith *et al.* (2009c), only one class of CWD was tracked in FLAMES (>6 mm).

The biological decomposition rate of CWD was estimated by assuming a mean annual input based on the stand structures of *Eucalyptus* open forests and woodland at Kapalga, in Kakadu National Park and from the Three Parks study, and calculating the decomposition rate required to produce the mean stock of CWD of 6.2 t ha^{-1} observed in this vegetation structure (Table 4 in Russell-Smith *et al.* 2009b) assuming the fire regime before the commencement of the WALFA project (Russell-Smith *et al.* 2013). A biological decomposition rate for CWD of 0.071 was estimated.

The role of termites in carbon dynamics in FLAMES is implicit in fine and CWD fuel dynamics as a contributor to biological decomposition rates; in carbon stock in live trees because the allometric relationships account for tree hollowing due mainly to termites; and in the probability of survival of trees, where large old trees become increasingly sensitive to fire due, in part, to hollowing caused by termites. Net emissions of methane by termites are relatively small (Jamali *et al.* 2013), and are relevant to estimations of non-CO_2 emissions budgets but not to estimations of carbon stock change.

Model runs

We ran the FLAMES model for climate stations representative of the Tiwi-Cobourg, Cape York Peninsula, Darwin Coastal, Pine Creek and Daly Basin bioregions. These bioregions have extensive savannas with a high tree biomass and are subject to frequent fires. For each bioregion, we ran the model using a 200-year spin-up and various settings listed in Table 13.1.

Validation of the FLAMES model output

Fire severity

The proportions of early and late dry season fire of low, moderate and high severity from the FLAMES modelling are similar to those determined from aerial assessments across northern Australia by Edwards *et al.* (Chapter 8), but both contrast with the earlier estimates based on ground-based observations at the Three Parks plots (Fig. 13.5). These ground-based observations show a strong tendency for early dry season fires to be of low severity. For the early dry season fires in the FLAMES modelling, a threshold of Grassland Fire Danger Index of 7.5 was applied to restrict fires to occasions with more benign fire weather. In order to reproduce the tendency for low severity fires in the Three Parks plots, a much lower threshold of fire danger would need to be applied.

Stand structure

We compared the tree stand characteristics from the Three Parks data with those from FLAMES runs with similar fire frequencies. Murphy *et al.* (2010) report that 46%, 40% and 56% of Kakadu, Nitmiluk and Litchfield national parks are burnt annually. In Fig. 13.6, the mean stand structure of all sandstone woodland or lowland woodland plots in the Three Parks dataset are compared with FLAMES runs for the Darwin Coastal and Daly Basin Bioregions, taking a mean of runs with fire frequencies of 0.4 yr^{-1} and 0.6 yr^{-1}. The number

Table 13.1. Parameterisation of the FLAMES model

Parameter	Value
Simulation length	Data presented are from the 200th year of the simulation
Tree stand	*Eucalyptus miniata/Eucalyptus tetrodonta* stand with nominal tropical grass
Soil	Kapalga kandosol, 1.5 m depth, 65% sand, 20% silt and 15% clay
Locations representing bioregions	Darwin Coastal (Darwin airport rainfall and climate records, using Batchelor airport wind speeds) Tiwi/Cobourg (Pirlangimpi records) Pine Creek (Pine Creek records) Daly Basin (Katherine Town Council records) Cape York (Coen records)
Rainfall	Stochastic realisation of local rainfall
Fire frequency	0.2, 0.3, 0.4, 0.5 and 0.6 yr^{-1}
Fire Style	Elliptical burns
Fire timing	0%, 20%, 40%, 60%, 80% and 100% in the early dry season (monthly probability evenly divided between May–Jul for early fires and Aug–Oct for late fires) All afternoon fires (1500 h)
Maximum Grass Fire Danger Index (GFDI) for prescribed burning	7.5
CWD & fine fuel decomposition rates (k_F and k_C)	0.8, 0.071
Number of replicate runs	20

of trees across the range of size classes shows remarkable agreement between the Three Parks data and the FLAMES outputs.

We fitted exponential curves to the stand structure data in Fig. 13.6. From the fitted parameters, we calculated the total basal area, total carbon stock and annual creation of dead tree carbon. The Three Parks data had slightly lower values of total basal area than the FLAMES runs, but the simulated carbon stock of the live trees in the Three Parks data was between the two FLAMES simulations (Table 13.2). The values for live tree carbon stock are similar to those for eucalypt open forest given by Hutley *et al.* (2011). The annual creation of dead tree carbon was slightly lower in the Three Parks data than in the FLAMES data.

Utility of the FLAMES model

We are confident that the FLAMES model is able to satisfactorily reproduce the fire behaviour seen in the latest analyses of high rainfall savannas using aerial observations (Edwards *et al.*, Chapter 8) and the structure of tree stands seen in the Three Parks data. Given that these attributes are emergent properties of the FLAMES model as outcomes of process rules, the similarity to real-world data is remarkable. It means that FLAMES can be used with

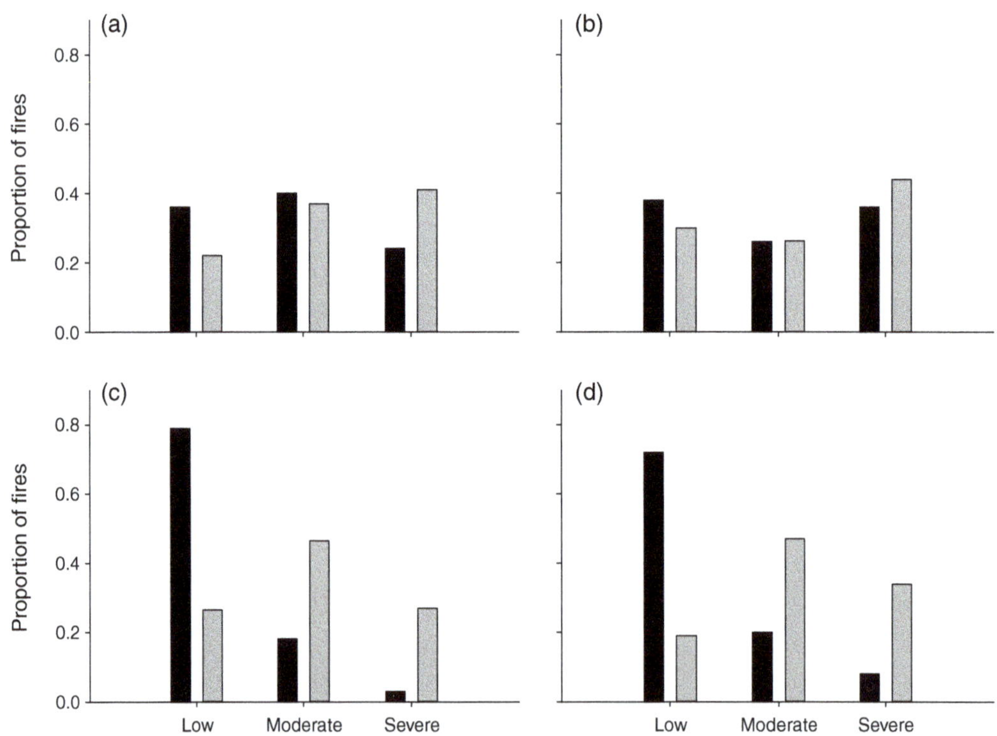

Figure 13.5. The proportion of early dry season (black bars) and late dry season fires (grey bars) in different fire severity classes as given for high rainfall savannas: (a) for aerial assessments reported by Edwards *et al.* (Chapter 8); (b) as simulated by FLAMES modelling in this chapter; (c) as used by Murphy *et al.* (2009); and (d) as given in (Russell-Smith *et al.* 2009c) and invoked in the savanna burning emissions abatement methodology (Commonwealth of Australia 2013).

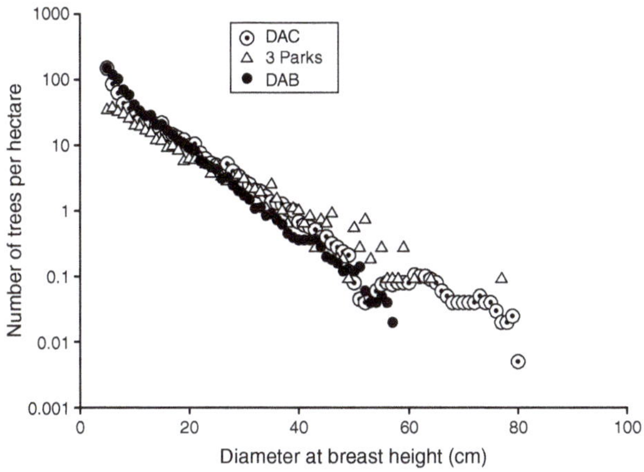

Figure 13.6. The mean stand structure of all plots from the lowland woodland and sandstone woodland of the Three Parks study compared with the mean stand structure from FLAMES runs for the Darwin Coastal (DAC) and Daly Basin (DAB) bioregions with mean fire frequencies of 0.6 and 0.4 fires yr^{-1}, respectively.

greater confidence to assess long-term variation in stand attributes with changes in fire regime.

Impact of fire on carbon as predicted by the FLAMES model

In these analyses, we first compare the attributes of the simulated savanna plots in equilibrium with various fire regimes. In these analyses, the responses are examined with respect to fire regime, and do not take account of the time period to respond to changes in fire regime. The response time is examined separately, by running the model for 200 years with a baseline fire regime observed in the WALFA project, and then for 50 years applying the with-project regime achieved in the WALFA project area (Russell-Smith *et al.* 2013).

Stand structure and tree mortality

The decline in the number of trees per hectare as the size class increases (Fig. 13.6) reflects that as a cohort of trees grows, a proportion of them die each year. With increasing fire frequency, death is more likely to be caused by fire and less likely to be caused by water

Table 13.2. Comparison of structural attributes of Three Parks plots with those of FLAMES outputs for the Darwin Coastal (DAC) and Daly Basin (DAB) bioregions

Parameter	Three Parks	FLAMES output	
		DAC	DAB
Fire frequency (yr^{-1})	0.40–0.56	0.50 (mean of 0.40 and 0.60)	0.40
Total basal area (m^2 ha^{-1})	7.7	9.0	8.7
Live tree carbon stock (t ha^{-1})	29	32	26
Annual production of dead tree carbon stock (t ha^{-1})	0.7	0.9	1.0

stress. This is reflected in the data on the numbers of trees dying from these two causes in the FLAMES output (Fig. 13.7).

Total basal area

Across all bioregions, equilibrium total basal area increased by 0.47–0.94 m^2 ha^{-1} for a decrease in fire frequency from 0.6 yr^{-1} to 0.2 yr^{-1} (Fig. 13.8). Overall, the rate of decline in equilibrium total basal area with increasing fire frequency (Fig. 13.8) was ~0.17 m^2 ha^{-1} for a 0.1 yr^{-1} increase in frequency. This change was remarkably similar to the relationship between observed total basal area and fire frequency in the data presented for 158 sites across the Top End of the Northern Territory (Lehmann *et al.* 2014) (0.14 m^2 ha^{-1} for a 0.1 yr^{-1} increase in frequency). Total basal area was found to decrease slightly with a decreasing proportion of early dry season fires (data not presented).

Total above-ground live tree carbon stock

In contrast to total basal area, which decreased with increasing fire frequency, the equilibrium above-ground live tree carbon stock increased with increasing fire frequency by ~0.3 t C ha^{-1} (Cape York Peninsula) to 0.8 t C ha^{-1} (Tiwi-Cobourg) with an increase in fire frequency of 0.1 yr^{-1} (Fig. 13.9).

The response of live above-ground carbon stock to the proportion of early dry season fires varied across bioregions, with the impacts being slight, and the direction of change being variable (data not presented). It is likely that the variation reflects chance effects of realisations of stand structure, and also patterns of fire intensities, which will vary with the climate of each bioregion.

Total above-ground carbon stock

Across all bioregions, the total above-ground carbon stock increased by 3.6 to 6.8 t C ha^{-1} with a decrease in fire frequency from 0.6 yr^{-1} to 0.2 yr^{-1} (Fig. 13.10). Overall the rate of decline in total carbon stock with increasing fire frequency was ~1.2 t C ha^{-1} for a 0.1 yr^{-1} increase in frequency. The total above-ground carbon stock was found to decrease slightly with a decreasing proportion of early fires (data not presented).

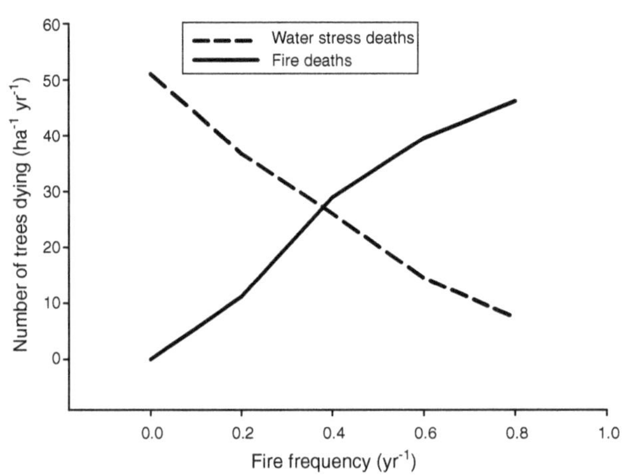

Figure 13.7. The effect of fire frequency on the mean number of tree stems dying of water stress or by fire across all tree sizes for the Daly Basin, Darwin Coastal and Cape York bioregion locations.

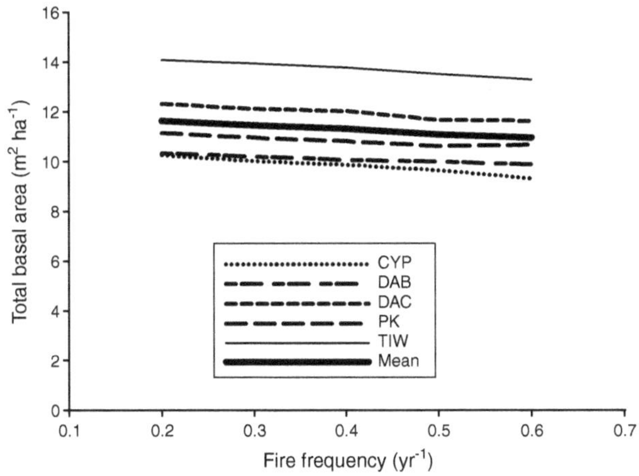

Figure 13.8. The effect of fire frequency on total basal area of eucalypt open forest as simulated by the FLAMES model for the Cape York (CYP), Daly Basin (DAB), Darwin Coastal (DAC), Pine Creek (PK) and Tiwi-Cobourg (TIW) bioregions and for the mean of all data. The equation represents a linear fit to the mean data.

The increase in total above-ground carbon stock with decreasing fire frequency derives from a strong increase in coarse woody debris, weaker increase in fine fuel (grass and tree litter) and, counter-intuitively, a weak decrease in live tree carbon stock with decreasing fire frequency (Fig. 13.11).

Sequestration rates

To investigate the time to reach equilibrium, we ran 50 iterations of the WALFA baseline fire regime over 200 years, then changed regime to the WALFA project fire regime and ran that for 50 years. The mean total basal area reached 0.3 m^2 ha^{-1} more than the baseline value in ~20 years, while increases in fine fuel accumulation drove a rapid increase in total above-ground carbon stock, which reached values close to maximum within 10 years (Fig. 13.12).

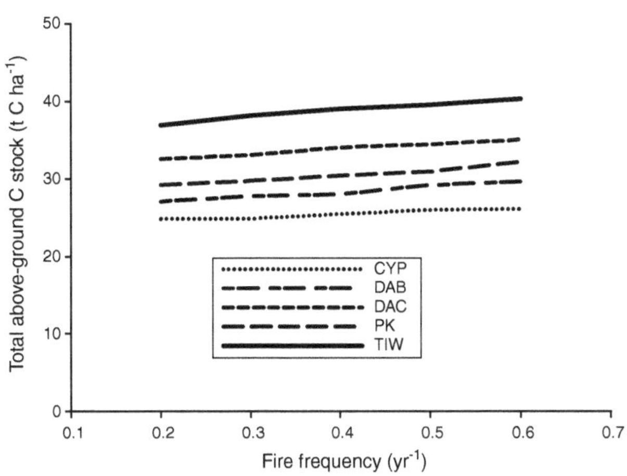

Figure 13.9. The effect of fire frequency on total live trees above ground in five bioregions simulated by the FLAMES model.

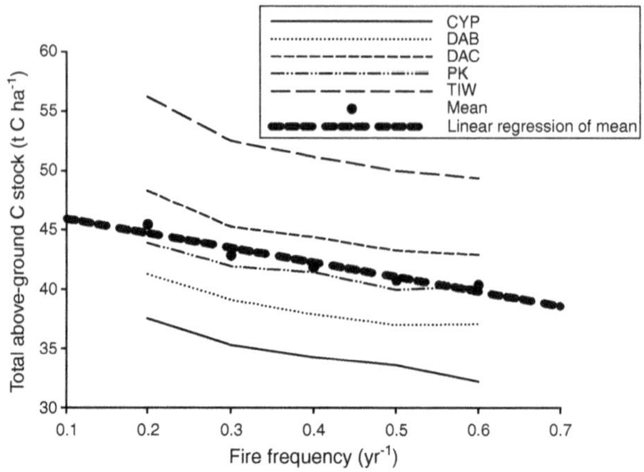

Figure 13.10. The effect of fire frequency on total above-ground carbon stock of eucalypt open forest as simulated by the FLAMES model for the bioregions given in Fig. 13.6 and for the mean of all data. The equation to the linear regression through the mean is $Y = 47 - 12.2x$ $R^2 = 0.91$.

Over the period of the runs of the project regime, the mean sequestration rate of total above-ground carbon in the years when the system was a net carbon sink was 1 t ha^{-1} yr^{-1}, and this occurred in 76% of all years. The mean source strength in the 24% of years when it was a source was 3.2 t ha^{-1} yr^{-1}. A synthesis of several field studies of carbon sequestration rates in eucalypt open forests of the Northern Territory found an average sequestration rate of 1 t ha^{-1} yr^{-1} (Williams *et al.* 2004), consistent with the high likelihood of observing that number shown by the FLAMES modelling. As argued by Liedloff and Cook (2011) and Hutley and Beringer (2011), relatively uncommon major disturbance events such as extreme fire and wind events can cause rapid carbon losses, negating the periods when the tropical savannas are a carbon sink.

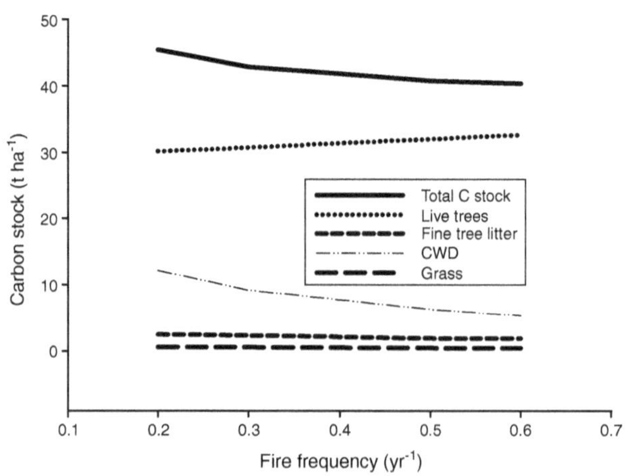

Figure 13.11. The effect of fire frequency on total above-ground carbon stock and the component fractions in live trees, fine tree litter, coarse woody debris and grass averaged across all bioregions.

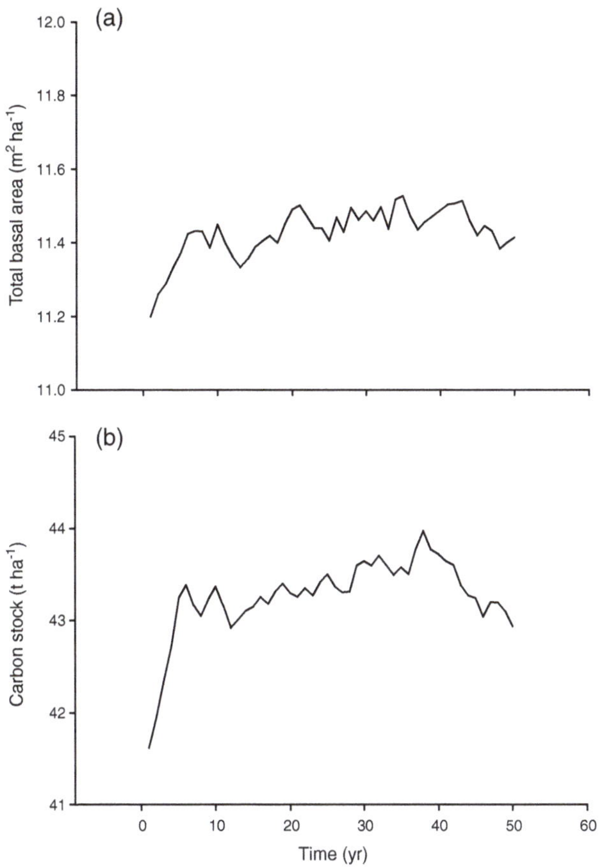

Figure 13.12. Changes in (a) total basal area and (b) total above-ground carbon stock over time following a change in management from the WALFA baseline to the WALFA with-project fire regime.

Generalisation of the predictions

Applying the statistical modelling of the Three Parks analysis to a suite of possible frequencies of fire and proportions of early fire enabled the development of an exact mathematical generalisation of the analysis that provides a predicted carbon stock at equilibrium with any likely fire regime (Fig. 13.13). General equations to predict the asymptotic live tree carbon stock (C_{LT}) for any combination of fire frequency (F) and proportion of early dry season fires (PE) were developed for the sandstone woodland and lowland woodland:

$$C_{LT} = ae^{(1.834PE - 2.61)F}$$

where a is live tree carbon stock in the absence of fire and has values of 45.4 t ha^{-1} for sandstone woodland and 52.1 t ha^{-1} for lowland woodland.

Application of this algorithm to the change in fire regime achieved by the WALFA project indicated that it would lead to an increase in total above-ground carbon stock of 10.5 t ha^{-1} for the sandstone woodland and 12.03 t ha^{-1} for the lowland woodland (Fig. 13.13). Based on the modelled rate of change in total carbon stock over a century

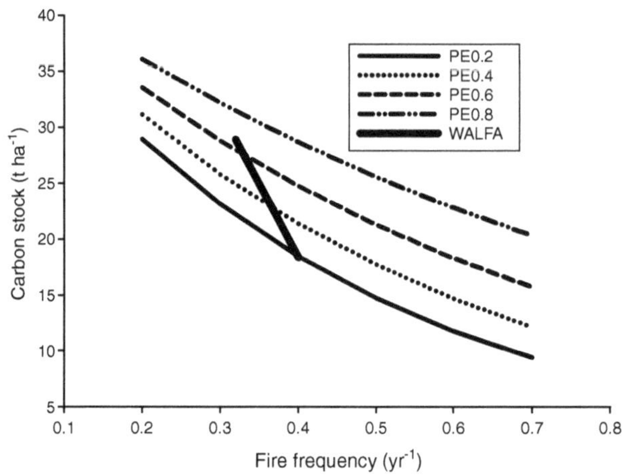

Figure 13.13. Generalisation of the results of the Three Parks analysis showing the effect of fire frequency on the asymptotic carbon stock of live trees in the sandstone woodland for various proportions of early dry season fires. Also shown is the impact of the change in fire management achieved in the WALFA project.

(Fig. 13.4), it is reasonable to assume that this modelled change would occur over a 100-year period, giving annual sequestration rates of 0.105 t C ha^{-1} yr^{-1} and 0.120 t C ha^{-1} yr^{-1} respectively.

For the FLAMES outputs, there is variation in the responses across the bioregions, and this would be due to different seasonal variations in fire weather, specifically wind, temperature and relative humidity, as well as due to different rainfall regimes. Taking the spatial variation in these properties into account would lead to the development of a Tier 3 methodology (*sensu* IPCC 2006) that would be inconsistent with the emissions abatement methodology and accounting procedures. Applying a mean relationship across a range of climate stations representative of the high rainfall savannas is consistent with the Tier 2

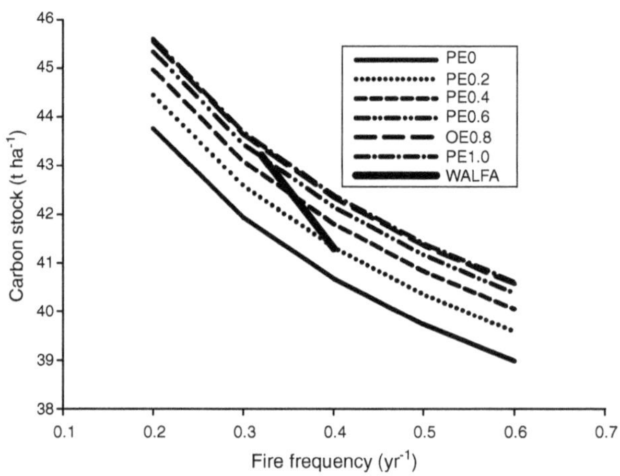

Figure 13.14. Generalisation of the FLAMES predictions showing the effect of fire frequency on the total above ground carbon stock of eucalypt open forest for various proportions of early dry season fires. Also shown is the impact of the change in fire management achieved in the WALFA project.

approach to quantifying savanna burning emissions. A generalisation of the FLAMES modelling to predict equilibrium total above-ground carbon stock (C_{AG}) in eucalypt open forest for possible combinations of frequencies of fire and proportions of early fire is shown in Fig. 13.14. The general equation for these relationships is:

$$C_{AG} = (-1.7PE^2 + 3.2PE + 37)F^{-0.11}$$

Application of this algorithm to the change in fire regime achieved by the WALFA project indicated that it would lead to an increase in total above-ground carbon stock of 1.93 t ha^{-1} (Fig. 13.14). Based on the modelled rate of change in total carbon stock over time (Fig. 13.13), it is reasonable to assume that this change would occur over a 25-year period, giving an annual sequestration rate of 0.077 t C ha^{-1} yr^{-1}.

Reconciliation of the results

There is a need to reconcile two major anomalies in projections of changes in carbon stock as a result of changes in fire regimes. First, it is assumed in the emissions abatement methodology (Commonwealth of Australia 2013: Table 2), that coarse woody debris (CWD: coarse and heavy fuel) loads do not change with variation in fire regime. However, the FLAMES analysis clearly shows that fire frequency should have a substantial effect on carbon stock in CWD. The Three Parks analysis does not account for carbon in either fine or coarse fuels. Second, the statistical analysis of the Three Parks data indicates that live tree carbon stock decreases strongly with increasing fire frequency, while the FLAMES analysis predicts that live tree carbon stock will increase with increasingly frequent fires (with intensity and severity tending to decrease with increasing frequency). Given that FLAMES appears to underestimate the seasonal switching in fire severity observed in the Three Parks plots (i.e. the strong tendency for early dry season fires to be of low severity), FLAMES may also underestimate the positive effects of a regime of frequent, very low intensity fire on tree biomass.

Coarse woody debris dynamics

The absolute insensitivity of CWD to fire regimes assumed in the savanna burning emissions abatement methodology is an outcome of the inability to detect a trend in CWD load from sampling of sites with varying fire histories as determined by satellite imagery (Russell-Smith *et al.* 2009c). Since CWD is primarily removed by only two processes, combustion or biological decomposition (including termite consumption), this assumption implies that biological decomposition rates for CWD are equivalent to prevailing rates of combustion. This implication is clearly at odds with available data on combustion and decomposition of CWD (Mackensen *et al.* 2003; Fensham 2005; Woldendorp and Keenan 2005). The stochastic nature of tree death and branch fall coupled with the high spatial variability in CWD density at typical sampling scales suggest that it would be very difficult to detect a response of CWD load to fire frequency in the field. However that does not mean that CWD loads do not respond to changing fire regimes at landscape scales. Modelling based on inferred biological decomposition rates and observed combustion rates clearly shows a strong impact of fire regime on CWD. Elsewhere in Australian woodlands, where fire frequency is much lower, the CWD loads are much greater than observed in northern Australia (Woldendorp and Keenan 2005; Russell-Smith *et al.* 2009c), fire regimes have strong effects on CWD loads (Aponte *et al.* 2014). There is no basis to reject the responsiveness of CWD to fire regime in northern Australian savannas simply because it cannot be detected by field sampling.

The effect of fire regimes on live tree carbon stocks: comparison with other evidence of fire impacts on vegetation

The simulations based on statistical modelling of the Three Parks data indicate that live tree carbon stocks are close to being in equilibrium with the current fire regime they face. However, under the baseline regime of the WALFA project area, carbon stocks were projected to decline and under the with-project WALFA regime to increase, such that there is an almost 50% difference in stock from baseline to with-project regime over 100 years. The magnitude of this projection contrasts with many other studies of the long-term effects of fire on eucalypt-dominated savannas. For example Russell-Smith *et al.* (2003) found that the effects on eucalypt trees from 23 years of no fire, annual early fires, annual late fires or biennial fires at Munmarlary, Kakadu National Park, was 'slight but significant'. They concluded that the trees had 'apparent insensitivity to fire regime' and that the response was 'in the main, muted over the range of applied fire treatments confirm(ing) the relative stability of eucalypt open forest and woodland under low- to moderate-intensity, frequent burning'. These conclusions were consistent with those of Bowman *et al.* (1988), who also analysed results of the Munmarlary fire experiment and found that with the exception of a fire-free treatment, varying fire regimes resulted in negligible floristic or structural differences in savanna vegetation. The analyses of Lehmann *et al.* (2014) similarly concluded that Australian savannas were little affected by fire regimes. Russell-Smith *et al.* (2003) described the 'stark contrast [between their finding of a muted response to fire] to the magnitude of responses observed for single, very intense fires by Lonsdale and Braithwaite (1991) and Williams *et al.* (1999)'. Further, there is little evidence that the current live tree carbon stock is declining substantially under current fire regimes; rather, the contrary has been observed (Williams *et al.* 2004, 2005a; Hutley and Beringer 2011).

The intensity of fires is of critical importance to the interpretation of impacts reported in the literature. The average fire intensities at the Kapalga fire experiment were much greater than those recorded in the Munmarlary experiment. The annual early dry season fires at Kapalga averaged 2.1 MW m^{-1} and the annual late dry season fires averaged 7.7 MW m^{-1} (Williams *et al.* 2003b). In the late plots, fire intensity was greatest at 15 MW m^{-1} in the first year of the experiment, after several years of fire protection, and ranged from 4 MW m^{-1} to 7 MW m^{-1} thereafter. Thus even the annual early fires would have been borderline between moderate and high fire severity *sensu* Russell-Smith and Edwards (2006b), while all the late fires were of high severity. The severity of the fire regime during the Kapalga experiment was consistent with the much larger plot size at Kapalga and the use of fronting fires that led to maximum rate of spread throughout the plots, compared with a mix of fronting, flanking and backing fires that would have resulted if point-source ignitions had been used (Catchpole *et al.* 1992; Cheney *et al.* 1998; Cook and Corbett 2003; Liedloff and Cook 2007). Williams *et al.* (2003b) found that the total basal area of trees remained more or less unchanged over a 6-year period under both unburnt and early-burn regimes with moderate–high severity fires, but declined by ~20% under annual late fronting fires. However, most of that decline under the regime of annual late fires occurred in the 2 years following the initial very intense fires of the experiment. There was little change during the remaining period of high severity fires. Thus the results of Kapalga are consistent with the findings of Lonsdale and Braithwaite (1991) that single very intense fires can have substantial impacts on total basal area, but are also consistent with findings from Munmarlary that fire regimes, with the exception of fire exclusion over long time periods, have a negligible impact on total basal area of trees.

The eucalypt decline syndrome

Close *et al.* (2009) proposed a model of 'premature tree decline' whereby an absence of fire hastens the mortality of overstorey eucalypts in some forests. This conceptual model was proposed to explain a syndrome affecting diverse eucalypt-dominated forests and woodlands. Competition for water between mid-stratum trees and mature overstorey trees is believed to be an important contributor. This process is a key component of the FLAMES model. Increasing fire frequency at the scale of individual trees in the model (i.e. taking account of patchiness) caused a decline in trees of intermediate size (5–40 cm DBH), but an increase in old trees (>40 cm DBH). This is entirely consistent with the proposed model of Close *et al.* (2009) describing eucalypt decline under infrequent fires and fire exclusion. Since carbon stock of individual trees increases with DBH to the power of 2.55, the increased dominance of old trees with increasing fire frequency will have a disproportionate impact on total carbon stock.

The issues of statistical extrapolation

The statistical extrapolation of observed short-term (~5 year) responses to a century or more risks not taking account of other mechanisms that might either dampen or exaggerate the trends. Using the statistical models developed from the Three Parks dataset to simulate long-term (100-year) change suggests a 48% increase in tree carbon stock in response to reducing fire frequency from 0.4 yr^{-1} to 0.32 yr^{-1}. This stands in marked contrast to the muted effect of experimental fire regimes comparing annual burning with no fires at all, and the lack of strong landscape scale responses to fire regimes (Bowman *et al.* 1988; Williams *et al.* 1999; Russell-Smith *et al.* 2003; Lehmann *et al.* 2014). Being a statistical model, homeostatic mechanisms, such as the release of juveniles observed when mature trees are killed (Fensham and Bowman 1992; Prior and Eamus 2000) and water availability providing ultimate limits to tree carrying capacity, cannot be taken into account. Similarly, the demographic shifts resulting from processes that favour or disfavour a dominance by large trees cannot be taken into account, despite such shifts having significant consequences for carbon storage given that large trees store carbon more efficiently than small trees. Such processes are critical to multi-decadal extrapolation.

The statistical approach of Murphy *et al.* (2009) and Murphy *et al.* (2010) was applied to 5-year blocks of the FLAMES simulation data. Similar to the Three Parks analysis, the results suggested that increasing frequencies of moderate–high severity fires reduce tree carbon stocks in the short term (5 years). A simple extrapolation of these short-term results to 100 years suggested that increasing fire frequency would *reduce* live tree carbon stocks – the opposite of FLAMES' long-term predictions. Hence, short-term statistical relationships between fire regimes and tree carbon stocks do not necessarily indicate the long-term trajectory of tree carbon stocks. By not considering important demographic processes, the statistical approach can lead to misleading conclusions.

Bark loss and tree growth rates

The published estimates of growth rate of trees in Australia's tropical savannas are invariably based on increments in DBH over bark. Given that substantial amounts of bark can be burnt off during fires, this has profound implications for interpretation of growth rate differences. For example Murphy *et al.* (2010) analysed differences in D increments measured over 10 years in sites with different ambient fire regimes. They reported reductions in D increments of ~0.5 to 1.4 mm yr^{-1} as a result of increasingly severe fire regimes. Although measurements were conducted to avoid the confounding effects of stem loss, they did not take account of bark combustion. In the context of reported losses of bark from bloodwood and stringybark eucalypts in southern Australia of 14 to 22 mm from single fires (Table 13.3), these differences in growth rate of less than 1.4 mm yr^{-1} may be an artefact of reliance on over-bark measurements and may not reflect true differences in growth. Fire intensities greater than ~3 to 4 MW m^{-1} can cause more than 10 mm of bark loss in *Eucalyptus marginata*, with greater losses (up to 30 mm) at lower intensities in many cases (Gould *et al.* 2007). Prior, Brook *et al.* (2006) discuss the difficulty in interpreting the effects of fire on growth rates when rates of bark loss are unknown, but nevertheless they concluded that severe fires do reduce growth rates of both saplings and adult trees.

This potential issue is not restricted to the Three Parks dataset. All northern Australian analyses of fire effects on tree biomass will be affected to some extent as they have all relied upon over-bark measurements of DBH. For example, Williams *et al.* (1999) reported a 27% decrease in live tree basal area through tree death following five high severity fires in 5 years at Kapalga. If 1 mm of bark was lost during each severe fire, bark loss alone would account for an apparent 10% decrease in basal area (assuming a stand structure typical of the Three Parks dataset: Fig. 13.6).

Apparently conflicting results from earlier FLAMES modelling

The FLAMES model has been used in various studies to model the outcome of continuing the regime of annual fronting or biennial fires at Kapalga for multiple decades (Cook and

Table 13.3. The loss of bark from eucalypt species during single fires

Species	Bark type	Bark loss expressed as DBH reduction (mm)*	Time since previous fire (yr)	Source
Eucalyptus marginata	Stringy–fibrous	14.8	7	Burrows *et al.* 2001
Eucalyptus marginata	Stringy–fibrous	1	3	Gould *et al.* 2007
Eucalyptus marginata	Stringy–fibrous	10.8–22.1	4–19	Gould *et al.* 2007
Corymbia calophylla	Tessellated short fibre bloodwood	21.8	7	Burrows *et al.* 2001
Eucalyptus macrorhyncha	Stringybark	12.26 (flank fire) – 18.3 (head fire)	?	Gill *et al.* 1986
Eucalyptus rossii	Gumbark	4 (flank fire) – 9.3 (head fire)	?	Gill *et al.* 1986
Eucalyptus obliqua and *E. rubida*	Stringy (*E. o.*); gumbark (*E. r.*)	16 (overwood trees)	Mostly >32	Chatto *et al.* 2003

* taken as twice the reported loss in bark thickness.

Liedloff 2000; Andersen *et al.* 2003; Liedloff and Cook 2007). It has shown strong declines in total basal area of trees. However, the fronting fires assumed in these scenarios ensure that fire intensity is the maximum possible for the fuel and weather conditions. Changing the assumed fire spread to include typical proportions of fronting, flanking and backing fires, as would occur in most non-experimental large-scale fires (Catchpole *et al.* 1992), markedly reduces the response (Liedloff and Cook 2007), consistent with the results of the present study. In some modelling, the allometry of Cook *et al.* (2005) has been used to estimate carbon budgets (e.g. Liedloff and Cook 2011), but these results have been superseded by the present study in which the more robust allometric approach of Williams *et al.* (2005b) has been adopted.

Conclusions

The 1996 guidelines for greenhouse gas accounting (IPCC 1996), on which the savanna burning emissions abatement methodology is based, assume that live vegetation in savannas does not respond to changes in fire regime. The two approaches to predicting the impacts of fire on carbon stocks in Australian mesic savannas produced divergent results. The Three Parks analysis predicting a strong increase in live tree carbon stock with reductions in fire frequency and increases in the proportion of early fires, and this increase continued for at least a century; it was driven by a physiological impact of fire on growth rate. In contrast, the FLAMES analysis predicted a decrease live tree carbon stock for similar changes in fire regime; this was driven by more frequent fires creating stands more dominated by large trees, which contain a much higher carbon stock than an equivalent basal area of smaller trees. The findings of the Three Parks analyses appear at odds with conclusion of other workers that Australian savanna trees are remarkably little affected by fire. The findings from the FLAMES analyses have not been recorded at all in Australian savannas, but most workers have used total basal area as their index of tree stands rather than carbon stock *per se*. A conceptual model has proposed similar impacts of fire exclusion in southern Australian woodlands (Close *et al.* 2009), but there is little support in the literature for the findings from the FLAMES analysis with regard to live tree carbon stock in northern Australia despite FLAMES being based on processes measured in the north. Accordingly we conclude that at present there is no robust basis to depart from the assumption in the 1996 guidelines that live tree carbon stocks are not affected by changing fire regimes.

In contrast to the ambiguous findings about the responses of live tree carbon stock to fire regimes, the dynamics of fine and coarse fuels is straightforward. It is clear that carbon stocks in both fine and coarse fuels increase with decreasing fire frequency. The FLAMES modelling provides an indication of the magnitude of the expected changes in fuel carbon stocks, but the findings are confounded with the as-yet uncertain dynamics of live trees. We recommend an approach to quantifying the impact of fire regime on fine and coarse fuels based on direct process-based modelling using the approach of Olson (1963).

Future work is required to reconcile the contrasting results of the FLAMES and Three Parks results. Is there evidence that increasing fire frequency will lead to greater carbon stocks through increasing the dominance of large trees? Is the observed decrease in tree growth rates after severe fires in the Three Parks study be an artefact of bark combustion, and, if not, could it continue to affect carbon stocks for many decades? How can changes in carbon stock in fuels be incorporated into the national greenhouse gas accounts?

References

Andersen AN, Cook GD, Williams RJ (2003) Synthesis: fire ecology and adaptive conservation management. In *Fire in Tropical Savannas: The Kapalga Experiment*. pp. 153–164. Springer: New York.

Aponte C, Tolhurst KG, Bennett LT (2014) Repeated prescribed fires decrease stocks and change attributes of coarse woody debris in a temperate eucalypt forest. *Ecological Applications* **24**(5), 976–989. doi:10.1890/13-1426.1

Beringer J, Hutley LB, Tapper NJ, Cernusak LA (2007) Savanna fires and their impact on net ecosystem productivity in North Australia. *Global Change Biology* **13**, 990–1004. doi:10.1111/j.1365-2486.2007.01334.x

Bowman DMJS, Wilson BA, Hooper RJ (1988) Response of eucalyptus forest and woodland to four fire regimes at Munmarlary, Northern Territory, Australia. *Journal of Ecology* **76**(1), 215–232. doi:10.2307/2260465

Burrows ND, Ward BG, Robinson AD (2001) Bark as fuel in a moderate intensity jarrah fire. *CaLMScience* **3**(4), 405–409.

Byram GM (1959) Combustion of forest fuels. In *Forest Fire Control and Use*. (Ed. KP Davies) pp. 61–89. McGraw-Hill: New York.

Catchpole WR, Alexander ME, Gill AM (1992) Elliptical-fire perimeter and area-intensity distributions. *Canadian Journal of Forest Research* **22**, 968–972. doi:10.1139/x92-129

Chatto K, Bell T, Kellas J (2003) 'Effects of repeated low-intensity fire on tree growth and bark in a mixed eucalypt foothill forest in south-eastern Australia.' Victorian Government Department of Sustainability and Environment: Melbourne.

Cheney NP, Gould JS, Catchpole WR (1998) Prediction of fire spread in grasslands. *International Journal of Wildland Fire* **8**, 1–13. doi:10.1071/WF9980001

Close DC, Davidson NJ, Johnson DW, Abrams MD, Hart SC, Lunt ID, Archibald RD, Horton B, Adams MA (2009) Premature decline of *Eucalyptus* and altered ecosystem processes in the absence of fire in some Australian forests. *Botanical Review* **75**(2), 191–202 [In English]. doi:10.1007/s12229-009-9027-y

Commonwealth of Australia (2013) Carbon Credits (Carbon Farming Initiative) Reduction of Greenhouse Gas Emissions through Early Dry Season Savanna Burning—1.1) Methodology Determination 2013. www.comlaw.gov.au/Series/F2013L01165

Cook GD (2003) Fuel dynamics, nutrients and atmospheric chemistry. In *Fire in Tropical Savannas: The Kapalga Fire Experiment*. (Eds AN Andersen, GD Cook, RJ Williams) pp. 47–58. Springer: New York.

Cook GD, Corbett LK (2003) Kapalga and the fire experiment. In *Fire in Tropical Savannas: The Kapalga Fire Experiment*. (Eds AN Andersen, GD Cook, RJ Williams) pp. 15–32. Springer: New York.

Cook G, Liedloff A (2000) Simulating the effects of the last hundred years of fire management and rainfall variability in north Australia. In *Australian Rangeland Society Centenary Symposium*, 2000, Broken Hill. (Eds S Nicolson and J Noble) pp. 55–59.

Cook GD, Liedloff AC (2001) Simulating fire ecology and fire management in north Australia. In 'International Congress of Modelling and Simulation: MODSIM 2001', 2001, Canberra, pp. 961–966.

Cook GD, Liedloff AC, Edwards A (2000) Applying a stand model of savanna trees at landscape and regional scales. 4th International Conference on Integrating GIS and Environmental Modeling: Problems, Perspectives and Research Needs. Banff, Alberta, Canada, 2–8 September 2000.

Cook GD, Williams RJ, Hutley LB, O'Grady AP, Liedloff AC (2002) Variation in vegetative water use in the savannas of the North Australian Tropical Transect. *Journal of Vegetation Science* **13**, 413–418. doi:10.1111/j.1654-1103.2002.tb02065.x

Cook GD, Liedloff AC, Eager RW, Chen X, Williams RJ, O'Grady AP, Hutley LB (2005) The estimation of carbon budgets of frequently burnt tree stands in savannas of northern Australia using allometric analysis and isotopic discrimination. *Australian Journal of Botany* **53**, 621–630. doi:10.1071/BT04150

Enquist BJ, West GB, Charnov EL, Brown JH (1999) Allometric scaling of production and life-history variation in vascular plants. *Nature* **401**, 907–911. doi:10.1038/44819

Fensham RJ (2005) Monitoring standing dead wood for carbon accounting in tropical savanna. *Australian Journal of Botany* **53**(7), 631–638. doi:10.1071/BT04109

Fensham RJ, Bowman DMJS (1992) Stand structure and the influence of overwood on regeneration in tropical eucalypt forest on Melville Island. *Australian Journal of Botany* **40**, 335–352. doi:10.1071/BT9920335

Fensham RJ, Fairfax RJ (2007) Drought-related tree death of savanna eucalypts: Species susceptibility, soil conditions and root architecture. *Journal of Vegetation Science* **18**(1), 71–80. doi:10.1111/j.1654-1103.2007.tb02517.x

Fensham RJ, Fairfax RJ, Archer SR (2005) Rainfall, land use and woody vegetation cover change in semi-arid Australian savannas. *Journal of Ecology* **93**, 596–606. doi:10.1111/j.1365-2745.2005.00998.x

Fensham RJ, Fairfax RJ, Ward DP (2009) Drought-induced tree death in savanna. *Global Change Biology* **15**, 380–387. doi:10.1111/j.1365-2486.2008.01718.x

Gill AM, Cheney NP, Walker J, Tunstall BR (1986) Bark losses from two eucalypt species following fires of different intensities. *Australian Forest Research* **16**(1), 1–8.

Gould JS, McCaw WL, Cheney NP, Ellis PF, Knight IK, Sullivan AL (2007) 'Project Vesta – Fire in dry eucalypt forest: fuel structure, fuel dynamics and fire behaviour.' Ensis-CSIRO, and WA Department of Environment and Conservation: Canberra and Perth.

Hutley LB, Beringer J (2011) Disturbance and climatic drivers of carbon dynamics of a north Australian tropical savanna. In *Ecosystem Function in Savannas: Measurement and Modeling at Landscape to Global Scales*. (Eds MJ Hill, NP Hanan) pp. 57–76. CRC Press: Boca Raton, FL.

Hutley LB, Beringer J, Isaac PR, Hacker JM, Cernusak LA (2011) A sub-continental scale living laboratory: spatial patterns of savanna vegetation over a rainfall gradient in northern Australia. *Agricultural and Forest Meteorology* **151**(11), 1417–1428. doi:10.1016/j.agrformet.2011.03.002

IPCC (1996) 'Revised 1996 IPCC guidelines for national greenhouse gas inventories.' Intergovernmental Panel on Climate Change: Exeter, UK.

IPCC (2006) '2006 IPCC guidelines for National Greenhouse Gas Inventories Volume 4 Agriculture, Forestry and Other Land Use.' Institute for Global Environmental Strategies: Hayama, Japan.

Jamali H, Livesley SJ, Hutley LB, Fest B, Arndt SK (2013) The relationships between termite mound CH_4/CO_2 emissions and internal concentration ratios are species specific. *Biogeosciences* **10**(4), 2229–2240. doi:10.5194/bg-10-2229-2013

Lehmann CER, Anderson TM, Sankaran M, Higgins SI, Archibald S *et al.* (2014) Savanna vegetation–fire–climate relationships differ among continents. *Science* **343**(6170), 548–552. doi:10.1126/science.1247355

Liedloff AC, Cook GD (2007) Modelling the effects of rainfall variability and fire on tree populations in an Australian tropical savanna using the FLAMES simulation model. *Ecological Modelling* **201**, 269–282. doi:10.1016/j.ecolmodel.2006.09.013

Liedloff AC, Cook GD (2011) The interaction of fire and rainfall variability on tree structure and carbon fluxes in savannas: Application of the FLAMES model. In *Ecosystem Function in Savannas: Measurement and Modeling at Landscape to Global Scales.* (Eds MJ Hill, NP Hanan) pp. 293–308. CRC Press: Boca Raton, FL.

Lonsdale WM, Braithwaite RW (1991) Assessing the effects of fire on vegetation in tropical savannas. *Australian Journal of Ecology* **16**(3), 363–374. doi:10.1111/j.1442-9993.1991.tb01064.x

Mackensen J, Bauhus J, Webber E (2003) Decomposition rates of coarse woody debris – A review with particular emphasis on Australian tree species. *Australian Journal of Botany* **51**(1), 27–37. doi:10.1071/BT02014

Murphy B, Russell-Smith J, Watt FA, Cook GD (2009) Fire management and woody biomass carbon stocks in mesic savannas. In *Culture, Ecology and Economy of Fire Management in North Australian Savannas: Rekindling the* Wurrk *Tradition.* (Eds J Russell-Smith, P Whitehead, P Cooke) pp. 361–378. CSIRO Publishing: Melbourne.

Murphy BP, Russell-Smith J, Prior LD (2010) Frequent fires reduce tree growth in northern Australian savannas: implications for tree demography and carbon sequestration. *Global Change Biology* **16**(1), 331–343. doi:10.1111/j.1365-2486.2009.01933.x

Olson JS (1963) Energy storage and the balance of producers and decomposers in ecological systems. *Ecology* **44**, 322–331. doi:10.2307/1932179

Prior LD, Eamus D (2000) Seasonal changes in hydraulic conductance, xylem embolism and leaf area in *Eucalyptus tetrodonta* and *Eucalyptus miniata* saplings in a north Australian savanna. *Plant, Cell & Environment* **23**, 955–965. doi:10.1046/j.1365-3040.2000.00612.x

Prior LD, Eamus D, Duff GA (1997) Seasonal and diurnal patterns of carbon assimilation, stomatal conductance and leaf water potential in *Eucalyptus tetrodonta* saplings in a wet–dry savanna in Northern Australia. *Australian Journal of Botany* **45**, 241–258. doi:10.1071/BT96017

Prior LD, Brook BW, Williams RJ, Werner PA, Bradshaw CJA, Bowman DMJS (2006) Environmental and allometric drivers of tree growth rates in a north Australian savanna. *Forest Ecology and Management* **234**(1–3), 164–180. doi:10.1016/j.foreco.2006.06.034

Prior LD, Murphy BP, Russell-Smith J (2009) Environmental and demographic correlates of tree recruitment and mortality in north Australian savannas. *Forest Ecology and Management* **257**(1), 66–74. doi:10.1016/j.foreco.2008.08.015

Richards SA (2005) Testing ecological theory using the information-theoretic approach: examples and cautionary results. *Ecology* **86**, 2805–2814. doi:10.1890/05-0074

Russell-Smith J, Edwards AC (2006a) Seasonality and fire severity in savanna landscapes of monsoonal northern Australia. *International Journal of Wildland Fire* **15**, 541–550. doi:10.1071/WF05111

Russell-Smith J, Edwards AC (2006b) Seasonality and fire severity in savanna landscapes of monsoonal northern Australia. *International Journal of Wildland Fire* **15**(4), 541–550. doi:10.1071/WF05111

Russell-Smith J, Whitehead PJ, Cook GD, Hoare JL (2003) Response of *Eucalyptus*-dominated savanna to frequent fires: lessons from Munmarlary, 1973–1996. *Ecological Monographs* **73**, 349–375. doi:10.1890/01-4021

Russell-Smith J, Edwards AC, Woinarski JCZ, McCartney J, Kerin S, Winderlich S, Murphy BP, Watt FA (2009a) Fire and biodiversity monitoring for conservation managers: a 10-year assessment of the 'Three Parks' (Kakadu, Litchfield and Nitmiluk) program. In *Culture, Ecology and Economy of Fire Management in North Australian Savannas: Rekindling the* Wurrk *Tradition.* (Eds J Russell-Smith, PJ Whitehead, PM Cooke) pp. 257–285. CSIRO Publishing: Melbourne.

Russell-Smith J, Murphy BP, Meyer CP, Cook GD, Maier S, Edwards AC, Schatz J, Brocklehurst P (2009b) Improving estimates of savanna burning emissions for greenhouse accounting in northern Australia: limitations, challenges, applications. In *Culture, Ecology and Economy of Fire Management in North Australian Savannas: Rekindling the* Wurrk *Tradition.* (Eds J Russell-Smith, P Whitehead, P Cooke) pp. 329–360. CSIRO Publishing: Melbourne.

Russell-Smith J, Murphy BP, Meyer MCP, Cook GD, Maier S, Edwards AC, Schatz J, Brocklehurst P (2009c) Improving estimates of savanna burning emissions for greenhouse accounting in northern Australia: limitations, challenges, applications. *International Journal of Wildland Fire* **18**, 1–18. doi:10.1071/WF08009

Russell-Smith J, Cook GD, Cooke PM, Edwards AC, Lendrum M, Meyer CP, Whitehead PJ (2013) Managing fire regimes in north Australian savannas: applying Aboriginal approaches to contemporary global problems. *Frontiers in Ecology and the Environment* **11**(s1), e55–e63. doi:10.1890/120251

Russell-Smith J, Edwards A, Woinarski J, Fisher A, Murphy B, Lawes M, Crase B, Thurgate N (2014) North Australian tropical savannas: the Three Parks Savanna Fire Effects Plot Network. In *Biodiversity and Environmental Change: Monitoring, Challenges and Direction.* (Eds D Lindenmayer, E Burns, N Thurgate, A Lowe) pp. 335–378. CSIRO Publishing: Melbourne.

West GB, Brown JH, Enquist BJ (1997) A general model for the origin of allometric scaling laws in biology. *Science* **276**, 122–126. doi:10.1126/science.276.5309.122

West GB, Brown JH, Enquist BJ (1999) A general model for the structure and allometry of plant vascular systems. *Nature* **400**, 664–667. doi:10.1038/23251

Williams RJ, Cook GD, Gill AM, Moore PHR (1999) Fire regime, fire intensity and tree survival in a tropical savanna in northern Australia. *Australian Journal of Ecology* **24**, 50–59. doi:10.1046/j.1442-9993.1999.00946.x

Williams RJ, Gill AM, Moore PHR (2003a) Fire behavior. In *Fire in Tropical Savannas: The Kapalga Experiment.* (Eds AN Andersen, GD Cook, RJ Williams) pp. 33–46. Springer: New York.

Williams RJ, Muller WJ, Wahren C-H, Setterfield SA, Cusack J (2003b) Vegetation. In *Fire in Tropical Savannas: The Kapalga Experiment.* (Eds AN Andersen, GD Cook, RJ Williams) pp. 79–106. Springer: New York.

Williams RJ, Hutley LB, Cook GD, Russell-Smith J, Edwards A, Chen X (2004) Assessing the carbon sequestration potential of mesic savannas in the Northern Territory, Australia: approaches, uncertainties and potential impacts of fire. *Functional Plant Biology* **31**, 415–422. doi:10.1071/FP03215

Williams RJ, Carter J, Duff GA, Woinarski JCZ, Cook GD, Farrer SL (2005a) Carbon accounting, land management, science and policy uncertainty in Australian savanna landscapes: introduction and overview. *Australian Journal of Botany* **53**, 583–588. doi:10.1071/BT05181

Williams RJ, Zerihun A, Montagu KD, Huffman M, Hutley LB, Chen X (2005b) Allometry for estimating aboveground tree biomass in tropical and subtropical eucalypt woodlands: towards general predictive equations. *Australian Journal of Botany* **53**(7), 607–619. doi:10.1071/BT04149

Woldendorp G, Keenan RJ (2005) Coarse woody debris in Australian forest ecosystems: A review. *Austral Ecology* **30**(8), 834–843. doi:10.1111/j.1442-9993.2005.01526.x

14

Towards a methodology for increased carbon sequestration in dead fuels through implementation of less severe fire regimes in savannas

Garry D Cook, Adam C Liedloff and Brett P Murphy

Summary

An approach to developing a methodology for carbon sequestration in dead fuels through the implementation of less severe fire regimes in savannas of northern Australia is described. The approach is intended to be complementary to the existing emissions abatement methodology. It accounts for carbon stock changes in above-ground fine and coarse fuels as a result of changing the frequency of fires and the proportion of early dry season fires.

Introduction

This chapter outlines an approach to develop a methodology for carbon sequestration under Australia's Carbon Farming Policy (CFP) framework. The proposed methodology would apply to projects aiming to increase the above-ground carbon stock through the implementation of less severe fire regimes across the tropical savannas in Australia that receive more than 1000 mm long-term average annual rainfall. It is intended to complement the 'Reduction of Greenhouse Gas Emissions through Early Dry Season Savanna Burning' carbon farming methodology (Commonwealth of Australia 2012). As such, it adopts the wording and procedures in that methodology where possible. Hereafter, that methodology will be referred to as the Savanna Burning Emissions Abatement Methodology. The main difference between this proposed methodology and the Savanna Burning Emissions Abatement Methodology is that this is a carbon sequestration methodology, not an emissions abatement methodology. Thus, in order to meet the permanency requirements of the CFP, the project fire regime resulting in the deemed sequestration would need to be maintained for 25 years. The annual carbon sequestration will be assumed to apply after 3 years of demonstrating that a new fire regime can be maintained. The amount of annual sequestration will be the difference between the stock at equilibrium with the baseline regime, and with the project regime divided by the time deemed to achieve that equilibrium and expressed as CO_2-equivalents (CO_2-e). The methodology applies to open forest and woodland savannas in high rainfall areas (>1000 mm).

The proposed methodology is based on modelling of the dynamics of above-ground fine and coarse fuel dynamics in eucalypt open forest using the FLAMES simulation model. It does not use the model itself, but a generalised algorithm based on simulation results that have been published (Cook *et al.*, Chapter 13).

Methodology scope

The project must apply strategic early dry season fire management in the project area. It must not use other indirect methods for reducing the fire area, such as the introduction of cattle to a project area. It cannot reduce the fire area within the project area by inducing an increase in late dry season fires or overall fire frequency in areas bordering the project area. Monthly fire maps must be produced using the same approach as the Savanna Burning Emissions Abatement Methodology.

Emissions credits obtained under the project will only apply to those parts of the project area containing open forest (with mixed tussock and hummock grassy fuels) or woodland (either hummock, or mixed tussock and hummock, grassy fuels). These vegetation types are identical to those defined in the revised Savanna Burning Emissions Abatement Methodology and must be mapped using the approach outlined in that methodology.

The greenhouse gases assessed in the proposed methodology are the deemed CO_2-equivalents of the carbon stock in above-ground vegetation during the baseline period, and during the project crediting period. The carbon budget must also account for the emissions of CO_2, CH_4 and N_2O from fuel use to establish and maintain the project (e.g. for helicopters and other energy-consuming equipment or drip torches) unless they have been accounted for under an accepted Savanna Burning Emissions Abatement project.

Calculation of carbon stock for the baseline period and the project period

From the analyses of Cook, Meyer and Liedloff (unpublished), the equilibrium total above-ground mass of fine and coarse fuel (F_F and F_C respectively: t ha^{-1}) for an area of open forest can be calculated for fire frequencies (Freq: yr^{-1}) and proportions of early dry season fires (P_E) using the following equations:

$$F_C = (1.33 + 3.63)\text{Freq}^{(0.081P_E - 0.54)} \quad (1)$$

$$F_F = (1.4P_E - 7.52)\text{Freq} + 0.21P_E + 8.6 \quad (2)$$

The corresponding equations for both woodland mixed and woodland hummock are:

$$F_C = (0.73P_E + 2)\text{Freq}^{(0.081P_E - 0.54)} \quad (3)$$

$$F_F = (0.92P_E - 4.8)\text{Freq} + 0.14P_E + 5.47 \quad (4)$$

The values of Freq and P_E for the baseline period and project period can be substituted into these equations to calculate the steady-state total above-ground mass in dead fuels. The definition of the baseline is the same as that in the Savanna Burning Emissions Abatement Methodology.

The total above-ground carbon stock will be calculated for the baseline regime and the project regime. The difference in stock is deemed to be the impact of the change in management at steady-state. The carbon contents (CC) for both fine and coarse fuel are deemed to be 0.46, the same as those for all fuels in the Savanna Burning Emissions Abatement Methodology.

Based on modelling of fine fuel dynamics, the change in steady-state fuel loads in response to a change in fire regime occurs within about a decade. Thus, a conservative estimate of the rate of change can be calculated by assuming that the carbon stock change occurs uniformly over a 25-year crediting period relevant to native forest protection projects. The annual carbon sequestration (S_F) in fine fuel for either forest or woodland sites expressed as CO_2-equivalents (t CO_2-e ha^{-1} yr^{-1}) can therefore be calculated as:

$$S_F = \frac{44 \times CC(F_F(\text{project}) - F_F(\text{baseline}))}{12 \times 25} \quad (5)$$

The calculations for coarse fuel are similar. Total sequestration is estimated by adding the fine and coarse sequestration rates for forest and woodland and multiplying by the areas (ha) of each vegetation type. Only 95% of this sequestration rate will be available for crediting purposes, which allows a 5% risk buffer.

Project eligibility

A project would be eligible if it encompasses open forest or woodland savannas receiving more than 1000 mm long-term annual average rainfall, as defined in the savanna burning emissions abatement methodology (Commonwealth of Australia 2012). The proponents would need to demonstrate that a new fire regime can be sustained and will achieve carbon sequestration through changes in fuel loads. Maintenance of the project regime for 3 years should be sufficient to demonstrate this capability. The technologies to achieve such changes in fire regimes as described in the Savanna Burning Emissions Abatement Methodology are applicable to this proposal.

Permanence and the time to reach equilibrium

Unlike forestry plantations, this proposed methodology alters the management of a multi-age stand of trees and the associated fine fuel and coarse woody debris (CWD). Furthermore, in any large project area (i.e. >100 km^2), fires are likely to occur on some of the project area every year. Thus the carbon stock will fluctuate. The sequestration will be the difference in mean carbon stocks under baseline and project fire regimes. However, in any one year carbon stocks may be greater or smaller depending on the vagaries of fire behaviour and the fate of individual live and dead trees. Spatial and temporal variability in stock would make measuring and verifying carbon stocks and ascribing their change to management difficult, if not insurmountable. The observed lack of trend in heavy and coarse fuels with time since fire (Russell-Smith *et al.* 2009) is an expression of these difficulties. Nevertheless, it is logically inconceivable for CWD to not respond to changing fire regimes. Thus the approach taken here is to understand and quantify the mechanisms leading to variation in the components of total carbon stock and incorporate them into a process-based fuel dynamics model rather than to aim to measure changes in carbon stock in the field.

Modelling fuel dynamics under the baseline and with-project fire regimes in the WALFA project area indicate that the changes in total fine and coarse fuel loads with the change in regime occur within a decade (Fig. 14.1). A 10-year baseline period to define the fire regime should therefore be adequate to provide a robust baseline carbon stock in fuels. Producing an average sequestration rate over a 25-year crediting period after a change in fire regime would be a conservative approach that would underestimate the potential sequestration rate. For open forest savannas within the WALFA project area, this would result in a deemed carbon sequestration rate of 0.19 t CO_2-e ha^{-1} yr^{-1}.

Greenhouse gas assessment boundary

The greenhouse gas assessment boundary is the boundary for assessing the changes in greenhouse gas emissions that occur as a result of the project and are within the control of the project proponent. In Table 14.1, we identify, for the fire abatement activity, all emissions sources and sinks directly or indirectly affected by the activity.

Key uncertainties

The estimation of carbon stock change in fuels excludes several key processes of carbon dynamics (Table 14.1). These are excluded due to insufficient data to enable their robust quantification. However, for all of these, exclusion results in a conservative estimate of total sequestration because they would provide an additional source of stored carbon.

The carbon sequestration algorithm depends strongly on the effect of fire regime on CWD dynamics. This in turn is highly dependent on the CWD decomposition rate. This was calculated as the value that was consistent with the sum of the coarse and heavy fuel in eucalypt open forest and woodlands (Table 2 of Commonwealth of Australia 2012), the

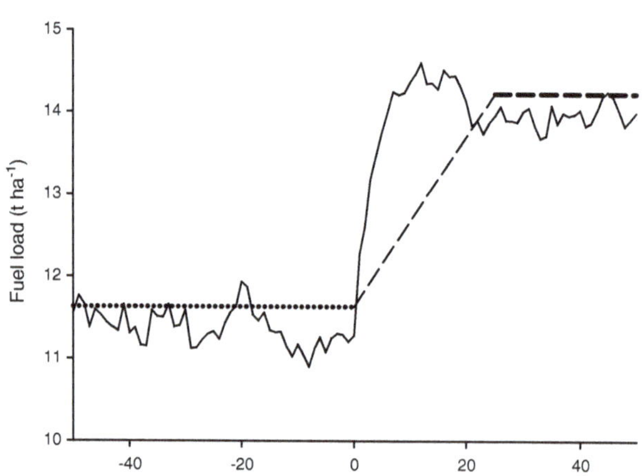

Figure 14.1. Modelled changes in biomass in total fine and coarse fuel in open forests and woodlands subject to the baseline and with-project fire regimes of the WALFA project area. The dotted line represents the steady-state fuel load of the baseline period, while the dashed line represents the fuel load in the with-project period with a deemed sequestration period of 25 years.

Table 14.1. Justification of greenhouse gas assessment boundary for the proposed savanna burning carbon sequestration methodology

Source or sink of carbon	Predicted impact of decreasing fire frequency and increasing proportion of early dry season fires	Included?	References
Live tree stock above-ground	No effect	No	Chapter 12
Coarse woody debris	Increase	Yes	Cook, Meyer and Liedloff (unpublished)
Fine fuel (including tree litter and grass)	Increase	Yes	Russell-Smith *et al.* (2009); Cook, Meyer and Liedloff (unpublished)
Live below-ground vegetation	No effect: likely to respond similarly to live above-ground tree stock	No	
Shrubby fuel	Likely to increase, but unquantified and excluded	No	Murphy *et al.* (2009); Russell-Smith *et al.* (2010)
Below-ground coarse woody debris	Unknown – likely to increase similarly to above-ground CWD	No	
Soil carbon	Likely to peak at a fire frequency of one in five years.	No	Richards *et al.* (2011)
Termite activity	Negligible additional impact because termite activity is latently included in the estimation of decomposition rates	Yes	Cook *et al.* (2005); Jamali *et al.* (2013); Cook, Meyer and Liedloff (unpublished)
CH_4 and N_2O emissions from combustion of fine fuel, shrubs and CWD	Ignored because these are already accounted for in the complementary savanna burning emissions abatement methodology	No	Commonwealth of Australia (2012)
Emissions of CO_2 and other greenhouse gases from fire management activities using vehicles	Relatively small emissions from fire management activities accounted as per the complementary savanna burning emissions abatement methodology	Yes	Commonwealth of Australia (2012)

baseline fire regime of the WALFA project area (Russell-Smith *et al.* 2013), the burning efficiency of heavy fuel and fire patchiness (Russell-Smith *et al.* 2009) and the annual production of CWD from trees that had been killed, as derived from the stand structure following the approach of Cook, Meyer and Liedloff (unpublished). Each of these measures has its own uncertainty, but there are also additional processes affecting CWD dynamics that have not been considered.

The most important process that has not been considered is the loss of branches from live trees and their subsequent input to the CWD pool. The inverse modelling to calculate the CWD decomposition rate in the present study assumed all CWD inputs derived from the death of whole trees, and thus underestimated the rate of creation. This assumption was made due to the lack of any data on branch fall rates from live trees. The underestimation of CWD input rates would lead to an underestimation of CWD decomposition rates. However, the effects of these underestimations should not affect the equilibrium value of CWD after long periods without fire, but only the rate at which the system approaches equilibrium. Given that the time for a change in fire regime to cause a change in carbon stock is substantially less than the 25 years over which we recommend the rate of change be calculated (Fig. 14.1), this should have negligible impact on the deemed sequestration rates.

The FLAMES model currently treats CWD as one pool. The burning efficiency of that pool was parameterised the using values for the heavy fraction (≥5 cm diameter) defined by Russell-Smith *et al.* (2009a). Although data are provided on stocks and burning efficiency of the coarse fraction (≥0.6 cm–<5 cm diameter) of total CWD, there are no data available to estimate the inputs of the coarse and heavy fractions. Treating CWD as one pool, parameterisation using the heavy fraction is likely to lead to an underestimation of the potential impact of a more benign fire regime on total CWD stocks. Thus, this is also a conservative assumption.

Shrub dynamics have been ignored in our estimate of carbon sequestration produced here. They comprise ~12% of available fuels in the open forest systems (Commonwealth of Australia 2012). It is likely that shrub carbon stocks will increase under a more benign fire regime and thus their exclusion is a conservative assumption.

Conclusion

An algorithm has been produced that enables a deemed estimation of the equilibrium carbon stock of open forest and woodland communities in Australian tropical savannas with more than 1000 mm mean annual rainfall. The algorithm requires input of the mean fire frequency and the mean proportion of early dry season (pre-August) fires. The carbon sequestration resulting from changing to a more benign fire regime is achieved well within a 25-year crediting period, so dividing the deemed change in carbon stock by 25 years produces a conservative estimate of the annual rate of sequestration. There are several key uncertainties and exclusions from the estimated sequestration, but these all lead to a conservative estimate of sequestration. The algorithms are a logical extrapolation of the assumptions in the Savanna Burning Emissions Abatement Methodology.

Application of the carbon sequestration algorithm to the change in fire regime achieved by the WALFA project (Russell-Smith *et al.* 2013) indicates a carbon sequestration rate in total fine and coarse fuel in open forest and woodland savannas of 0.19 t CO_2-e ha^{-1} yr^{-1} for 25 years.

References

Commonwealth of Australia (2012) Carbon Farming (Reduction of Greenhouse Gas Emissions through Early Dry Season Savanna Burning) Methodology Determination 2012. Federal Register of Legislative Instruments. Vol. F2012L01499, p. 37. Australian Government: Canberra.

Cook GD, Liedloff AC, Eager RW, Chen X, Williams RJ, O'Grady AP, Hutley LB (2005) The estimation of carbon budgets of frequently burnt tree stands in savannas of northern

Australia using allometric analysis and isotopic discrimination. *Australian Journal of Botany* **53**, 621–630. doi:10.1071/BT04150

Jamali H, Livesley SJ, Hutley LB, Fest B, Arndt SK (2013) The relationships between termite mound CH_4/CO_2 emissions and internal concentration ratios are species specific. *Biogeosciences* **10**(4), 2229–2240. doi:10.5194/bg-10-2229-2013

Murphy B, Russell-Smith J, Watt FA, Cook GD (2009) Fire management and woody biomass carbon stocks in mesic savannas. In *Culture, Ecology and Economy of Fire Management in North Australian Savannas: Rekindling the* Wurrk *Tradition*. (Eds J Russell-Smith, P Whitehead, P Cooke) pp. 361–378. CSIRO Publishing: Melbourne.

Richards AE, Cook GD, Lynch BT (2011) Optimal fire regimes for soil carbon storage in tropical savannas of northern Australia. *Ecosystems* **14**, 503–518. doi:10.1007/s10021-011-9428-8

Russell-Smith J, Murphy BP, Meyer MCP, Cook GD, Maier S, Edwards AC, Schatz J, Brocklehurst P (2009) Improving estimates of savanna burning emissions for greenhouse accounting in northern Australia: limitations, challenges, applications. *International Journal of Wildland Fire* **18**, 1–18. doi:10.1071/WF08009

Russell-Smith J, Price OF, Murphy BP (2010) Managing the matrix: decadal responses of eucalypt-dominated savanna to ambient fire regimes. *Ecological Applications* **20**(6), 1615–1632. doi:10.1890/09-1553.1

Russell-Smith J, Cook GD, Cooke PM, Edwards AC, Lendrum M, Meyer CP, Whitehead PJ (2013) Managing fire regimes in north Australian savannas: applying Aboriginal approaches to contemporary global problems. *Frontiers in Ecology and the Environment* **11**(s1), e55–e63. doi:10.1890/120251

15

Where to from here?

Jeremy Russell-Smith, Mila Bristow, Peter Brocklehurst, Garry D Cook, Nick Cuff, Andrew C Edwards, Rohan Fisher, Lindsay B Hutley, Peter Jacklyn, Glenn James, Sarah Legge, Adam C Liedloff, Dominique Lynch, Stefan Maier, CP (Mick) Meyer, Catherine Monagle, Brett P Murphy, Sofia Oliveira, Peter J Whitehead and Cameron P Yates

Summary

This final chapter provides an overview and discussion of various significant issues and ongoing challenges that have been identified in, or arise from, discussion in the earlier chapters. Our coverage addresses a variety of key topical areas, including development and presentation of integral mapping products, ongoing abatement and sequestration methodology challenges, and challenges associated with project implementation, governance, and supportive policy settings. We identify several critical issues and sources of uncertainty that should be the focus of future research effort.

Introduction

Previous chapters: (1) set out a rationale for developing carbon offset methodologies for application in fire-prone northern Australian settings, particularly for the benefit of Indigenous landholders, but also for land managers generally in economically and ecologically marginal pastoral enterprise settings; and (2) provide a detailed description of research undertaken towards the development of two new formal carbon accounting methodologies. The first new methodology is an extension of the higher rainfall (>1000 mm per annum [p.a.]) savanna burning abatement methodology to lower rainfall (600–1000 mm p.a) conditions, and the second is a methodology addressing carbon sequestration in woody biomass components due to enhanced fire management under higher rainfall conditions.

The purpose of the present chapter is to highlight significant issues relating to the ongoing development and application of savanna burning projects, principally in fire-prone northern Australian settings, but, by extension, in savanna systems globally. To that end, this chapter identifies key savanna burning challenges in five broad categories:

1. mapping of fire extent, severity and intensity, fuel types and fuel curing, and related challenges associated with maintaining imagery data continuity, and distribution and communication of data products

2. abatement methodology issues, including assessing woody fuel dynamics, emission factors, significance of termite emissions, and reconciliation of 'top down vs. bottom up' accounting approaches
3. ongoing sequestration methodology issues, including challenges associated with reconciling modelling approaches for predicting fire regime effects on sequestration in woody biomass, modelling woody fuel dynamics, use of remote sensing for measuring biomass change, and measuring and predicting carbon sequestration in soils
4. implementation challenges, including the critical importance for developing appropriate project governance arrangements, especially in cross-cultural, multi-partner settings
5. current national and international policy challenges.

Mapping issues

Fire extent mapping

Fire extent mapping is currently a semi-automated process undertaken by experienced spatial scientists (Fisher and Edwards, Chapter 3). While efficiencies in image processing may help reduce the time taken to undertake fire mapping (Chapter 3), available automated fire mapping products are relatively unreliable, especially when mapping small early dry season burnt areas. This has significant implications for operational fire management. We suggest that, for an operational product to be adopted by fire managers, a human operator is required in the mapping process to react to user requirements. Those undertaking the fire mapping report that ~15% of their time is spent liaising with end-users. This high level of user engagement has many benefits. For example, crowd-sourced validation/feedback provides regular checks of the reliability of the mapping product, as well as a communal sense of ownership of the data.

Imagery continuity and requirements

Chapter 3 describes a simple and robust method for mapping burnt areas across northern Australia. The semi-automated approach is shown to be both *quantitatively* and qualitatively better than other publicly available products. While other automated burnt area mapping products provide important broad-scale fire information, they are generally used as a descriptive rather than an operational tool for fire managers. Thus, the role of the burnt area data as provided through the North Australian Fire Information (NAFI) website has a function significantly different from that provided by other sources.

The key issue in maintaining the burnt area mapping is the ongoing availability and pre-processing of base imagery. The current satellites carrying MODIS sensors, while still performing well, are already several years past their planned operational life span (2011 Terra, 2012 Aqua). The failure of the MODIS sensor on the Terra satellite in particular could significantly reduce the frequency of operational fire mapping provided through NAFI. While still providing high quality imagery, the afternoon overpass of the MODIS sensor on the Aqua satellite increases the likelihood of cloud cover interfering with burnt area mapping. NASA's MODIS continuity plan uses the Joint Polar Satellite System (JPPS). Currently operational through the Suomi satellite platform, the Visible Infrared Imaging Radiometer Suite (VIIRS) sensor is providing quality continental-scale imagery at a MODIS-equivalent resolution (400 m). The next-generation JPPS is expected to be launched in 2017.

A second alternative will be the Ocean and Land Colour Instrument (OLCI) and the Sea and Land Surface Temperature Radiometer (SLSTR) on the Sentinel-3 series of satellites. It will be an operational system from the European Space Agency (ESA) and the European Organisation for the Exploitation of Meteorological Satellites (EUMETSAT). The first satellite is scheduled to be launched in 2015. One major difference is the reduced spatial resolution in the bands used for burnt area mapping: VIIRS (340 m) and OLCI (300 m) compared to MODIS (250 m).

Additionally, the pre-processing and serving of the MODIS imagery required for mapping is currently unfunded and thus also very vulnerable to discontinuation. Due to the current high operational dependency of northern Australian land managers on these data, further resilience needs to be built into data processing and storage. Opportunities to institutionalise both these processes and future datasets need to be explored. The semi-automated mapping method is flexible and lends itself to rapid adaptation to alternative sensors. Further operational redundancy is provided by the manual process being built on the use of readily available software and a fairly simple set of processing skills. This allows for mapping and maintenance capacity to be shared among several operators.

Maintaining NAFI services

Chapter 5 highlights three important characteristics of the NAFI service that will affect its future. First, the service has engaged effectively with northern fire managers, working with them to develop tools and data that enabled better fire management. In this way, the service is able to build the capacity of northern fire managers to undertake projects of national and international significance. Second, the service was seen to provide a simple web-based map that has become a *lingua franca* for fire management in the north, with all the consequent benefits this has for collaboration. Third, the centralisation of the service at a single location in Darwin was seen as being vulnerable to network and other information technology system failures. This is a concern given the widespread use of the site by the emergency services in landscape-scale fire incident contexts. Any model for sustaining the NAFI service into the future will need to reduce this vulnerability without sacrificing the first two characteristics, which confer significant benefits.

A way forward starts with observing that the NAFI service also comprises two components requiring different skills, knowledge and capacities. One component is centred on engagement and working with fire managers so they can effectively use satellite data, particularly burnt area data, to achieve specific goals. This component is associated with managing natural and cultural resources and emission reduction, and requires skills in remote sensing, mapping, user engagement, web development and northern fire management, plus a knowledge of country. Many elements of this aspect of the NAFI service are not very sensitive to network and hardware failure because they involve occasional downloads of data to users' own software. It should be feasible to sustain this component by drawing on the current capacities of the NAFI service, particularly if funding is linked with emerging offset markets.

The second component involves maintaining the NAFI map viewer with its regularly updated display of hotspots and fire scars. This is the basic service that NAFI started out with – it is very widely used and is relied on by a range of fire managers, including the emergency services in the northern wildfire season. This component has the potential to be significantly affected by network or hardware failure during the fire season, given the need to continuously update the hotpot data and deliver it using a map viewer application. In the future it would be prudent to work with organisations like Landgate WA and Geoscience Australia to support the display of a NAFI-style map viewer on a range of platforms to reduce vulnerability to network and hardware failures.

Fire severity mapping

To date, there has been limited implementation of the fire severity mapping product. Overall, the mapping accuracy varies from ~60 to 70%, far below the mapping accuracy achieved for fire extent mapping (~90%) and the level of accuracy required for vegetation fuel types mapping (>80%). Adding validation data collected since the original 2011 calibration dataset did not markedly improve the accuracy, suggesting that it is not possible to improve the current algorithm using only near-infrared images. The proportional reflectance in the short wave infrared (SWIR) highlights the occurrence of plant moisture (Dennison *et al.* 2005; Peterson *et al.* 2008) and similarly non-photosynthetic vegetation with MODIS channel 6 (Edwards *et al.* 2013). To date, it has not been possible to incorporate this SWIR band into an algorithm due to its malfunction on Aqua, and there is the issue of the lower resolution (500 m) of the SWIR bands.

As noted in Chapter 8, addition of reliable fire severity mapping in subsequent iterations of savanna burning methodologies will negate the requirement to use seasonally based estimates of fire severity to derive burning efficiency factors (BEFs; refer Chapter 6). Rather, fire severity may be derived directly using spatially explicit remotely sensed data with values for each fire-affected pixel. Such fire severity mapping products will also have very significant ecological value, including as input into modelling fire effects scenarios (Russell-Smith *et al.* 2010, 2012; Edwards *et al.* unpublished).

Curing data for fire management operations

Although not directly applicable to savanna burning methodologies, the status of grass fuel curing (i.e. the proportion of dead, relative to live, grass biomass) determined from remote sensing has significant potential for informing fire management operations. Curing data describe three phases: uncured (green) grass at the end of the wet season which will not burn; grasses cured sufficiently to carry low-intensity patchy fires; and fully cured grasses which are most likely to support higher intensity, and potentially very extensive, fires. To date, national curing research has established useful relationships between ground-layer curing and satellite-derived indices only for open grassland conditions in south-eastern Australia (Newnham *et al.* 2010). The majority of Australian landscape-scale fire occurs under tree overstoreys in monsoonal regions of northern and central Australia. A program addressing remote sensing of curing in these landscapes has recently commenced. When operationally developed, it is proposed that curing data will be available through the NAFI website.

Potential for direct measurement of fire intensity

Currently, fire severity is used to determine the burning efficiency and infer the proportion of biomass consumed with respect to three fire severity classes (see Chapter 6). The relationship describing biomass burnt with respect to fire intensity (Wooster *et al.* 2003) is much stronger and more direct than that achievable with fire severity (Russell-Smith *et al.* 2009). Using MODIS imagery, fire intensity can be determined directly from the energy released, the so-called 'fire radiative power' (FRP) (Oliveira *et al.* 2015). Currently, FRP is measured by the MODIS sensors, with a 6-hour overpass interval. Geostationary satellites with sensors capable of thermal infrared measurement would provide near-continuous fire intensity measurements for fire-affected pixels and hence allow for more reliable continuous measurement of biomass consumption – albeit at coarser spatial resolutions than achievable with MODIS (Roberts *et al.* 2005), and with a consequent reduction in the reliability of detecting smaller, less intense fires (FRP <100 MW).

Vegetation fuels mapping

In Chapter 4, it is suggested there is the potential to derive fuel-type mapping from existing vegetation mapping sources. Our experience of this process, and that of others (Adamo *et al.* 2014; Comber *et al.* 2004), is that there is still substantial field calibration required to deliver reliable vegetation mapping products. For example, Queensland's Regional Ecosystems mapping provides descriptions for five strata in each unit (Neldner *et al.* 2005). Desktop re-combination of these strata in the absence of calibration may, for example, result in the same vegetation being described either as open forest or as woodland. Hence, re-interpreting the data without recourse to calibration data is not effective. Applying a standard fuels mapping method, using MODIS imagery (Edwards and Russell-Smith 2009), has required extensive field-sourced data for both calibration and validation in an iterative process when applied across the higher rainfall section of Western Australia's Kimberley region. Our experience demonstrates that there is no fit-for-purpose map that can be derived for the whole region without substantial field assessment.

Savanna burning abatement issues

Fuel dynamics and accounting

Chapters 6, 7 and 14 address fuel load dynamics. Rather than presenting competing descriptions of fuel dynamics they are complementary, and in combination they open pathways to the next stages of national and project-based accounting. Carbon accounting for national greenhouse gas reporting and management is primarily concerned with increasing the precision of emission estimates. Increases in absolute accuracy occasionally occur when new knowledge shows that previous estimates are in error due to missing sources or poorly quantified processes. For the most part, however, aggregated national emission estimates have proven to be relatively accurate, from the commencement of national reporting. Most of the improvements to inventory estimates have come about by reducing uncertainties in parameter values, thus increasing the precision of the emission estimates. The impetus for this has been the development of emission reductions strategies through fire management. These require methodologies that explicitly account for the management mechanism, and for the abatement to be reported in the national aggregate; the complexity of algorithms in the national methodology must increase to the same level as the project methodologies. However, it is also important that the complexity of the methodology matches, but does not exceed, the detail of the available activity data; to do otherwise increases the cost of the accounting without compensating returns in precision.

The first major improvement for savanna fire accounting was the inclusion of an additional emission source, coarse woody debris (CWD), which improved overall accuracy of the accounts, and fire seasonality, which allowed fire management activities to be accounted explicitly. Described by Russell-Smith *et al.* (2009), the improved algorithm required description of fire seasonality, vegetation class and fuel size class, and the evaluation of all parameters for those strata. It was possible to evaluate all parameters by direct measurement. This avoided the need to formulate and verify models of fire and fuel dynamics, which could have increased, rather than decreased, uncertainty in the emission estimates. However, this restricted the application of the methodology to the high rainfall (>1000 mm p.a.) regions where the parameters were measured. Chapter 6 extends the range of application of the methodology to the low rainfall (<1000 mm p.a.) savanna woodlands, and by increasing the number and geographic spread of independent measurements reduces the risk of spatial bias. Most importantly, it provides an extensive database to constrain models of fuel accumulation and ecosystem production. However, the emissions abatement methodology accounts only for the direct emissions of non-CO_2 greenhouse gases to the atmosphere from the fire events. Accounting for sequestration of carbon into dead fine and coarse biomass requires some further elaboration, which was provided in Chapter 14.

Emission factors

There has been more than a century of research on growth and decay of woody biomass in forests, and it shows; the ecological processes are reasonably well understood, the remotely sensed products are robust, and, in general, the estimates of stocks and fluxes are adequately constrained. This is not the case for combustion processes; described in Chapter 9, the knowledge of emission factors continues to rely on field measurements because the combustion processes cannot yet be defined and simulated accurately in the laboratory. The challenge for the future is to develop general methods for predicting modified combustion efficiency [MCE], particularly using remotely sensed products; however, there are no solutions on the horizon. To be useful at the precision required, the remotely sensed measurements of total carbon emission or CO and CO_2 concentrations would need to be simultaneous and coincident, and techniques for deconvoluting mean plume MCE values to MCE for the different fuel size classes will need to be developed. The issue for inorganic species, particularly nitrous oxide (N_2O), is more challenging (Chapter 9). Perhaps the most surprising finding is the lack of research focus on emission factors for this species; almost no new published measurements were made between 1994 and 2008, when work commenced in northern Australia, in contrast to the extensive studies of the organic trace species. Hence, at the moment, we remain reliant on empirical measurements.

Significance of termite emissions

In addition to methane (CH_4) produced by fires, a currently unaccounted for source of CH_4 flux is derived from the substantial termite biota that occur across the savanna and rangelands of northern Australia (Dawes-Gromadzki 2005). Termites produce significant CH_4 fluxes as part of a fermentation process during digestion of organic matter. Termites are found in four nesting habits: epigeal (mounds above ground), hypogeal (subterranean nests), nests within woody debris, and arboreal nests. Termite mounds are a strong source of CH_4 relative to the soil, which tends to be a net sink for methane in these savannas (Jamali *et al.* 2011b). Jamali *et al.* (2011b) estimated CH_4 emissions from epigeal and arboreal termites at a high rainfall site near Darwin at +8 kg CO_2-e ha^{-1} yr^{-1}. These emissions were modest when compared to CH_4 emission from savanna fires. Comparison with the West Arnhem Land Fire Abatement (WALFA) project's pre- and post-management implementation suggests that termite methane emissions are equivalent to between about 7 and 11% of annual fire emissions. If the natural soil CH_4 sink is included, the net CH_4 emission from soils plus termite mounds is a sink of 30.1 kg CO_2-e ha^{-1} yr^{-1} (Jamali *et al.* 2011b) and we can conclude fire dominates savanna emissions.

There is significant uncertainty associated with these scaled estimates, given the paucity of sites where mound and arboreal fluxes per mound area have been quantified. There is between 5 and 25-fold difference in mound fluxes from wet to dry season, depending on species. Seasonal differences must therefore be quantified to scale on an annual basis with any confidence. In Australian savannas, there have been few studies where these fluxes have been quantified, and all studies have been undertaken in high rainfall (~1000–1700 mm p.a.) areas (Jamali *et al.* 2011a). The sites used by Jamali and colleagues were all frequently burnt (frequency ~0.7 fires yr^{-1}) and it is unknown what the impact of a shift in fire regime from late dry season (LDS) to early dry season (EDS) would do to termite abundance and subsequent emissions. Reductions in fire frequency may result in fuel accumulation, providing both wood and grass substrate to support higher levels of termite emissions.

There are no data describing fluxes in lower rainfall areas (<1000 mm p.a.), where anecdotal observations suggest termite mound abundance can be significantly higher than those observed in high rainfall areas. Grazing and fire interactions and termite abundance in lower rainfall areas are also not well understood. While emissions per mound volume and/or area would likely be lower in low rainfall areas, this may be offset by a far higher mound density and basal area. Hence, despite some available data in high rainfall areas, there is still significant uncertainty when scaling to the entire 1.9 million km^2 savanna biome. Further work is needed to examine fluxes and mound densities in these regions to improve confidence in the notion that termite and soil emissions are trivial when compared to fire-derived emissions.

Top down vs. bottom up

All of the methodologies presented in this volume are 'bottom up' accounts which estimate emissions from estimated flux rates or integrals of flux rates. The non-CO_2 greenhouse gas account of the low rainfall emissions abatement methodology is an example of the former; the sequestration methodology is an example of the latter. The accuracy of these estimates depends directly on the accuracy with which the algorithm describes the emission process, and on the accuracy of each of the many parameters defined in the algorithm. The uncertainties in both model and parameter values are often substantial, and therefore so too are the emission estimates. Intergovernmental Panel on Climate Change (IPCC) good practice requires that emission estimates are verified by independent methods. Chapter 11 discusses the potential for applying 'top down' methods, which estimate the emission rate from the smoke concentrations observed in the atmosphere, to verify or independently estimate emissions. The results are encouraging. Regionally and seasonally, the observed smoke concentrations are consistent with the concentrations we would expect if 'bottom up' emission estimates are correct. Directly estimating emissions from either surface observations of smoke concentrations or satellite observations is more problematic. Although good, the atmospheric transport models are far from perfect, and at present the remotely sensed observations of aerosol optical depth (AOD) over northern Australia are neither sufficiently accurate nor sufficiently extensive to provide an alternative to 'bottom up accounts'. The situation could change quickly, however, as new remotely sensed products are developed, and as the accuracy and coverage of currently available products (such as fire radiative power [FRP]) improves.

Carbon sequestration

Sequestration in living biomass

Though short-term changes in tree biomass in response to fire are relatively well established, we have only a limited capacity to predict the levels at which tree biomass will equilibrate under long-term fire management, and this poses a major obstacle to development of projects that use fire management to sequester carbon into live tree biomass.

Chapter 12 suggests that biomass of the eucalypt overstorey is typically close to carrying capacity, determined by soil water availability, yet the non-eucalypt midstorey tends to be much more sensitive to fire and is typically well below carrying capacity (Cook *et al.*, Chapter 13; Russell-Smith *et al.* 2003; Woinarski *et al.* 2004). A key source of uncertainty is the extent to which the eucalypt overstorey and the non-eucalypt midstorey compete for water. If competition for soil water is strong, then it seems unlikely that the biomass of the non-eucalypt midstorey can increase significantly in the long term without being offset by a reduction in overstorey biomass. However, if the eucalypt overstorey and non-eucalypt midstorey access water from different parts of the soil profile (i.e. midstorey from shallow layers, overstorey from deep layers), or at different times of the year (i.e. the predominantly deciduous midstorey during the wet season only, the evergreen overstorey year-round), then changes in the biomass of one layer could be expected to have little impact on biomass of the other layer.

A conspicuous outcome of Chapter 13 is the discrepancy between the long-term predictions of the Three Parks statistical model and the FLAMES process model. The causes of this discrepancy have yet to be identified, though at least two factors may have been important:

1. The FLAMES model runs only included eucalypts, which tend to be highly fire-resistant, while the Three Parks analysis included non-eucalypts (typically representing ~30% of tree basal area), which tend to be more fire-sensitive and so could be expected to respond more strongly to a reduction in fire frequency if the stands are below carrying capacity.
2. Predictions of the Three Parks model do not take into account potential fire-driven changes in stand structure (e.g. the possible shift towards domination by small trees under a regime of low intensity fires, as suggested by Cook *et al.* Chapter 13), nor the role of homeostatic mechanisms such as water limitation during the long dry season.

Future research should focus on exploring changes in the non-eucalypt midstorey using process models such as FLAMES and validating the somewhat counter-intuitive predictions made in Chapter 13 – that fire management is likely to decrease tree biomass due to increasing domination by small trees. Continued research in this area requires ongoing maintenance of long-term monitoring such as the Three Parks program.

Role of remote sensing for measuring woody biomass change

There are several possible approaches to measuring above-ground woody biomass using remote sensing. The most accurate approach (often exceeding the accuracy of non-destructive field sampling) uses airborne lidar, though this is relatively expensive for larger areas. Other approaches utilise data from satellite-based radar or passive optical sensors. While these approaches deliver reasonable accuracies in case studies focusing on single dates and small areas, their performance with respect to monitoring bio-sequestration efforts in natural environments over larger areas and under varying environmental conditions have yet to be evaluated.

One promising method to estimate the amount of above ground biomass is to place sensors on unmanned aerial vehicles (UAV). This approach overcomes spatial, spectral, and temporal limitations of satellite sensors. It provides a low-cost approach for repeated data collection, and is more flexible in terms of revisit time than any other remote sensing data collection method. UAV equipped with high spatial resolution thermal and spectral sensors can quantify vegetation structure parameters such as leaf area index (Hunt *et al.* 2008), tree stem volume (Salo *et al.* 2012), and biomass (Beeri *et al.* 2005; Hunt *et al.* 2005; Pölönen *et al.* 2012a, b). As technology develops, innovative applications involving both UAV and lidar technologies may provide increasingly accurate and precise data for predicting biomass change.

Sequestration in dead fuel fractions

Quantifying sequestration into the dead woody biomass pool is a long-term process; it requires a precision that is significantly greater than is possible from direct measurements of fuel loads at the landscape scale. Average fuel loads are determined not by the temporal and spatial sequence of fire events, but by fire regimes. Chapter 14 describes this issue and the accounting mechanism. The fire regime integrates numerous fire events across both time and space, and cannot be measured at a single point in time and space. Therefore, accounting for trends in fuel loads, which are strongly affected by fire events, requires a landscape model of fuel dynamics. Cook *et al.* (unpublished) develop a fuel accumulation model described by annual fuel input and decay rate from the local (m^2) scale to the landscape (100 km^2) scale, and show how this model can be parameterised in terms of fire regime. This moves us from direct measurements of fuels and combustion parameters at specified locations in the landscape, to measurements of model parameters for defined fire regimes and regions. Chapter 7 describes how annual fuel input and decay rate vary with annual rainfall, and with vegetation structure, specifically total basal area (TBA) and foliage projected cover (FPC). Robust ecological functional relationships between rainfall, soil type, TBA and FPC can be described for the Australian savannas. Highly prospective remotely sensed products such as fractional cover are already available through the AUSCOVER project (www.auscover.org.au) with others in development, including a TBA product using data from the ALOS satellite. These spatial datasets, combined with the functional relationships for annual litterfall, should allow us to estimate fuel load dynamics throughout the Australian savannas and therefore to implement a carbon sequestration methodology for dead biomass at both project and national scales.

Reality is what is measured on the ground; biomass stocks, in the case of sequestration methodologies, and atmospheric concentrations of smoke constituents in the case of emissions (Chapter 11). Temporal and spatial variability limits the precision with which we can measure the large area averages that are required for inventory accounting, but through the averaging inherent in regression analysis, ecological functional relationships, such as fuel accumulation curve models defined by measured parameters, resolve the problem. However, it needs to be remembered that models are simplifications of reality and that functional relationships developed from decadal datasets may not capture the influences of disturbance events that occur on longer time scales.

Cyclones and severe storm events, for example, can clear large swaths of forest, and in model terms these events transfer large fractions of live biomass instantly to the dead organic matter pools. At the regional scale does this invalidate the application of Olson curves? Hutley *et al.* (2014) analysed the impacts of the most intense tropical cyclone ever recorded in the Australian region (Cyclone Monica), and concluded that while short-term impacts were extreme, on decadal timescales they were less significant; fire frequency and fuel consumption increased transiently, increasing the decay constant of the CWD pool, and annual carbon fluxes had returned to baseline levels in approximately a decade, suggesting that fuel input rates would recover to the area mean in 10–20 years. Also, such events are infrequent and, averaged across the total area of Australian savannas, their relative impact is likely to be small. Regionally averaged fuel accumulation curves, therefore, are unlikely to be invalidated by infrequent disturbance events; however, a detailed analysis to test the question would be prudent.

Sequestration in soils

There is some evidence that increases in tree biomass due to fire management will be accompanied by increases in soil organic carbon (Coetsee *et al.* 2010). Such changes in soil carbon stocks are likely to be very important, given that soil organic carbon is by far the largest pool in these savannas, representing ~75% of ecosystem carbon (Chen *et al.* 2003). Richards *et al.* (2011) used the CENTURY soil carbon model to predict that soil carbon stock increases with decreasing fire frequency, due to higher soil mineralisation rates, lower nitrogen losses from fire, and less leaching of mineral nitrogen. Greater nitrogen availability boosts productivity of woody plants, providing greater inputs of carbon to the soil. However, these authors were unable to detect an effect of fire management on measured soil carbon stocks after 5 years, and suggest that it would take about a century of strongly contrasting fire regimes for changes in soil carbon stocks to be detectable. Empirically demonstrating responses of soil carbon stocks to management is extremely difficult, due to the very high spatial variability of soil carbon stocks (Allen *et al.* 2010) and the slow response times. Accurately accounting for sequestration in soils will require the development of low-cost methods for monitoring soil carbon stocks, and validation of predictive models such as CENTURY for northern Australia.

Implementation

Operational and regulatory – is it worth it?

Without active intervention, contemporary fire regimes across much of northern Australia will further damage habitats and wildlife, reduce the long-term sustainability of native pastures, and degrade cultural and aesthetic values. However, fire management across these vast landscapes requires recurrent resourcing. To varying extents (and with varying degrees of success) ongoing fire management (with recurrent funding) is restricted to individual (and therefore relatively small) parcels of government-owned land. Various state and Commonwealth government grant schemes have also intermittently supported larger-scale fire management programs on non-government land, but these projects are ephemeral. One of the significant potential benefits of savanna burning emission abatement is the opportunity to construct large-scale projects with a longer-term and self-supporting funding model, especially over pastoral and Indigenous tenures. Such large-scale projects are likely to lead to the greatest biodiversity, production and sociocultural benefits over time. Nevertheless, there are some obstacles to capitalising on the opportunity. First, the regulatory and administrative burden of managing a savanna burning emissions reduction project is complex and formidable, requiring a fair amount of technical and industry understanding; this puts the opportunity out of the reach of most northern landholders and managers unless they have access to the appropriate expertise. At the moment, expertise is not readily available. Second, the focus on abatement and carbon credits has the potential to distort the focus of fire management away from primary aims (e.g. of biodiversity conservation, pastoral production), and can also threaten the cohesion of participating entities because actions and outcomes may now have economic consequences that are shared unevenly across the group of participants. Very clear project objectives and governance structures may help alleviate this risk.

Linkages with pastoral management, biodiversity and stewardship

The generation of carbon credits by adopting fire management practices that reduce greenhouse gas emissions and increase carbon sequestration presents a potential opportunity for northern Australian pastoral land managers to diversify their livelihoods (Walsh *et al.* 2014). However, and especially on productive pastoral lands, savanna burning activities need to be considered in the broader context of pastoral production, particularly interactions with woody thickening and grazing management (Cowley *et al.* 2014). To date, there has been little systematic exploration of integrating the potential contributions of savanna burning and related ecosystem services values (e.g. biodiversity benefits, soil and water conservation) as part of the pastoral enterprise.

Fire management opportunities in fire-prone central Australian rangelands

The current methodologies address fire and associated greenhouse gas emissions abatement and carbon sequestration management challenges only under monsoonal climatic conditions as far south as the mean annual 600 mm rainfall isohyet for emissions abatement, and 1000 mm for sequestration. We recognise that this is a somewhat arbitrary climatic management envelope (Whitehead *et al.*, Chapter 2), but predicated on the observation that rainfall is sufficiently reliable to re-set fuel loads annually and allow for the implementation of a mitigating savanna burning fire regime.

However, we recognise also that economic opportunities for supporting strategic fire management are desperately required also under more erratic rainfall event-driven fire regimes in the central Australian rangelands, particularly over very extensive regions dominated by highly flammable spinifex (*Triodia* spp.) grasslands (Allan and Southgate 2002). To that end, we note that the first stage of a feasibility study assessing the potential for implementing an analogous strategic prescribed burning program in central Australian rangelands, based essentially on Indigenous fire management practices, has just been completed (Burrows 2014). That assessment indicates that while there is a significant opportunity for implementing such a program to deliver both abatement and sequestration outcomes, a major challenge involves the vast operational scale required to deliver economic returns. Nonetheless, as for savanna burning projects, such a program would provide very substantial community and biodiversity co-benefits.

Governance challenges for Indigenous enterprises

Customary connection to ancestral lands remains a core feature of 'caring for country' obligations and thereby project governance interests in contemporary Aboriginal land management. Socioeconomic disadvantage, especially in remote regions, limits capacity to manage biophysical, social and customary economic assets at large scales, so opportunities that increase that capacity are highly valued – particularly where strong synergies with supporting agency demands are apparent, as in the current case. Although local knowledge, social organisation, demography and land (where relevant tenure exists) contexts are very rich, the dearth of financial and administrative capacity means development of most savanna burning projects has been dependent on multiple grants, creating complex reporting obligations and uncertainty about capacity to complete project development. As a result, local struggles for political and economic independence have been compromised (see also Thiele 1982).

The operational scale issue is exacerbated by poor recognition of sociocultural demands (Gerritsen 2007; Morphy 2010) and protocols faced by estate-based groups needing (whether wanting or not) to up-scale local governance arrangements. One impact of grant and other resource dependence is the imposition of prescribed outcomes (often driven by pan-Aboriginal or other policy imperatives) and external measures of success. Despite stark differences in 'cultures of governance' (Smith 2008: 76) in this cross-cultural space, best practice endeavours can develop a level of integration of knowledge and technologies (Altman *et al.* 2007) colloquially referred to as the 'two tool box' approach.

Bolstered by the stand-out performance and resilience of the WALFA project and the hard-won first abatement methodology, the Australian Government has committed to expanding enterprise opportunities with a suite of related methodologies aimed at encouraging Indigenous participation in the carbon market. Despite this new impetus for economic engagement from the top down, partnership arrangements like the WALFA project remain challenged to formalise governance arrangements working from the bottom up through a single entity capable of engaging with a carbon buyer, whether through a formal market, a government auction system, or any other centralised body. Other carbon projects from Queensland to the Kimberley face similar issues.

Reconciling local groups' demands for a prosperous cultural future, in which relative independence is a key measure of success, while simultaneously accepting the discipline of coordinated action to access commercial benefits that may appear unrelated to performance of customary obligations, is no trivial task. Where 'healthy country–healthy people' meets fire and carbon farming, governance arrangements must integrate inherently localised decision-making, high quality planning, excellent training, highly coordinated field operations, mechanisms for conflict resolution, and business management arrangements in an enterprise subject to erratically variable incomes and difficult-to-manage cash flows. The conceptual leap to the oft-romanticised self-sustaining enterprise challenges such groups to establish, and deliver against, their own measures of success and capacity (Hunt and Smith 2007).

Project scale, recognition of difference in 'cultures of governance' (Smith 2008), and management of key resources (human, financial, institutional) are governance issues also for governments in the formative development stages. Flexibility and creative and innovative ways to meet Indigenous requests for assistance (Marika *et al.* 2009) are demanded if this unique socioeconomic opportunity for reliable strong Indigenous futures is to be realised.

Policy issues

Emissions Reduction Fund (ERF)

The Emissions Reduction Fund (ERF) is the centrepiece of the current Australian Government's Direct Action Plan to reduce national emissions by 5% by 2020. Put simply, the ERF will buy emission reductions from individuals or corporations who demonstrate that they have taken actions going beyond normal business to reduce emissions. The ERF will use public funds to buy 'lowest cost' credits from bidders competing in 'reverse' auctions. Challenges with the ERF for producers of land sector credits include:

1. excluding effective but not necessarily lowest cost projects – forcing all bidders to compete against each other in all auctions pits producers drawing incomes only from sale of credits against, for example, energy efficiency projects that reduce costs and so boost long-term profitability, independent of income from credits

2. no recognition of co-benefits – additional environmental (e.g. biodiversity) and social (e.g. remote area employment) benefits of land-based projects are not considered
3. access to one short contract – 5 years is not long enough to recover investments needed for some types of projects, especially in land management, or to secure commercial loans
4. penalties for undersupply – providers unable to meet projected outputs may be forced to buy credits to make up shortfalls, disadvantaging projects like savanna burning, where substantial year-to-year variation is normal
5. price uncertainty – the ERF will not connect to existing markets and prices are likely to vary from auction to auction, depending on the array of bidders choosing to compete
6. stifling innovation – only emissions included in Australia's national accounts are eligible, ensuring that all public expenditure goes to meeting the 5% reduction target, but in the process excluding additional options for real benefits for the atmosphere and Australian environments.

These features and others work against high participation rates. Simple solutions to problems proposed in many submissions to government,[1] like banding auctions by sector, were rejected in the White Paper setting out the government's intentions (Commonwealth of Australia 2014). An exposure draft of an implementing Bill provides for the Regulator to use 'other procurement processes' additional to reverse auctions. And very large projects (>250 000 t CO_2-e year^{-1}) may be able to get longer contracts and more than one crediting period. There are a few other positives like modification of the CFI's unrealistic 100-year permanence criterion for sequestration, with projects being able to choose 25-year commitments subject to a 20% reduction in price. Much of the detail on process will be specified in regulations, which may be influenced by feedback from consultations on the draft Bill.

Full impact will be clearer when new and amending legislation is enacted, regulations prepared and provisions tested through experience. However, on balance the already difficult challenges for remote Indigenous groups with limited resources to build projects will be increased. If participation from the land sector in general and Indigenous land managers in particular is a serious goal, a few simple changes could make a big difference:

Longer or repeat contracts: Insistence on manifestly inadequate time to make a return on investment is hard to understand. But in combination with provisions for longer crediting periods, it implies that policy-makers believe that projects will transition to other buyers at the end of 5 years. Is the ERF seen as a temporary aberration pending restoration of a more durable mechanism? We suggest that this strangeness is resolved by providing for repeat or longer contracts when the Minister or Regulator determines that it is in the public interest. This would provide options to consider other benefits (employment in otherwise employment-free locations) in ways that don't threaten the procedural 'integrity' of reverse auctions.

Remove penalties for under-delivery: Savanna fire projects require a long baseline to take account of inherent year-to-year variation. Punishing suppliers for variation in weather implies fundamental lack of interest in or misunderstanding of land sector credits. We suggest dropping these punitive provisions and rely instead on a capacity to terminate contracts for *repeated* manifestly inadequate performance.

Immediate payment for over-delivery: It is proposed to 'allow' additional (beyond contract) credits to be purchased on completion of contracts. Given immediate penalties for under-supply it appears unreasonable to require successful providers to wait up to 5 years. We suggest that the Regulator be required to purchase oversupply at the contracted price at the time of supply, perhaps within agreed limits up to (say) 25% in excess of contracted amounts. This would offer providers incentive to maximise output and the government a buffer against risk of under-supply in other projects. Given general assessment that meeting the targeted reduction in national emissions through ERF will be challenging, bringing forward some supply should not create unmanageable budgetary difficulties, and could actually help with orderly management of year-to-year funding allocations. Producers operating inherently time-variable projects can make risk-averse bids yet still have incentives to maintain annual outputs.

Regulator discretion on purchase methods: Principles proposed for exercising discretion in purchase methods stop at cheap credits. We suggest that, as with contract numbers and periods, the Regulator might be offered guidance in law about what other aspects of the public interest might be taken into account for identifying and managing alternative arrangements for securing supply. Issues might include: reduced net demands on the Commonwealth budget taking account of downstream savings; cost-effective management of urgent environmental issues; and maintenance of employment in disadvantaged settings. Arrangements might include invitations to nominated regions or specified abatement sectors to participate in auctions or tenders of smaller scale with longer contract periods matched to payback periods. An open and transparent process for identifying and selecting opportunities would obviously be required. In our view, Regulator decisions on additional aspects of the public interest requires no more subjective judgments than existing expectations that they will reliably predict real net cost.

Some propositions to change the draft legislation move in the right direction but in the process highlight apparent inconsistencies in approach to contract periods, crediting periods, large projects and the like. Internal contradictions or asymmetries of approach remain. When the rationale for settings cannot be understood, confidence – to enter into what will be a complex and expensive process for new entrants – is greatly weakened.

If government wants to access reliable supply from new Indigenous projects between now and 2020, more than a better ERF will be needed, including access to socioeconomically oriented programs that build capability specifically for commercial projects. Support for ranger projects has been extraordinarily successful in building operational capability transferable to such projects, but gaps remain in essential business skills and infrastructure.

International challenges

Between 1997 and 2009, fires in savanna (incorporating grassland, open savanna and woodland) accounted for 60% of total global fire emissions. Recent studies estimate that around half of global fire carbon emissions come from Africa, with South America contributing between 15 and 27%, and Australia <10% (Russell-Smith *et al.* 2013). The magnitude of these figures, combined with the lessons of the Australian experience, suggest that across the fire dependent biomes[2] of southern Africa,[3] central and southern America,[4] and Asia[5] the potential is high for fire management to contribute significantly to emissions reductions, biodiversity protection, poverty reduction and other sustainable development goals.

As currently being explored by the international scientific and policy community,[6] similarities between fire dependent landscapes globally suggest that the Australian experience is likely to be broadly relevant in other contexts, when adapted for local conditions. This potential is, as yet, unrealised. It need not be so. The preconditions for implementation of fire management methodologies are known, concrete and achievable. They include: a supportive policy and regulatory environment; technical infrastructure for the measuring and verification of area burnt; the development of methodologies applicable in a given region; robust governance frameworks at the project implementation level; and sustainable financing mechanisms. All of these have their own complexities and require an initial investment of time, resources and capacity development. Despite requiring an initial investment, the Australian experience has demonstrated that fire management projects can ultimately become self-sustaining, and create sustainable livelihood opportunities for the local and Indigenous communities who implement them. It is the initial investment that the international community, alongside countries themselves, will need to make to realise the significant contribution to global goals possible through improved fire management.

Notes

1. NAILSMA's submission is available at www.nailsma.org.au/sites/default/files/NAILSMA_ERF%20submission_2014_21Feb.pdf.
2. Including but not restricted to tropical and sub-tropical savannas.
3. Noting that fire-dependent tropical and sub-tropical savannas occur across several countries of southern Africa including Angola, Botswana, the Democratic Republic of the Congo, Madagascar, Mozambique, Namibia, South Africa, Tanzania, Zambia and Zimbabwe.
4. Noting that fire dependent tropical and sub-tropical savannas occur across several countries of Central and South America including Bolivia, Belize, Brazil, Colombia, Guatemala, Guyana, Honduras, Mexico, Peru, Suriname and Venezuela.
5. Noting that fire dependent tropical and subtropical savannas occur across several countries of Asia and Oceania, including Australia, Cambodia, India, Indonesia, Myanmar, Papua New Guinea, Thailand and Timor Leste.
6. Noting for example the International Savanna Fire Management Initiative of the United Nations University – Institute for the Advanced Study of Sustainability (see www.unutki.org), the work program of the Food and Agriculture Organization of the United Nations (see www.fao.org/forestry/firemanagement/en), and various other projects globally including the GIZ funded project for the prevention, control and monitoring of fires in the Cerrado (see www.giz.de/en/worldwide/12505.html).

References

Adamo M, Tarantino C, Tomaselli V, Kosmidou V, Petrou Z, Manakos I, Lucas R, Mucher C, Veronico G, Marangi C, de Pasquale V, Blonda P (2014) Expert knowledge for translating land cover/use maps to General Habitat Categories (GHC). *Landscape Ecology* **29**, 1045–1067. doi:10.1007/s10980-014-0028-9

Allan GE, Southgate IR (2002) Fire regimes in the spinifex landscapes of Australia. In *Flammable Australia: The Fire Regimes and Biodiversity of a Continent*. (Eds RA Bradstock, JE Williams, AM Gill) pp. 145–176. Cambridge University Press: Cambridge, UK.

Allen DE, Pringle MJ, Page KL, Dalal RC (2010) A review of sampling designs for the measurement of soil organic carbon in Australian grazing lands. *The Rangeland Journal* **32**, 227–246. doi:10.1071/RJ09043

Altman JC, Buchanan GJ, Larson L (2007) 'The Environmental Significance of the Indigenous Estate: Natural Resource Management as Economic Development in Remote Australia.' Centre for Aboriginal Economic Policy Research, Discussion Paper No 286/2007. Australian National University Press: Canberra.

Beeri O, Phillips R, Carson P, Liebig M (2005) Alternate satellite models for estimation of sugarbeet residue nitrogen credit. *Agriculture, Ecosystems & Environment* **107**, 21–35. doi:10.1016/j.agee.2004.10.030

Burrows N (2014) 'Potential for Indigenous fire management in central Australia to reduce greenhouse gas emissions and increase biosequestration.' Western Australias Government Department of Parks and Wildlife: Perth.

Chen XY, Hutley LB, Eamus D (2003) Carbon balance of a tropical savanna of northern Australia. *Oecologia* **137**, 405–416. doi:10.1007/s00442-003-1358-5

Coetsee C, Bond WJ, February EC (2010) Frequent fire affects soil nitrogen and carbon in an African savanna by changing woody cover. *Oecologia* **162**, 1027–1034. doi:10.1007/s00442-009-1490-y

Comber AJ, Law ANR, Lishmna JR (2004) Application of knowledge for automated land cover change monitoring. *International Journal of Remote Sensing* **25**, 3177–3192. doi:10.1080/01431160310001657795

Commonwealth of Australia (2014) 'Emissions Reduction Fund White Paper.' Commonwealth of Australia: Canberra.

Cowley RA, Joyce K, Valencia M, Cowley T, Hearnden M, Pettit C, Dyer R (2014) Not too cool, not too often, and not too infrequent: getting fire frequency and timing just right for optimal management of woody cover and pasture composition in a grazed tropical savanna, Northern Territory, Australia. *The Rangeland Journal* **36**, 323–345.

Dawes-Gromadzki TZ (2005) Bugs beneath the surface: the functional significance of soil macroinvertebrates to landscape health in Australia's tropical savannas. *Insect Science* **12**, 307–312. doi:10.1111/j.1005-295X.2005.00037.x

Dennison PE, Roberts DA, Peterson SH, Rechel J (2005) Use of normalised difference water index for monitoring live fuel moisture. *International Journal of Remote Sensing* **26**, 1035–1042. doi:10.1080/0143116042000273998

Edwards AC, Russell-Smith J (2009) Ecological thresholds and the status of fire-sensitive vegetation in western Arnhem Land, northern Australia: implications for management. *International Journal of Wildland Fire* **18**, 127–146. doi:10.1071/WF08008

Edwards AC, Maier SW, Hutley LB, Williams RJ, Russell-Smith J (2013) Spectral analysis of fire severity in north Australian tropical savannas. *Remote Sensing of Environment* **136**, 56–65. doi:10.1016/j.rse.2013.04.013

Gerritsen R (2007) Indigenous employment in natural resource management in Australia: options and prospects. In *Investing in Indigenous Natural Resource Management*. (Eds MK Luckert, BM Campbell, JT Gorman, ST Garnett) pp. 79–85. Charles Darwin University Press: Darwin.

Hunt ER, Cavigelli M, Daughtry CST, McMurtrey JE, Walthall CL (2005) Evaluation of digital photography from model aircraft for remote sensing of crop biomass and nitrogen status. *Precision Agriculture* **6**, 359–378. doi:10.1007/s11119-005-2324-5

Hunt ER, Hively WD, Daughtry CS, McCarty GW, Fujikawa ST, Ng TL, Tranchitella M, Linden DS, Yoel DS (2008) Remote sensing of crop leaf area index using unmanned airborne vehicles. In *ASPRS Pecora 17 Conference Proceedings*, Bethesda, MD, USA: American Society for Photogrammetry and Remote Sensing. www.asprs.org/a/publications/proceedings/pecora17/0018.pdf. Accessed August 2014.

Hunt J, Smith D (2007) 'Indigenous Community Govenance Project: Year Two Research Findings.' Centre for Aboriginal Economic Policy Research, Working Paper No. 36. Australian National University: Canberra. www.anu.edu.au/caepr/Publications/WP/CAEPRWP36.pdf

Hutley LB, Evans BJ, Beringer J, Cook GD, Maier SW, Razon E (2014) Impacts of an extreme cyclone event on landscape-scale savanna fire, productivity and greenhouse gas emissions. *Environmental Research Letters* **8**. doi:10.1088/1748-9326/8/4/045023

Jamali H, Livesley SJ, Dawes T, Cook GD, Hutley LB, Arndt SK (2011a) Diurnal and seasonal variations in CH_4 flux from termite mounds in tropical savannas of the Northern Territory, Australia. *Agricultural and Forest Meteorology* **151**, 1471–1479. doi:10.1016/j.agrformet.2010.06.009

Jamali H, Livesley SJ, Dawes T, Cook GD, Hutley LB, Arndt SK (2011b) The importance of termites to the CH_4 balance of a tropical savanna woodland of northern Australia. *Ecosystems* **14**, 698–709. doi:10.1007/s10021-011-9439-5

Marika R, Yunupingu Y, Marika-Mununggiritj R, Muller S (2009) Leaching the poison – the importance of process and partnership in working with Yolngu. *Journal of Rural Studies* **25**, 404–413. doi:10.1016/j.jrurstud.2009.05.007

Morphy F (2010) (Im)mobility: regional population structures in Aboriginal Australia. *The Australian Journal of Social Issues* **45**, 363–382.

Neldner VJ, Wilson BA, Thompson EJ, Dillewaard HA (2005) Methodology for survey and mapping of regional ecosystems and vegetation communities in Queensland, ver. 3.1. Queensland Herbarium, Environmental Protection Agency: Brisbane.

Newnham GJ, Grant IF, Martin DN, Anderson SA (2010) 'Satellite based curing methods and mapping.' Final Report: Project A, 1. Bushfires Cooperative Research Centre: Melbourne.

Oliveira SLJ, Maier SW, Pereira JMC, Russell-Smith J (2015) Seasonal differences in fire activity and intensity in tropical savannas of northern Australia using satellite measurements of fire radiative power. *International Journal of Wildland Fire* **24**, 249–260.

Peterson SH, Roberts DA, Dennison PE (2008) Mapping live fuel moisture with MODIS data: a multiple regression approach. *Remote Sensing of Environment* **112**, 4272–4284. doi:10.1016/j.rse.2008.07.012

Pölönen I, Saari H, Kaivosoja J, Honkavaara E, Pesonen L (2012a) Hyperspectral imaging based biomass and nitrogen content estimations from light-weight UAV. In *SPIE Remote Sensing*. pp. 88870J–88870J. International Society for Optics and Photonics: Bellingham, WA, USA.

Pölönen I, Salo H, Saari H, Kaivosoja J, Pesonen L, Honkavaara E (2012b) Biomass estimator for NIR image with a few additional spectral band images taken from light UAS. In *SPIE Defense, Security, and Sensing*. pp. 836905–836905. International Society for Optics and Photonics: Bellingham, WA, USA.

Richards AE, Cook GD, Lynch BT (2011) Optimal fire regimes for soil carbon storage in tropical savannas of northern Australia. *Ecosystems* **14**, 503–518. doi:10.1007/s10021-011-9428-8

Roberts GM, Wooster J, Perry GLW, Drake N, Rebelo LM, Dipotso F (2005) Retrieval of biomass combustion rates and totals from fire radiative power observations: application to southern Africa using geostationary SEVIRI imagery. *Journal of Geophysical Research* **11**, D21111. doi:10.1029/2005JD006018

Russell-Smith J, Whitehead PJ, Cook GD, Hoare JL (2003) Response of *Eucalyptus*-dominated savanna to frequent fires: lessons from Munmarlary, 1973–1996. *Ecological Monographs* **73**, 349–375. doi:10.1890/01-4021

Russell-Smith J, Murphy BP, Meyer CP, Cook GD, Maier S, Edwards AC, Schatz J, Brocklehurst P (2009) Improving estimates of savanna burning emissions for greenhouse accounting in northern Australia: limitations, challenges, applications. *International Journal of Wildland Fire* **18**, 1–18. doi:10.1071/WF08009

Russell-Smith J, Price O, Murphy BP (2010) Managing the matrix: decadal responses of eucalypt-dominated mesic savanna to ambient fire regimes. *Ecological Applications* **20**, 1615–1632. doi:10.1890/09-1553.1

Russell-Smith J, Edwards AC, Price OF (2012) Simplifying the savanna: the trajectory of fire-sensitive vegetation mosaics in northern Australia. *Journal of Biogeography* **39**, 1303–1317. doi:10.1111/j.1365-2699.2012.02679.x

Russell-Smith J, Monagle C, Jacobsohn M, Beatty RL, Bilbao B, Millán A, Vessuri H, Sánchez-Rose I (2013) Can savanna burning projects deliver measurable greenhouse emissions reductions and sustainable livelihood opportunities in fire-prone settings? *Climatic Change*. doi:10.1007/s10584-013-0910-5

Salo H, Tirronen V, Pölönen I, Tuominen S, Balazs A, Heikkilä J, Saari H (2012) Methods for estimating forest stem volumes by tree species using digital surface model and CIR images taken from light UAS. In *SPIE Defense, Security, and Sensing*. pp. 83900G–83900G. International Society for Optics and Photonics: Bellingham, WA, USA.

Smith D (2008) Cultures of governance and the governance of culture: transforming and containing Indigenous institutions in West Arnhem Land. In *Contested Governance – Culture Power and Institutions in Indigenous Australia*. (Eds J Hunt, D Smith, S Garling, W Sanders) pp. 75–111. Research Monograph No. 29. Centre for Aboriginal Economic Policy Research. Australian National University Press: Canberra.

Thiele S (1982) 'Yugul, An Arnhem Land Cattle Station.' North Australian Research Unit Monograph P66. Australian National University: Canberra.

Walsh D, Russell-Smith J, Cowley R (2014) Fire and carbon management in a diversified north Australian rangelands economy: research, policy and implementation challenges. *The Rangeland Journal* **36**, 313–322.

Woinarski JCZ, Risler J, Kean L (2004) Response of vegetation and vertebrate fauna to 23 years of fire exclusion in a tropical *Eucalyptus* open forest, Northern Territory, Australia. *Austral Ecology* **29**, 156–176. doi:10.1111/j.1442-9993.2004.01333.x

Wooster MJ, Zhukov B, Oertel D (2003) Fire radiative energy for quantitative study of biomass burning – dDerivation from the Bird Experimental Satellite and comparison to MODIS fire products. *Remote Sensing of Environment* **86**, 83–107. doi:10.1016/S0034-4257(03)00070-1

Index

www.ingramcontent.com/pod-product-compliance
Lightning Source LLC
LaVergne TN
LVHW070907210926
848482LV00017B/1012
9780643108516